Hands-On Software Architecture with Java

Learn key architectural techniques and strategies to design efficient and elegant Java applications

Giuseppe Bonocore

BIRMINGHAM—MUMBAI

Hands-On Software Architecture with Java

Copyright © 2022 Packt Publishing

Group Product Manager: Aaron Lazar

Publishing Product Manager: Richa Tripathi

Senior Editor: Nisha Cleetus

Content Development Editor: Rosal Colaco

Technical Editor: Maran Fernandes

Copy Editor: Safis Editing

Project Coordinator: Manisha Singh

Proofreader: Safis Editing

Indexer: Hemangini Bari

Production Designer: Joshua Misquitta

Marketing Coordinator: Pooja Yadav

First published: February 2022
Production reference: 1100222

Published by Packt Publishing Ltd.
Livery Place
35 Livery Street
Birmingham
B3 2PB, UK.

ISBN 978-1-80020-730-1

www.packt.com

Contributors

About the author

Giuseppe Bonocore is a solution architect dealing with application development, Java technology, JBoss middleware, and Kubernetes projects since 2014. He has more than 10 years of experience in open source software, in different roles. His professional experience includes Red Hat, Accenture, and Docomo Digital, covering many technical leadership roles and deploying huge open source projects across Europe.

I want to thank the people who have been close to me and supported me,
especially my wife.

About the reviewer

Andres Sacco is a technical lead at MercadoLibre and has experience in different languages, such as Java, PHP, and Node.js. In his previous job, he helped to find alternative ways to optimize the transference of data between microservices, which helps to reduce the cost of infrastructure by 55%. Also, he has dictated some internal courses about new technologies, and he has written some articles on Medium.

Stefano Violetta is a creative backend developer with over 14 years of expertise in software development and architecture, managing all stages of the development cycle; he has worked in many different companies, from start-ups to tech giants such as eBay. He likes putting together well-written code that helps to create advanced applications that are fit for purpose, functionally correct, and meet the user's precise needs. On a personal level, he possesses really strong interpersonal skills, being respectful and collaborative. He lives (and works) in the suburbs of Milan, Italy, with his wife and two kids. When he isn't dealing with software, he likes to read and watch movies.

Table of Contents

2

Software Requirements – Collecting, Documenting, Managing

3

Common Architecture Design Techniques

4

Best Practices for Design and Development

5

Exploring the Most Common Development Models

Section 2: Software Architecture Patterns

6

Exploring Essential Java Architectural Patterns

7

Exploring Middleware and Frameworks

8
Designing Application Integration and Business Automation

9
Designing Cloud-Native Architectures

10

Implementing User Interaction

11

Dealing with Data

Section 3: Architectural Context

12
Cross-Cutting Concerns

13
Exploring the Software Life Cycle

14

Monitoring and Tracing Techniques

15

What's New in Java?

Preface

Despite being a technology born in 1995, Java is still alive and well.

Every now and then, an article pops up saying that Java is old and should be dismissed and replaced by other languages. But the reality is, Java is here to stay.

There are many reasons for that, but the most important is that it just *works*: it solves common issues in the software development world so well.

Java technology is the main topic of this book. However, in each of the chapters, we will have the opportunity to talk about many different ideas, and I think that most of them go beyond the Java programming language and are likely to be useful in other situations too.

Indeed, in this book, I've tried to distill concepts around many aspects of software development, particularly development in the enterprise world, in big and complex businesses and projects. The goal is to give you insights by focusing on the most important topics. Of course, given the breadth and complexity of the topics, it would be impossible to take a deep dive into every aspect. But you will be provided with some good starting points, and you can easily find more resources if you want further details.

Following more or less the timeline of a typical software project, we will start with the fundamentals of software architecture, from requirement gathering to modeling architecture basics. We will also look at the most common development models, including, of course, DevOps.

In the second section of the book, we will explore some common software architecture patterns. This will include Java architecture patterns, as well as middlewares (both for traditional and cloud-native approaches) and other essential parts of software architecture, such as integration, user interfaces, and data stores.

In the third and final section of the book, we will cover some additional topics, including cross-cutting concerns (such as security, monitoring, and tracing) as well as some considerations around software life cycle management. Finally, we will have a quick look at the latest version of the Java technology.

Who this book is for

This book is for Java software engineers who want to become software architects and learn the basic concepts that a modern software architect needs to know. The book is also for software architects, technical leaders, engineering managers, vice presidents of software engineering, and CTOs looking to extend their knowledge and stay up to date with the latest developments in the field of software architecture.

No previous knowledge is required, and even if you are already familiar with the Java language and the basic concepts of software development, you will still benefit from this book's recap of the different architecture-related topics.

What this book covers

Chapter 1, Designing Software Architectures in Java – Methods and Styles, introduces the approach toward the examples that we will take throughout this book. We will introduce a number of different scenarios and some real-world examples, in order to clarify abstract concepts and shift our point of view toward implementation.

Chapter 2, Software Requirements – Collecting, Documenting, Managing, explains some techniques for requirement gathering and some tools to document and track them.

Chapter 3, Common Architecture Design Techniques, covers the most commonly used architecture definition formats and the goals they aim to achieve. We will look at an example application, described using different architecture diagrams. Moreover, we will walk through some examples of modeling use cases using BPMN and a business rule using DMN.

Chapter 4, Best Practices for Design and Development, is where we will have a look at the different methods that can be used to help us with both our understanding of the overall solution and the implementation of it.

Chapter 5, Exploring the Most Common Development Models, is where we will have an overview of the most common software development models and their implications, including more traditional and historical ones (such as waterfall) as well as more modern approaches such as agile and DevOps.

Chapter 6, Exploring Essential Java Architectural Patterns, looks at architectural patterns. There are some architecture patterns that are so common that they have become more or less standard. While sometimes being overused, these architectures must be considered as basic building blocks that we need to know about in order to solve common architectural problems.

Chapter 7, Exploring Middleware and Frameworks, is where we will see how to use middleware and frameworks, understanding their role in designing and building our architecture.

Chapter 8, Designing Application Integration and Business Automation, is where, as a follow-up to the previous chapter, we will see two typical middleware implementations. Indeed, application integration and business automation are two commonly used middleware functions, used to build efficient and reusable enterprise architectures.

Chapter 9, Designing Cloud-Native Architectures, is where we will have a look at what a cloud-native application is, what the recommended practices are, and how to enhance existing applications to better suit a cloud-enabled world.

Chapter 10, Implementing User Interaction, is where we will detail the omnichannel approach by having a look at the different entry points for customer interaction.

Chapter 11, Dealing with Data, is where we will have a look at the different kinds of data persistence and how and when to mix them together.

Chapter 12, Cross-Cutting Concerns, is where we will summarize the most important cross-cutting topics to be taken into account, including identity management, security, and resilience.

Chapter 13, Exploring Software Life Cycle, will discuss all the ancillary concepts of software development projects, such as source code management, testing, and releasing. This will include some interesting concepts, such as **Continuous Integration and Continuous Delivery/Deployment** (also known as **CI/CD**).

Chapter 14, Monitoring and Tracing Techniques, will explore concepts related to the visibility and maintenance of applications running in production. This includes things such as log management, metric collection, and application performance management.

Chapter 15, What's New in Java?, will focus on the latest Java release (17) as well as a bit of the history of the language (including versioning schemes) and the ecosystem of Java vendors.

To get the most out of this book

The code samples provided with this book are generic enough to be run with the most recent Java versions, provided by any vendor. All the most common operating systems (Windows, macOS, and Linux) will work. The build and dependency management tool used is **Maven**.

The suggested configuration is **Java OpenJDK 11** and **Apache Maven 3.6**. For the React examples, **Node.js 8.1** and **React 17** were used.

Software/hardware covered in the book	Operating system requirements
Maven 3.6	Windows, macOS, or Linux
Java OpenJDK 11	Windows, macOS, or Linux
Node.js 8.1	Windows, macOS, or Linux

If you are using the digital version of this book, we advise you to type the code yourself or access the code from the book's GitHub repository (a link is available in the next section). Doing so will help you avoid any potential errors related to the copying and pasting of code.

Download the example code files

You can download the example code files for this book from GitHub at `https://github.com/PacktPublishing/Hands-On-Software-Architecture-with-Java`. If there's an update to the code, it will be updated in the GitHub repository.

We have other code bundles from our rich catalog of books and videos available at `https://github.com/PacktPublishing/`. Check them out!

Download the color images

We also provide a PDF file that has color images of the screenshots and diagrams used in this book. You can download it here: `https://static.packt-cdn.com/downloads/9781800207301_ColorImages.pdf`.

Conventions used

There are a number of text conventions used throughout this book.

`Code in text`: Indicates code words in text, database table names, folder names, filenames, file extensions, pathnames, dummy URLs, user input, and Twitter handles. Here is an example: "Each test method is identified by the `@Test` annotation."

A block of code is set as follows:

```
...
@Test
public void testConstructor()
    {
        Assertions.assertEquals(this.hello.getWho(),
        "default");
    }
...
```

Any command-line input or output is written as follows:

```
mvn io.quarkus:quarkus-maven plugin:1.12.2.Final :create
```

Bold: Indicates a new term, an important word, or words that you see onscreen. For instance, words in menus or dialog boxes appear in **bold**. Here is an example: "The following diagram shows you a comparison of **IaaS**, **PaaS**, and **SaaS**."

> **Tips or Important Notes**
> Appear like this.

Get in touch

Feedback from our readers is always welcome.

General feedback: If you have questions about any aspect of this book, email us at customercare@packtpub.com and mention the book title in the subject of your message.

Errata: Although we have taken every care to ensure the accuracy of our content, mistakes do happen. If you have found a mistake in this book, we would be grateful if you would report this to us. Please visit www.packtpub.com/support/errata and fill in the form.

Piracy: If you come across any illegal copies of our works in any form on the internet, we would be grateful if you would provide us with the location address or website name. Please contact us at copyright@packt.com with a link to the material.

If you are interested in becoming an author: If there is a topic that you have expertise in and you are interested in either writing or contributing to a book, please visit `authors.packtpub.com`.

Share your thoughts

Once you've read *Hands-On Software Architecture with Java*, we'd love to hear your thoughts! Scan the QR code below to go straight to the Amazon review page for this book and share your feedback.

https://packt.link/r/1-800-20730-1

Your review is important to us and the tech community and will help us make sure we're delivering excellent quality content.

Section 1: Fundamentals of Software Architectures

In this section, you will gain all the foundations needed for defining and understanding complex software architectures.

We will start with what software architecture is, the different kinds of it, and the importance of properly defining it. We will then step into the first phases of a software development project, including requirement collection and architecture design.

The focus will then be on best practices for software design and development. Last but not least, we will have an overview of the most common development models, such as waterfall, Agile, and DevOps.

This section comprises the following chapters:

- *Chapter 1, Designing Software Architectures in Java – Methods and Styles*
- *Chapter 2, Software Requirements – Collecting, Documenting, Managing*
- *Chapter 3, Common Architecture Design Techniques*
- *Chapter 4, Best Practices for Design and Development*
- *Chapter 5, Exploring the Most Common Development Models*

1
Designing Software Architectures in Java – Methods and Styles

In this chapter, we will focus on some core concepts that we can use as a base to build on in the upcoming chapters. We will explore different ways to represent the software architecture, paying attention to the intended audience and their specific point of view. Additionally, we will elaborate on the importance of a proper architectural design and its role in the software development life cycle. Following this, we will move on to the Java ecosystem, which is the core topic of this book, to discover why it's a good choice for implementing a complete enterprise application.

In particular, we will cover the following topics:

- The importance of software architecture
- Different types of architecture design – from doodling on paper to more accurate modeling

- Other kinds of architectural diagrams
- The changing role of Java in cloud-native applications
- Case studies and examples
- Software components diagram

By the end of this chapter, you should have a clear view of why design is a critical part of the software development process and what the main types of architecture schemas are. Additionally, you will become familiar with the role of Java technology in modern application development.

These skills are crucial for implementing functional and elegant software solutions. It will also be a good basis for personal development and career enhancement.

The importance of software architecture

Often, software development is all about cost and time. No one knows exactly why, but the software industry is almost always associated with tight deadlines, insufficient resources, and long hours. Under this kind of pressure, it's common to question the importance of everything that is not strictly *coding*. Testing is a common victim of this, along with documentation and, of course, design. But of course, these phases are essential for the success of a project. While we will quickly touch on most of those aspects, architecture design is the core of this book, and I believe that by understanding the practices and goals, the need for it will become clear to everybody.

In this section, we will discover what the fundamental objects of a properly designed architecture are. Highlighting those simple but crucial points is useful in raising awareness about the importance of this phase. If you start advocating those good practices in your team, the quality of your software deliverables will increase.

The objectives of architecture design in the software life cycle

The ultimate goal of this book is not to define the architecture *per se*; there are plenty of papers and interesting things available on that matter, including the awesome work of Martin Fowler. Nevertheless, there are a couple of considerations that we need to bear in mind.

The architecture should support the crucial decisions within our software project. However, the architecture itself is actually a loose concept, often including different plans (such as physical, logical, network, and more) and points of view (such as users, business logic, machine-to-machine interactions, and more).

Let's take the most overused metaphor as an example: a software project is like a building. And similarly to a construction project, we require many different points of view, with different levels of detail, ranging from general overviews to detailed calculations and the bills of materials. A general overview is useful to give us an idea of where we are and where we want to go. In addition to this, it is an essential tool for being sure we are on the right path. However, a system overview doesn't provide enough details for teams such as networking, security, sysops, and, ultimately, the developers that require a more substantiated and quantitative view to drive their day-to-day decisions.

The main goals of designing a proper software architecture include the following:

- Prospecting a **birds-eye view** to project sponsors and investors. While it is not a good practice to drive a business discussion (for example, an elevator pitch) toward technical elements too soon, a higher level of management, venture capitalists, and the like are becoming increasingly curious about technical details, so a high-level overview of the application components can be crucial for winning this kind of discussion.

- Defining a **shared lingo** for components of our solution, which is crucial for collaborating across the team.

- Providing **guidance for technological choices** since putting our design decisions on paper will clarify important traits of our application. *Will data be central? Do we need to focus on multiple geographies? Are user interactions the most common use case?* Some of those reasonings will change over time. However, correctly designing our application will drive some crucial technology choices, in terms of choosing components and stacks to rely on.

- Splitting **roles and responsibilities**. While a proper project plan, a statement of work, or a **Responsible**, **Accountable**, **Consulted**, **Informed** (**RACI**) (which is a classical way to categorize who does what) table will be used for real project management, writing the software backbone down on paper is our first look at who we have to involve for proper project execution.

Indeed, the architecture is an excellent example of planning in advance. However, a proper software architecture should be much more than a technological datasheet.

Architecture, as with buildings, is more about the styles and guidelines to be followed all around the project. The final goal of a piece of software architecture is to find elegant solutions to the problems that will arise during the project plan. Ultimately, it will act as guidance throughout the project's life cycle.

The software architect – role and skills

As a role, the software architect is often identified as the more senior technical resource in the IT team. In fact, the job role of an architect is almost always seen as a career progression for developers, especially in enterprise environments. While not necessary, being good at coding is crucial for a complete comprehension of the overall functioning of the system.

There are several different other skills that are required to be a successful architect, including creativity, the ability to synthesize, and vision. However, above all, experience is what it takes to become an architect.

This includes firsthand experience on many different projects, solving real-world issues: what a proper software design looks like and how the design has evolved. This skillset is very useful to have in the background of the architect.

Additionally, it's vital to have a huge library of solutions to choose from in order to avoid reinventing the wheel. While we love to think that our problem is very unique, it's very unlikely to be so.

This leads us to the approach that we will use in this book: we will not focus on just one aspect or technology to drill down on, but we will take a *horizontal* approach, discussing a number of different topics and offering ideas on how to approach potential problems. We hope to act as a handbook to support you when making real-world choices.

Is architecture design still relevant in modern development?

There will be a couple of chapters dedicated to discussing **Microservices**, **DevOps**, and the cloud-native avalanche, but it's safe to assume that in one form or another, you will have plenty of opportunities to hear something about them.

As you might have gathered, most of these concepts are not really new. The **Agile Manifesto**, which is a seminal work detailing some of the practices commonly used in modern development techniques, was published in 2001, yet most of the common-sense principles it contains are misinterpreted. When I was working in IT consulting back in 2008, a common joke among development teams was "*Yes, we do agile. We skip documentation and testing.*"

Of course, that's just an opinion based on personal experience. There are plenty of teams who do not underestimate the importance of proper planning and documentation and are doing wonderfully while working with Agile. Yet, in some cases, less structured development methodologies have been taken as an excuse to skip some crucial steps of the development life cycle.

As we will elaborate, in *Chapter 5*, *Exploring the Most Common Development Models*, Agile is much more than slimming down boring phases of the project. Indeed, testing and documentation are still very relevant, and Agile is no excuse to skip that.

There are plenty of reflections you can take in terms of how to adapt your design techniques to DevOps, Agile, and more, and we will discuss this topic later in this book. However, one thing is certain: *architecture matters*. *Design is very relevant*. We have to spend the correct amount of time planning our choices, revisiting them when needed, and generally, sticking with some well-defined guiding principles. The alternative is poor quality deliverables or no deliverables at all.

Now, let's take a look at what the first phases of software design usually look like.

Different types of architecture design – from doodling on paper to more accurate modeling

When we start to shape the architecture of a new application, the result is often familiar.

I would say that across different geographies, industries, and application types, some elements are common. The architectural sketches are usually made of boxes and lines, with labels, arrows, and similar artifacts. That's an intuitive way to shape our thoughts on paper.

However, in the following section, we will go through *different* ways of expressing those concepts. This will make us aware of available styles and techniques and will make our diagram clearer and, ultimately, easier to share and understand.

But first, let's find out what the characteristics of architectural sketching actually are.

Sketching the main architectural components

As we discussed earlier, there are a number of different components that are recurrent in a high-level architectural sketch. Let's examine them one by one:

- **Boxes**: These represent the software components. They can refer to one complete application or specific subcomponents (such as packages, modules, or similar things).

- **Lines**: These describe the relationships between the boxes. Those links imply some sort of communication, commonly in the form of APIs. The lines can also represent inheritance or a grouping of some sort. A direction (that is, *an arrow*) can also be specified.

- **Layers**: These are a dotted or dashed line, grouping components and their relationships. They are used to identify logical slices of the architecture (such as the frontend, backend, and more), the grouping of subcomponents (for example, validation and business logic), network segments (such as the intranet and DMZ), physical data centers, and more.

- **Actors**: Simulating the interactions of users within the systems, actors are usually represented as stickmen, sitting on top of some components (usually frontends or UIs of some sort). It is not uncommon to observe different channels represented, in the form of laptops or mobile phones, depending on the industry and type of application (for example, ATMs, branch offices, and physical industrial plants).

Now, let's view an example sketch:

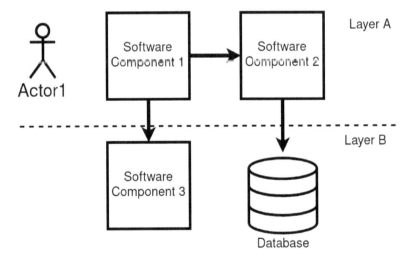

Figure 1.1 – The common components on a first architectural sketch

As we've already mentioned, the quick and dirty representation shown in this diagram is useful since it's an easy way to start thinking about how our application should look. However, on a closer look, there are some common inaccuracies:

- The software components (that is, our boxes) might be represented with different levels of zoom: sometimes representing applications, sometimes features, and sometimes software modules. This is inconsistent and could generate confusion.

- Some components are specialized (for example, databases), while others are not. As noted in the preceding point, this leads to an inhomogeneous view.

- In some parts of the diagram, we are representing use cases or information flows (for example, with the actors), while elsewhere, we are drawing a static picture of the components.

- Some points of view don't cope well with others because we might be representing network firewalls but not referencing any other networking setup.

Now that we've learned what a naïve representation looks like and what its limits are, let's take a look at some other types of diagrams and how they represent alternative points of view.

Other kinds of architectural diagrams

As we discovered in the previous section, the first sketches of a piece of architecture often end up as an intuitive and naïve view, lacking essential details. In this section, we will look at an overview of different types of architectural diagrams. This will help us to pick the right diagram for the right situation, defining a clearer view of our architecture. So, let's dig into some details.

Common types of architectural diagrams

In order to define a clearer and more detailed view of what our software will look like, it's essential to start picking layers and points of view to represent. This will naturally lead us to focus on more tailored designs. While not exhaustive, a list of possible architectural diagrams includes the following:

- **Software components**: This kind of schema includes different software modules (such as applications or other components) and the interaction between them (for example, read from, write to, listen, and more). One particular instance of this diagram could include protocols and formats of communication between those components, becoming close to a complete API documentation:

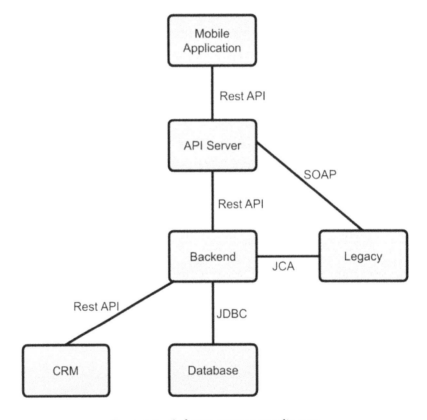

Figure 1.2 – Software components diagram

- **Network architecture**: This is a pretty common design type and is often considered the more *scientific* and detailed one. It includes data such as network segments (**DMZ** and **INTRANET**), **Firewall**, IP addressing, and more:

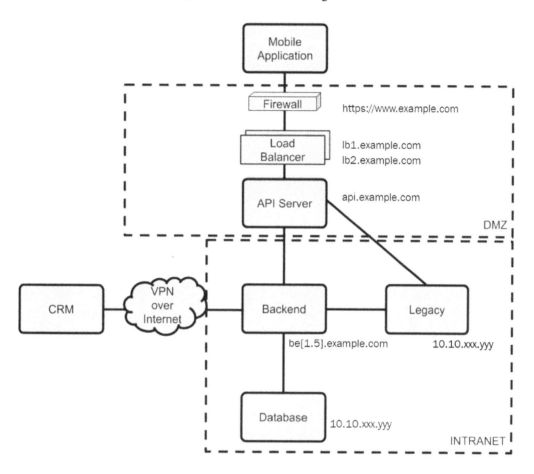

Figure 1.3 – Network architecture diagram

- **Physical architecture**: This is a mapping of software modules into server deployments. Usually, it's complete with information about the server hardware and model. In the case of a multiple datacenter setup (which is pretty common in enterprise environments), it can also contain details about racks and rooms. Storage is another relatively common component. Depending on the implementation, this architecture might include information about virtualization technology (for example, the mapping of VMS to the physical servers that are hosting it). Additionally, it could, where relevant, include references to cloud or container deployments:

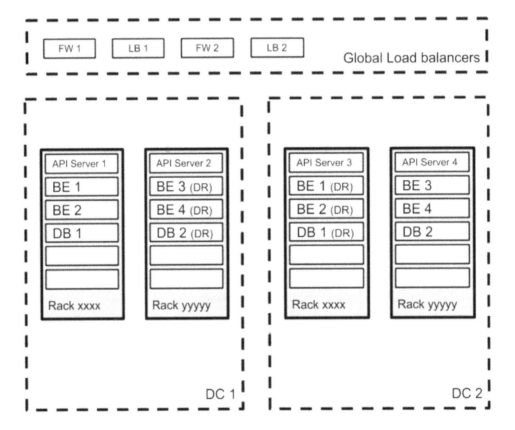

Figure 1.4 – Physical architecture diagram

These are the very basic points of view in an architecture diagram and an essential starting point when detailing the design of your application. Diving further into the application specification life, other kinds of diagrams, often derivatives of those, could be elaborated (for example, cloud deployment diagrams, software modules, and more) depending on your specific needs. In the next section, we will focus on Java technology, which is the other fundamental topic of this book and crucial for completing our architectural view of modern applications.

The changing role of Java in cloud-native applications

Now that we've briefly touched on the various kinds of designs and diagrams of an application, let's focus on the other fundamental topic of this book: the Java language.

It's not uncommon to hear that Java is dead. However, if you are reading this book, you probably agree that this is far from the truth.

Of course, the panorama of software development languages for enterprise applications is now wider and more complicated than the golden age of Java; nevertheless, the language is still alive and widespread, especially in some areas.

In this section, we will explore the usage of Java technology in the enterprise software landscape. Then, we will take a quick glance at the history of **Java Enterprise Edition** (**JEE**). This will be a good foundation to understand existing enterprise architectures and model modern, cloud-native applications based on this technology.

Now, let's examine why Java technology is still thriving.

Why Java technology is still relevant today

The most important reason for Java's popularity is probably the availability of skill. There are plenty of experts on this language, as many polls and studies show (for example, PYPL and Tiobe). Another crucial point is the relevance of the ecosystem, in terms of the quantity and quality of libraries, resources, and tooling available for the Java platform.

Rewriting complex applications (including their dependencies) from Java to another language could probably take years, and, long story short, there might be no reason to do that. Java just works, and it's an incredibly productive platform. It might be slow and resource-intensive in some scenarios, but this is balanced by its stability. The language has been battle-tested, is feature-rich, and essentially, covers all the use cases required in an enterprise, such as transactionality, integration with legacy environments, and manageability.

Now, let's take a look at where and how Java technology is used in enterprise environments. This can be very useful to understand existing scenarios and fit new applications into existing application landscapes.

Java usage in enterprise environments

In order to fit our Java application in the overall architecture, it's important to understand the typical context of a large enterprise, from a software architecture perspective.

Of course, the enterprise architecture depends a lot on the industry domain (for instance, banking, telecommunications, media, and more), geography, and the tenure of the organization, so my vision might be slightly biased toward the segment I have worked with for the longest (a large enterprise in the EMEA area). Still, I think we can summarize it as follows:

- **Legacy**: Big applications, usually running very core functions of the enterprise for many years (at least more than 10 and commonly more than 20). Needless to say, the technology here is not the most current (**Cobol** is widespread in this area, but it is not uncommon to see other things such as **PL SQL**, huge batch scripts, and even C/C++ code). However, the language is seldom an issue here. Of course, nowadays, those skills are very rare to find on the job market, but usually, the software *just works*. The point here is that most of the time, nobody exactly knows what the software does, as it's poorly documented and tested. Moreover, you usually don't have automated release procedures, so every time you perform a bugfix, you have to cross your fingers. Needless to say, a proper testing environment has never been utilized, so most of the things have to be tested in production.

- **Web (and mobile)**: This is another big chunk of the enterprise architecture. Usually, it is easier to govern than legacy but still very critical. Indeed, by design, these applications are heavily customer-facing, so you can't afford downtime or critical bugs. In terms of technologies, the situation here is more fragmented. Newer deployments are almost exclusively made of **Single-Page Applications** (**SPAs**) based on JavaScript (implemented with frameworks such as Angular, Vue, and React). Backends are REST services implemented in JavaScript (**Node.js**) or Java.

- **Business applications**: Often, the gap between web applications and business applications is very thin. Here, the rule of thumb is that business applications are less web-centric (even if they often have a web GUI), and usually, they are not customer exposed. The most common kind of business application is the management of internal back-office processes. It's hard to find a recurrent pattern in business applications since it's an area that contains very different things (such as CRMs, HR applications, branch office management, and more).

- **BigData**: Under various names and nuances (such as data warehouses, data lakes, and AI), BigData is commonly a very huge workload in terms of the resources required. Here, the technologies are often packaged software, while custom development is done using various languages, depending on the core engine chosen. The most common languages in this area are Java (Scala), **R** (which is decreasing in popularity), and **Python** (which is increasing in popularity). In some implementations, a big chunk of SQL is used to stitch calculations together.

- **Middlewares and infrastructure**: Here falls everything that glues the other apps together. The most common pattern here is the integration (synchronous or asynchronous). The keywords are ESB, SOA, and messaging. Other things such as **Single Sign-On** and identity providers can be included here.

As I mentioned, this is just a coarse-grained classification, useful as reference points regarding where our application will fit and which other actor our application will be interacting with.

Notice that the technologies mentioned are mostly *traditional* ones. With the emergence of modern paradigms (such as the cloud, microservices, and serverless), new languages and stacks are quickly gaining their place. Notable examples are Go in the microservice development area and Rust for system programming.

However, those technologies and approaches are often just evolutions (or brand-new applications) belonging to the same categories. Here, the most interesting exception is in the middleware area, where some approaches are decreasing in popularity (for example, SOA) in favor of lighter alternatives. We will discuss this in *Chapter 7, Exploring Middleware and Frameworks*.

Now that we've explored the widespread usage of Java in an enterprise context, let's take a look at its recent history.

JEE evolution and criticism

JEE, as we have learned, is still central in common enterprise applications. The heritage of this language is just great. The effort that has been done in terms of standardizing a set of APIs for common features (such as transactionality, web services, and persistence) is just amazing, and the cooperation between different vendors, to provide interoperability and reference implementation, has been a very successful one.

However, in the last couple of years, a different set of needs has emerged. The issue with JEE is that in order to preserve long-term stability and cross-vendor compatibility, the evolution of the technology is not very quick. With the emergence of cloud and more modular applications, features such as observability, modular packaging, and access to no SQL databases have become essential for modern applications. Of course, standards and committees have also had their moments, with developers starting to move away from vanilla implementations and using third-party libraries and non-standard approaches.

> **Important Note:**
>
> The objective of this book is not to recap the history and controversy of the JEE platform. However, organizational issues (culminating with the donation of the project to the Eclipse Foundation) and less frequent releases have contributed to the decrease in popularity of the platform.

The upcoming of the **Platform-as-a-Service** (**PaaS**) paradigm is another important event that is changing the landscape. Modern orchestration platforms (with Kubernetes as the most famous example), both in the cloud or on-premises, are moving toward a different approach. We will examine this in greater detail later, but essentially, the core concept is that for the sake of scalability and control, some of the typical features of the application server (for example, clustering and the service registry) are delegated to the platform itself. This has a strict liaison with the microservice approach and the benefits they bring. In the JEE world, this means that those features become duplicated.

Another point is about containerization. One of the focal points of container technology is immutability and its impacts in terms of stability and the quality of the applications. You package one application into a container and easily move it between different environments. Of course, this is, not in the same direction as JEE servers, which have been engineered to host multiple applications, managing hot deploys and live changes of configurations.

A further consideration regarding application servers is that they are, by design, optimized for transaction throughput (often at the expense of startup times), and their runtime is general-purpose (including libraries covering many different use cases). Conversely, the cloud-native approach is usually aimed at a faster startup time and a runtime that is as small as possible, bringing only the features needed by that particular application. This will be the focus of our next section.

Introducing cloud-native Java

Since the inception of the microservices concept, in the Java development community, the paradigm has increasingly shifted toward the *fat jar approach*. This concept is nothing new, as the first examples of *uber jars* (a synonym of the fat jar) have been around since the early 2000s, mainly in the desktop development area. The idea around them is pretty simple: instead of using dynamic loading of libraries at runtime, let's package them all together into an executable jar to simplify the distribution of our application. This is actually the opposite of the model of the application servers, which aim to create an environment as configurable as possible, supporting things such as hot deployment and the hot-swapping of libraries, privileging the uptime to immutability (and predictability).

In container-based and cloud-native applications, fat jar approaches have begun to be viewed as the perfect candidate for the implementation of cloud-native, microservices-oriented applications. This is for many different reasons:

- **Testability**: You can easily run and test the application in a local environment (it's enough to have a compatible **Java Virtual Machine** or **JVM**). Moreover, if the interface is properly defined and documented, it's easy to mock other components and simulate integration testing.

- **Ease of installation**: The handover of the application to ops groups (or to testers) is pretty easy. Usually, it's enough to have the .jar file and configuration (normally, on a text file or environment variable).

- **Stability across environments**: Since everything is self-contained, it's easy to avoid the *works-on-my-machine* effect. The development execution environment (usually, the developer machine) is designed pretty similarly to the production environment (aside from the configuration, which is usually well separated from the code, and of course, the external systems such as the databases). This behavior mirrors what is provided by containers, and it's probably one of the most important reasons for the adoption of this approach in the development of microservices.

There is one last important consideration to pay attention to: curiously enough, the all-in-one fat jar approach, in contrast with what I've just said, is theoretically conflicting with the optimization provided by the containerization.

Indeed, one of the benefits provided by every container technology is *layerization*. Put simply, every container is composed by starting with a *base image* and just adding what's needed. A pretty common scenario in the Java world is to create the application as a tower composed of the operating system plus the JVM plus dependencies plus the application artifact. Let's take a glance at what this looks like in the following diagram. In gray, you will see the base image, which doesn't change with a new release of the application. Indeed, a change to the application artifact means only redeploying the last layer on top of the underlying **Base Image**:

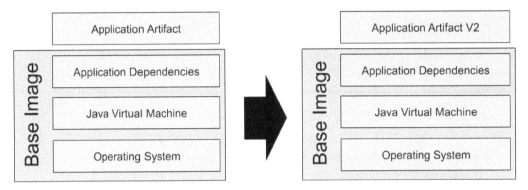

Figure 1.5 – Layering container images

As you can see in the preceding diagram, the release in this scenario is as light as simply replacing the **Application Artifact** layer (that is, the top layer).

By using the fat jar approach, you cannot implement this behavior. If you change something in your application but nothing in the dependencies, you have to rebuild the whole **Fat JAR** and put it on top of the JVM layer. You can observe what this look like in the following diagram:

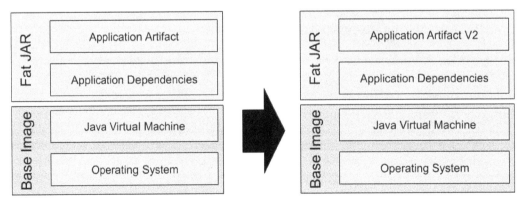

Figure 1.6 – Layering container images and fat jars

In this scenario, the release includes all of the application dependencies, other than the application by itself.

While this might appear to be a trivial issue, it could mean hundreds of megabytes copied back and forth into your environment, impacting the development and release time since most of the things composing the container cannot be cached by the container runtime.

Some ecosystems do a bit of experimentation in the field of *hollow jars* to essentially replicate an approach similar to the application server. Here, the composed (fat) jar is split between the application layer and the dependencies layer in order to avoid having to repackage/move everything each time. However, this approach is far from being widespread.

The Java microservices ecosystem

One last consideration goes to the ecosystem in the Java microservices world. As we were beginning to mention earlier, the approach here is to delegate more things to *the platform*. The service itself becomes simpler, having only the dependency that is required (to reduce the size and the resource footprint) and focusing only on the business logic.

However, some of the features delegated to the application server are still required. The service registry, clustering, and configuration are the simplest examples that come to mind.

Additionally, other, newer needs start to emerge:

- **HealthCheck** is the first need. Since there is no application server to ensure your application is up and running, and the application is implemented as more than one running artifact, you will end up having to monitor every single microservice and possibly restarting it (or doing something different) if it becomes unhealthy.

- **Visibility** is another need. I might want to visualize the network of connections and dependencies, the traffic flowing between components, and more.

- Last but not least: **resiliency**. This is often translated as the circuit breaker even if it's not the only pattern to help with that. If something in the chain of calls fails, you don't want the failure to cascade.

So, as we will discover in the upcoming chapters, a new ecosystem will be needed to survive outside the JEE world.

Microservices has been a groundbreaking innovation in the world of software architectures, and it has started a whole new trend in the world of so-called cloud-native architectures (which is the main topic of this book). With this in mind, I cannot avoid mentioning another very promising paradigm: **Serverless**.

Serverless borrows some concepts from microservices, such as standardization and horizontal scaling, and takes it to the extreme, by relieving the developer of any responsibility outside the code itself and delegating aspects such as packaging and deployment to an underlying platform. Serverless, as a trend, has become popular as a proprietary technology on cloud platforms, but it is increasingly used in hybrid cloud scenarios.

Java is not famous in the serverless world. The need for compilation and the weight added by the JVM has, traditionally, been seen as a showstopper in the serverless world. However, as we will explore further in *Chapter 9*, *Designing Cloud-Native Architectures*, Java technology is now also gaining some momentum in that area.

And now, in order to better clarify different architectural designs, we will examine some examples based on a reference case study.

Case studies and examples

Following up on the handbook approach, each time we face a complex concept, I will try to clarify it by providing case studies. Of course, while the cases are not real (for reasons you can imagine), the challenges closely resemble several first-hand experiences I've incurred in my professional history.

In this section, we will start from scratch by designing a piece of software architecture. Then, we will add details to portray a more precise view. This will help you to better understand the first steps in the design of a complex piece of architecture.

Case study – mobile payments

In this case study, we will simulate the architecture design of a mobile payment solution. As contextual background, let's suppose that a huge bank, in order to increase the service offering toward their customers and following some market research, wants to implement a mobile payment application. By definition, a mobile payment is a pretty broad term, and it includes many different use cases involving financial transactions completed using smartphones.

In this particular implementation, we will consider the use case of paying with your smartphone by charging you via your mobile phone bill.

Essentially, this means implementing a client-server architecture (with the clients implemented as a mobile application), interacting both with existing enterprise applications and external systems exposed by telecommunication operators. Now, let's now try to analyze some use cases related to this scenario and model it by using the different schemas we've discussed so far.

Whiteboarding the overall architecture

Beginning on white space, let's start whiteboarding the overall architecture. As we've learned, the first step is usually to sketch, at a high level, the relevant modules and the relationships between them. It's not important to be super detailed, nor to use a particular style. We are just brainstorming the first shapes on paper:

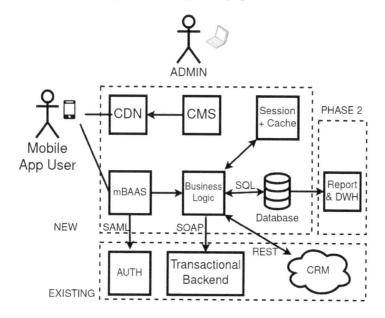

Figure 1.7 – Architecture whiteboarding

Here, we have drafted a birds-eye view of the use case. We now know where the transaction starts, where the data is saved, and how the user interacts with the system.

Additionally, we have identified the main components of the application:

- The mobile application (represented together with the user)
- The (**CDN**) to serve static resources to the application
- The (**CMS**) to configure content to be delivered to the app
- The backend (**mobile Backend as a Service** or **mBaaS**) to proxy requests and responses
- The business logic of the application
- **Session** and **Cache**, to store non-persistent data of the users

- **Database**, to store persistent data
- Other parts of the application: reporting and data warehousing, authentication, **Transactional Backend**, and **Customer Relationship Management** (**CRM**)

As expected, this kind of design has some intrinsic issues:

- You can observe mixed-use cases (both of the mobile user and the CMS administrator), which can be foreseen by the arrows between different components, but it's barely designed.
- There is a view in the project timeline regarding the implementation of components (reporting and data warehousing appear to be optional in the first phase of the project).

Some protocols in the interactions are named (for example, SOAP and REST), but it's not an API specification, nor a network schema. Anyway, even if it's not super detailed, this schema is a good starting point. It helps us to define the main application boundaries, it gives a high-level overview of the integration points, and overall, it's a good way to kick off a more detailed analysis. We will improve on this in the next section.

Software components diagram

In order to address some of the issues highlighted in the previous section, I've modeled the same system by focusing on software components. This does not follow any specific standard even if is pretty similar to the **C4** approach (where **C4** stands for **Context, Containers, Components, and Code**; we will discuss this further in later chapters):

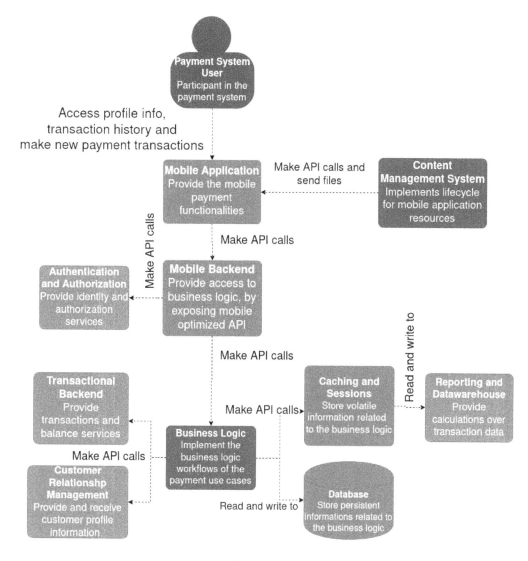

Figure 1.8 – Software components diagram

As you can see, this schema is more homogeneous and better organized than the first sketch. At a first glance, you can view what features are provided to the user. Additionally, it highlights how the system interacts with each other in a structured way (for example, using API calls, reads and writes, and more).

Compared to the first sketch, there are some considerations that we can observe:

- The components are almost the same as the other schema.

- The diagram is less focused on the use case, even if the user is still represented (together with a high-level recap of the features available to them).

- There is no view on the project phases. This helps you to focus on just one point of view (architectural components), making the schema less confusing.

- No protocols are named, only high-level interactions (such as reads, writes, and API calls).

- Some technical components are preserved (the database), while others are skipped since they have less impact on the functional view (for example, the CDN, which is probably more relevant on a network schema).

In this section, we learned how to approach the first design of our mobile payments application; first, with a more naïve view, then by trying to detail the view in a more structured way. In the upcoming chapters, we will discuss how to further clarify and enrich those views.

Summary

In this first chapter, we just scratched the surface on the two most essential topics of this book: the different types of architectural design and the relevance of Java technology in the enterprise world.

We have discovered what the first sketches of our software architecture look like and why they are relevant, even if they are not very detailed. Then, we moved on to different schemas (such as software components, the infrastructure, and the network) to get a glimpse of other schema styles, which is useful to address specific areas of interest. On the Java side, we made some considerations about the role of Java in the enterprise landscape and how the language is evolving to meet the challenges of modern cloud environments.

These concepts will be useful starting points for the two core concepts of this book. On the architectural side, we've grasped how complex and important it is to view, analyze, and design a proper architecture. From a technological point of view, we've learned how Java, the technology we will focus on for the rest of this book, is very widespread in the enterprise context and how it is still relevant for building modern, cloud-native applications.

In the next chapter, we will start working with requirements. Requirement gathering and specifications are essential in order to rework our architectural design, adding more details and ensuring the final product will meet customer expectations.

Further reading

- *Who Needs an Architect?* by Martin Fowler (`http://files.catwell.info/misc/mirror/2003-martin-fowler-who-needs-an-architect.pdf`)

- *Don't Put Fat Jars in Docker Images* by Philipp Hauer (`https://phauer.com/2019/no-fat-jar-in-docker-image`)

2
Software Requirements – Collecting, Documenting, Managing

Collecting requirements is arguably one of the most frustrating activities in software production for several reasons. Difficulties often arise because it is never completely clear who the owner is, as well as because architects cannot do a good design without certain requisites, and developers, of course, can't do a proper job without the designs.

However, it is fairly common practice for a development team to start doing something without a complete requirements collection job because there is no time. Indeed, what often happens, especially in regards to large and complex projects, is that the milestones are put in place before the project scope is completely defined. In this industry, since software is an intangible product (not like a building or a bridge), budget approval is usually a more *fluid* process. Therefore, it's not unusual to have a project approved before all the details (including requirements, feasibility, and architectural design) are fully defined. Needless to say, this is an inherently bad practice.

In this chapter, we will look at different techniques for requirements gathering and analysis in order to increase the quality of our software deliverables.

You will learn about the following:

- The different types of requirements: **functional** and **non-functional**
- What characteristics a requisite must have
- How to formalize requirements in standard formats
- How to collect requirements by using agile and interactive techniques

Once you have completed this chapter, you will be able to organize productive requirements gathering sessions and document them in a clear way. Being able to collect and properly document requisites can be a real gamechanger for your career in software development in several ways:

- The quality of the software you produce will be better, as you will focus on what's really needed and be able to prioritize well.
- You will have a better understanding of the language of business and the needs of your customers, and you will therefore implement features that better fit their needs.
- You will have the possibility to run informal and interactive sessions on requirements gathering. (As an example, see the *Event Storming* section.)
- You will have a primer about international standards in software requirements specifications, which may be a hard constraint in some environments (for example, when working for regulated industries such as government or healthcare).

Since requirements collection and management is a practice mostly unrelated to a specific programming language, this chapter doesn't directly reference Java technology.

Now, let's start exploring the discipline of software requirements engineering.

Introducing requirements engineering

From a purely metaphorical perspective, if an algorithm is similar to a food recipe, a software requirement is the order we place at a restaurant. But the similarity probably ends here. When we order our food, we pick a specific dish from a discrete list of options, possibly with some small amount of *fine tuning*.

Also, continuing with our example, the software requirement has a longer and more complex life cycle (think about the testing and evolution of the requirement itself), while the food order is very well timeboxed: the customer places the order and receives the food. In the worst case, the customer will dislike the food received (like a user acceptance test going wrong), but it's unusual to evolve or change the order. Otherwise, everything is okay when the customer is happy and the cook has done a great job (at least for that particular customer). Once again, unlike the software requirement life cycle, you will likely end up with bug fixes, enhancements, and so forth.

Requirements for software projects are complex and can be difficult to identify and communicate. Software requirements engineering is an unusual job. It requires a concerted effort by the customer, the architect, the product manager, and sometimes other various professionals. *But what does a technical requirement actually look like?*

Feature, Advantage, and Benefit

As we will see in a few sections, requirements collection involves many different professionals working together to shape what the finished product will look like. These professionals usually fall into two groups, business-aware and technology-aware. You should of course expect those two groups to have different visions and use different languages.

A good way to build common ground and facilitate understanding between these two groups is to use the **Feature, Advantage, and Benefit logical flow**.

This popular framework, sometimes referred to as **FAB**, is a marketing and sales methodology used to build messaging around a product. While it may not seem immediately relevant in the requirements gathering phase, it is worth looking at.

In the FAB framework, the following apply:

- A **Feature** is an inherent product characteristic, strictly related to what the product can do.

- The **Advantage** can be defined as what you achieve when using a particular Feature. It is common to have more than one Advantage linked to the same technical feature.

- The **Benefit** is the final reason why you would want to use the Feature. If you want, it's one further step of abstraction starting from advantages, and it is common to have more than one Benefit linked to the same feature.

Let's see an example of FAB, related to the mobile payment example that we are carrying over from the previous chapter:

- A **Feature** is the possibility of authorizing payments with biometric authentication (such as with your fingerprint or face ID). That's just the technical aspect, directly related to the way the application is implemented.

- The related **Advantage** is that you don't need to insert a PIN or password (and overall, you will need a simpler interaction with your device – possibly just one touch). That's what the feature will enable, in terms of usage of the application.

- The linked **Benefit** is that your payments will be faster and easier. But another benefit can be that your payments will also be safer (no one will steal your PIN or password). That's basically the reason why you may want to use this particular feature.

As you can imagine, a non-technical person (for example, a salesperson or the final customer) will probably think of each requirement in terms of benefits or advantages. And that's the right way to do it. However, having reasoning on the FAB flow could help in having a uniform point of view, and possibly *repositioning* desiderata into features and eventually requirements. We can look at a simple example regarding user experience.

Sticking with our mobile payments sample application, a requirement that business people may want to think about is the advantages that the usage of this solution will bring.

One simple example of a requirement could be to have a list of payments easily accessible in the app. A feature linked to that example would allow the customers to see their transaction list immediately after logging into the system.

In order to complete our flow, we should also think about the benefits, which in this case could be described as the ability to keep your expenses under control. However, this could also work the other way around. When reasoning with more tech-savvy stakeholders, it's easier to focus on product features.

You may come up with a feature such as *a user currently not provisioned in the system should be presented with a demo version of the application.*

The advantage here is having an easy way to try the application's functionalities. The benefit of this for customers is that they can try the application before signing up for an account. The benefit for the business is that they have free advertising to potentially draw in more customers.

You might now ask, *so what am I looking for, when doing requirements gathering, that is, searching for features?* There are no simple answers here.

My personal experience says that a feature may be directly considered a requirement, or, more often, be composed of more than one requirement. However, your mileage may vary depending on the type of product and the kind of requirements expressed.

One final thing to note about the FAB reasoning is that it will help with clustering requirements (by affinity to similar requirements or benefits), and with prioritizing them (depending on which benefit is the most important).

Now we have a simple process to link the technical qualities of our product to business impacts. However, we haven't yet defined exactly what a requirement is and what its intrinsic characteristics are. Let's explore what a requirement looks like.

Features and technical requirements

As we saw in the previous section, requirements are usually strictly related to the **features** of the system. Depending on who is posing the request, requirements can be specified with varying amounts of technical detail. A requirement may be as low-level as the definition of an API or other software interfaces, including arguments and quantitative input validation/outcome. Here is an example of what a detailed, technically specified requirement may look like:

When entering the account number (string, six characters), the system must return the profile information. Result code as int *(0 if operation is successful), name as* string, *and surname as* string *[...]. In the case of account in an invalid format, the system must return a result code identifying the reason of the fault, as per a mapping table to be defined.*

Often requirements are less technical, identifying more behavioral aspects of the system. In this case, drawing on the model we discussed in the previous section (*Feature, Advantage, and Benefit*), we are talking about something such as a feature or the related advantage.

An example here, for the same functionality as before, may look like this:

The user must have the possibility to access their profile, by entering the account number.

It's easy to understand that a non-technical requirement must be detailed in a quantitative and objective way before being handed over to development teams. *But what makes a requirement quantitative and objective?*

Types and characteristics of requirements

There are a number of characteristics that make a requirement effective, meaning easy to understand and respondent to the customer expectations in a non-ambiguous way.

From my personal point of view, in order to be effective, a requirement must be the following:

- **Consistent**: The requirement must not conflict with other requirements or existing functionalities unless this is intentional. If it is intentional (for example, we are removing old functionalities or fixing wrong behaviors), the new requirement must explicitly override older requirements, and it's probably an attention point since corner cases and conflicts are likely to happen.

- **Implementable**: This means, first of all, that the requirement should be feasible. If our system requires a direct brain interface to be implemented, this of course will not work (at least today). Implementable further means that the requirement must be achievable in the right amount of time and at the right cost. If it needs 100 years to be implemented, it's in theory feasible but probably impractical.

  Moreover, these points need to be considered within the context of the current project, since although it may be easy to implement something in one environment it may not be feasible in another. For example, if we were a start-up, we could probably launch a brand-new service on our app that would have little impact on the existing userbase. If we were a big enterprise, however, with a large customer base and consolidated access patterns, this may need to be evaluated more thoroughly.

- **Explicit**: There should be no room for interpretation in a software requirement. Ambiguity is likely to happen when the requirement is defined in natural language, given that a lot of unspoken data is taken erroneously for granted. For this reason, it is advised to use tables, flowcharts, interface mockups, or whatever schema can help clarify the natural language and avoid ambiguity. Also, straightforward wording, using defined quantities, imperative verbs, and no metaphors, is strongly advised.

- **Testable**: In the current development philosophies, heavily focused on experimentation and trial and error (we will see more on this in the upcoming chapters), a requirement must be translated in a software test case, even better if it can be fully automated. While it may be expected that the customer doesn't have any knowledge of software testing techniques, it must be possible to put testing scenarios on paper, including things such as tables of the expected outputs over a significant range of inputs.

  The QA department may, at a later stage, complement this specification with a wider range of cases, in order to test things such as input validation, expected failures (for example, in the case of inputs too large or malformed), and error handling. Security departments may dig into this too, by testing malicious inputs (for example, SQL injections).

This very last point leads us to think about the technical consequences of a requirement. As we were saying at the beginning of this chapter, requirements are commonly exposed as business features of the system (with a technical standpoint that can vary in the level of detail).

However, there are *implicit* requirements, which are not part of a specific business use case but are essential for the system to work properly.

To dig deeper into this concept, we must categorize the requirements into three fundamental types:

- **Functional requirements**: Describing the business features of the system, in terms of expected behavior and use cases to be covered. These are the usual business requirements impacting the use cases provided by the system to be implemented.

- **Non-functional requirements**: Usually not linked to any specific use case, these requirements are necessary for the system to work properly. Non-functional requirements are not usually expressed by the same users defining functional requirements. Those are usually about implicit aspects of the application, necessary to make things work. Examples of non-functional requirements include performance, security, and portability.

- **Constraints**: Implicit requirements are usually considered a *must* and are mandatory. These include external factors and things that need to be taken for granted, such as obeying laws and regulations and complying with standards (both internal and external to the company).

One example here could be the well-known **General Data Protection Regulation** (**GDPR**), the EU law about data protection and privacy, which you have to comply with if you operate in Europe. But you may also have to comply with the industry standards depending on the particular market in which you are operating (that's pretty common when working with banks and payments), or even standards enforced by the company you are working with. A common example here is the compatibility of the software (such as when it has to be compatible with a certain version of an operating system or a particular browser).

Now that we've seen the different types of requirements and their characteristics, let's have a look at the life cycle of software requirements.

The life cycle of a requirement

The specification of a requirement is usually not immediate. It starts with an idea of how the system should work to satisfy a use case, but it needs reworking and detailing in order to be documented. It must be checked against (or mixed with) non-functional requirements, and of course, may change as the project goes on. In other words, the life cycle of requirements can be summarized as follows. Each phase has an output, which is the input for the following one (the path could be non linear, as we will see):

- **Gathering**: Collection of use cases and desired system features, in an unstructured and raw format. This is done in various ways, including interviews, collective sketches, and brainstorming meetings, including both the customer and the internal team. **Event Storming** (which we will see soon) is a common structured way to conduct brainstorming meetings, but less structured techniques are commonly used here, such as using sticky notes to post ideas coming from both customers and internal teams. In this phase, the collection of data usually flows freely without too much elaboration, and people focus more on the creative process and less on the details and impact of the new features. The output for this phase is an unstructured list of requirements, which may be collected in an electronic form (a spreadsheet or text document), or even just a photograph of a wall with sticky notes.

- **Vetting**: As a natural follow-up, in this phase the requirements output from the previous phase is roughly analyzed and categorized. Contradicting and unfeasible topics must be addressed. It's not unusual to go back and forth between this phase and the previous one. The output here is still an unstructured list, similar to the one we got from the previous step. But we started to polish it, by removing duplicates, identifying the requirements that need more details, and so on.

- **Analysis**: In this phase, it's time to conduct a deeper analysis of the output from the previous phase. This includes identifying the impact of the implementation of every new feature, analyzing the completeness of the requirement (desired behavior on a significant list of inputs, corner cases, and validation), and the prioritization of the requirement. While not necessary, it is not unusual in this case to have a rough idea of the implementation costs of each requirement. The output from this step is a far more stable and polished list, basically a subset of the input we got. But we are still talking about the unstructured data (not having an ID or missing some details, for example), which is what we are going to address in the next phase.

- **Specification**: Given that we've completed the study of each requirement, it's now time to document it properly, capturing all the aspects explored so far. We may already have drafts and other data collected during the previous phases (for example, schemas on paper, whiteboard pictures, and so on) that just need to be transcribed and polished. The documentation redacted in this phase has to be accessible and updatable throughout the project. This is essential for tracking purposes. As an output of this phase, you will have each requirement checked and registered in a proper way, in a document or by using a tool. There are more details on this in the *Collecting requirements – formats and tools* section of this chapter.

- **Validation**: Since we got the formal documentation of each requirement as an output of the previous phase, it is a best practice to double-check with the customer whether the final rework covers their needs. It is not unusual for, after seeing the requirements on paper, a step back to the gathering phase to have to be made in order to refocus on some use cases or explore new scenarios that have been uncovered during the previous phases. The output of this phase has the same format as the output of the previous phase, but you can expect some changes in the content (such as priorities or adding/removing details and contents). Even if some rework is expected, this data can be considered as a good starting point for the development phase.

So, the requirement life cycle can be seen as a simple workflow. Some steps directly lead to the next, while sometimes you can loop around phases and step backward. Graphically, it may look like the following diagram:

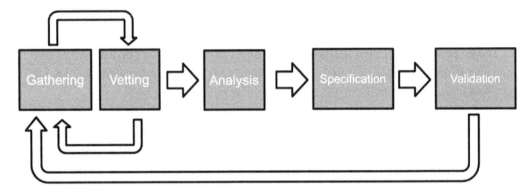

Figure 2.1 – Software requirements life cycle

As you can see in the previous diagram, software requirements specification is often more than a simple phase of the software life cycle. Indeed, since requirements shape the software itself, they may follow a workflow on their own, evolving and going through iterations.

As per the first step of this flow, let's have a look at requirements gathering.

Discovering and collecting requirements

The first step in the requirements life cycle is **gathering**. Elicitation is an implicit part of that. Before starting to vet, analyze, and ultimately document the requirements, you need to start the conversation and start ideas flowing.

To achieve this, you need to have the right people in the room. It may seem trivial, but often it is not clear who the source of requirements should be (for example, the business, a vague set of people including sales, executive management, project sponsors, and so on). *Even if you manage to have those people onboard, who else is relevant for requirement collection?*

There is no golden rule here, as it heavily depends on the project environment and team composition:

- You will need for sure some **senior technical resources**, usually lead architects. These people will help by giving initial high-level guidance on technical feasibility and ballpark effort estimations.

- Other useful participants are **enterprise architects** (or business architects), who could be able to evaluate the impact of the solution on the rest of the enterprise processes and technical architectures. These kinds of profiles are of course more useful in big and complex enterprises and can be less useful in other contexts (such as start-ups). As a further consideration, experienced people with this kind of background can suggest well-known solutions to problems, compared with similar applications already in use (or even reusing functionalities where possible).

- **Quality engineers** can be a good addition to the team. While they may be less experienced in technical solutions and existing applications, they can think about the suggested requirements in terms of test cases, narrowing them down and making them more specific, measurable, and testable.

- **Security specialists** can be very helpful. Thinking about security concerns early in the software life cycle can help to avoid surprises later on. While not exhaustive, a quick assessment of the security impacts of proposed requirements can be very useful, increasing the software quality and reducing the need to rework.

Now that we have all the required people in a room, let's look at a couple of exercises to break the ice and keep ideas flowing to nail down our requirements.

The first practice we will look at is the **lean canvas**. This exercise is widely used in the start-up movement, and it focuses on bringing the team together to identify what's important in your idea, and how it will stand out from the competition.

The lean canvas

The **lean canvas** is a kind of holistic approach to requirements, focusing on the product's key aspects, and the overall business context and sustainability.

Originating as a tool for start-ups, this methodology was developed by Ash Maurya (book author, entrepreneur, and CEO at LEANSTACK) as an evolution/simplification of the Business Model Canvas, which is a similar approach created by Alexander Osterwalder and more oriented to the business model behind the product. This method is based on a one-page template to gather solution requirements out of a business idea.

The template is made of nine segments, highlighting nine crucial aspects that the final product must have:

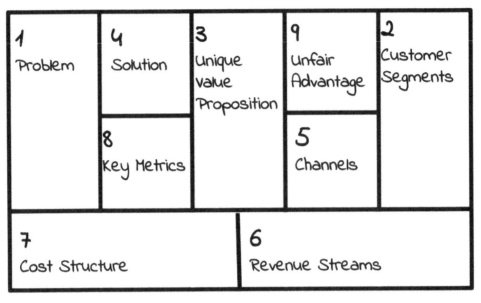

Figure 2.2 – The lean canvas scaffold

Note that the numbering of each segment reflects the order in which the sections should be filled out. Here is what each segment means:

1. **Problem**: *What issues will our customers solve by using our software product?*

2. **Customer Segments**: *Who is the ideal person to use our software product (that is, the person who has the problems that our product will solve)?*

3. **Unique Value Proposition**: *Why is our software product different from other potential alternatives solving similar problems?*

4. **Solution**: *How will our software product solve the problems in section 1?*

5. **Channels**: *How will we reach our target customer?* (This is strictly related to how we will market our software solution.)

6. **Revenue Streams**: *How we will make money out of our software solution?*

7. **Cost Structure**: *How much will it cost to build, advertise, and maintain our software solution?*

8. **Key Metrics**: *What are the key numbers that need to be used to monitor the health of the project?*

9. **Unfair Advantage**: *What's something that this project has that no one else can copy/buy?*

The idea is to fill each of these areas with one or more propositions about the product's characteristics. This is usually done as a team effort in an informal setting. The canvas is pictured on a whiteboard, and each participant (usually product owners, founders, and tech leads) contributes ideas by sticking Post-it notes in the relevant segments. A *postprocess* collective phase usually follows, grouping similar ideas, ditching the less relevant ideas, and prioritizing what's left in each segment.

As you can see, the focus here is shifted toward the feasibility of the overall project, instead of the detailed list of features and the specification. For this reason, this methodology is often used as a support for doing elevator pitches to potential investors. After this first phase, if the project looks promising and sustainable from the business model point of view, other techniques may be used to create more detailed requirement specifications, including the ones already discussed, and more that we will see in the next sections.

While the lean canvas is more oriented to the business model and how this maps into software features, in the next section we will explore Event Storming, which is a discovery practice usually more focused on the technical modeling of the solution.

Event Storming

Event Storming is an agile and interactive way to discover and design business processes and domains. It was described by Alberto Brandolini (IT consultant and founder of the Italian Domain Driven Design community) in a now-famous blog post, and since then has been widely used and perfected.

The nice thing about this practice is that it is very friendly and nicely supports brainstorming and cross-team collaboration.

To run an Event Storming session, you have to collect the right people from across various departments. It usually takes at least business and IT, but you can give various different flavors to this kind of workshop, inviting different profiles (for example, security, UX, testers) to focus on different points of view.

When you have the right mix of people in the room, you can use a tool to help them interact with each other. When using physical rooms (the workshop can also be run remotely), the best tool is a wall plus sticky notes.

The aim of the exercise is to design a business process from the user's point of view. *So how do you do that?*

1. You start describing **domain events** related to the user experience (for example, a recipient is selected). Those domain events are transcribed on a sticky note, traditionally orange, and posted to the wall respecting the temporal sequence.

2. You then focus on what has caused the domain event. If the cause is a user interaction (for example, the user picks a recipient from a list), it's known as a **command** and tracked as a blue sticky note, posted close to the related event.

3. You may then draft the **user** behind the command (for example, a customer of the bank). This means drafting a persona description of the user carrying out the command, tracking it on a yellow sticky note posted close to the command.

4. If domain events are generated from other domain events (for example, the selected recipient is added to the recently used contacts), they are simply posted close to each other.

5. If there are interactions with **external systems** (for example, the recipient is sent to a CRM system for identification), they are tracked as pink sticky notes and posted near to the related domain event.

Let's have a look at a simple example of Event Storming. The following is just a piece of a bigger use case; this subset concisely represents the access of a user to its transactions list. The use case is not relevant here, it's just an example to show the main components of this technique:

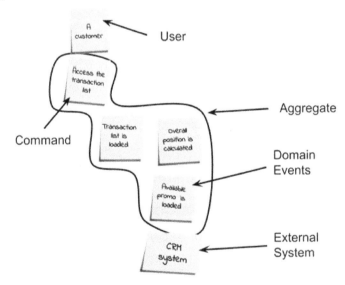

Figure 2.3 – The Event Storming components

In the diagram, you can see a small but complete subset of an Event Storming session, including stickies representing the different components (**User**, **Command**, and **Domain Events**) and the grouping representing the aggregates.

What do you achieve from this kind of representation?

- A shared understanding of the overall process.

- A clustering of events and commands, identifying the so-called **aggregates**. This concept is very important for the modeling of the solution, and we will come back to this in *Chapter 4*, *Best Practices for Design and Development*, when talking about Domain-Driven Design.

- The visual identification of bottlenecks and unclear links between states of the system.

It's important to note that this methodology is usually seen as a scaffold. You may want to customize it to fit your needs, tracking different entities, sketching simple user interfaces to define commands, and so on. Moreover, these kinds of sessions are usually iterative. Once you've reached a complete view, you can restart the session with a different audience to further enrich or polish this view, to focus on subdomains and so on.

In the following section, we will explore some alternative discovery practices.

More discovery practices

Requirements gathering and documentation is somewhat of a composite practice. You may find that after brainstorming sessions (for example, a lean canvas, Event Storming, or other comparable practices), other requirement engineering techniques may be needed to complete the vision and explore some scenarios that surfaced during the other sessions. Let's quickly explore these other tools so you can add them to your toolbox.

Questionnaires

Questions and answers are a very simple and concise way of capturing fixed points about a software project. If you are capable of compiling a comprehensive set of questions, you can present your questionnaire to the different stakeholders to collect answers and compare the different points of view.

The hard part is building such a list of questions. You may have some ideas from previous projects, but given that questions and answers are quite a *closed-path* exercise, it isn't particularly helpful if you are at the very beginning of the project. Indeed, it is not the best method to use if you are starting from a blank page, as it's not targeted at nurturing creative solutions and ideas. For this reason, I would suggest proceeding with this approach mostly to detail ideas and use cases that surfaced in other ways (for example, after running brainstorming sessions).

Mockups and proofs of concepts

An excellent way to clarify ideas is to directly test what the product will look like by playing with a subset of functionalities (even if fake or just stubbed). If you can start to build cheap prototypes, or even just mockups (fake interfaces with no real functionalities behind the scenes), you may be able to get non-technical stakeholders and final users on board sooner, as you give them the opportunity to interact with the product instead of having to imagine it.

This is particularly useful in UX design, and for showcasing different solutions. Moreover, in modern development, this technique can be evolved toward a shorter feedback loop (release early, release often), having the stakeholders test alpha releases of the product instead of mockups so they can gain an understanding of what the final product will look like and change the direction as soon as possible.

A/B testing

A further use for this concept is to have the final users test by themselves and drive the product evolution. This technique, known as **A/B testing**, is used in production by high-performing organizations and requires some technological support to be implemented. The principle is quite simple: you pick two (or more) alternative features, put them into production, and measure how they perform. In an evolutionary design, the best performing will survive, while the others will be discarded.

As you can imagine, the devil is in the details here. Implementing more alternatives and discarding some of them may be expensive, so often there are just minor differences between them (for example, the color or position of elements in the UI). Also, the performance must be measurable in an objective way, for example, in e-commerce you might measure the impact on purchases, or in advertising the conversions of banners and campaigns.

Business intelligence

Another tool to complete and flesh out the requirement list is **business intelligence**. This might mean sending surveys to potential customers, exploring competitor functionalities, and doing general market research. You should not expect to get a precise list of features and use cases by using only this technique, but it may be useful for completing your view about the project or coming up with some new ideas.

You may want to check whether your idea for the finished system resonates with final customers, how your system compares with competitors, or whether there are areas in which you could do better/be different. This tool may be used to validate your idea or gather some last pieces to complete the picture. Looking at *Figure 2.1*, this is something you may want to do during the validation phase.

Now, we have collected a wide set of requirements and points of view. Following the requirements life cycle that we saw at the beginning of this chapter, it is now time for requirements analysis.

Analyzing requirements

The discovery practices that we've seen so far mostly cover the gathering and vetting of requirements. We've basically elicited from the stakeholders details of the desired software functionalities and possibly started organizing them by clustering, removing duplicates, and resolving macroscopic conflicts.

In the analysis phase, we are going to further explore the implications of the requirements and complete our vision of what the finished product should look like. Take into account that product development is a fluid process, especially if you are using modern project management techniques (more on that in *Chapter 5*, *Exploring the Most Common Development Models*). For this reason, you should consider that most probably not every requirement defined will be implemented, and certainly not everything will be implemented in the same release – you could say we are shooting at a moving target. Moreover, it is highly likely that more requirements will be developed later on.

For this reason, requirements analysis will probably be performed each time, in an iterative approach. Let's start with the first aspect you should consider when analyzing the requirements.

Checking for coherence and feasibility

In the first section, we clearly stated that a requirement must be consistent and implementable. That is what we should look for in the analysis phase.

There is no specific approach for this. It's a kind of qualitative activity, going through requirements one by one and cross-checking them to ensure they are not conflicting with each other. With big and complex requirement sets, this activity may be seen as a *first pass*, as no explicit conflict may arise later during design and implementation. Similar considerations may be made with regard to feasibility. In this phase, it's important to catch the big issues and identify the requirements that seem to be unfeasible, however, more issues can arise during later phases.

If incoherent or unfeasible requirements are spotted, it's crucial to review them with the relevant stakeholders (usually business), in order to reconsider the related features, and make changes. From time to time, small changes to the requirement can make it feasible. A classic scenario is related to picking a subset of the data or making similar compromises. In our mobile payments example, it may not be feasible to show instantaneously the whole list of transactions updated in real time, however, it could be a good compromise to show just a subset of them (for example, last year) or have a small visualization delay (for example, a few seconds) when new transactions occur.

Checking for explicitness and testability

Continuing with requirements characteristics, it is now time to check the explicitness and testability of each requirement. This may be a little more systematic and quantitative compared to the previous section. Essentially, you should run through the requirements one by one and check whether each requirement is expressed in a defined way, making it easy to understand whether the implementation has been completed correctly. In other words, the requirement must be testable and it is best if it is testable in an objective and automatable way.

Testing for explicitness brings with it the concept of completeness. Once a requirement (and the related feature) is accepted, all the different paths must be covered in order to provide the product with predictable behavior in most foreseeable situations. While this may seem hard and complex, in most situations it's enough to play with possible input ranges and conditional branches to make sure all the possible paths are covered. Default cases are another important aspect to consider; if the software doesn't know how to react to particular conditions it's a good idea to define reasonable, standard answers to fall into.

Checking non-functional requirements and constraints

As the last step, it's important to run through the requirements list, looking for non-functional requirements and constraints. The topic here is broad and subjective. It's likely not possible (nor useful) to explicate all the non-functional requirements and constraints and put them on our list. Most of them are shared with existing projects, regulated by external parties, or simply not known.

However, there are areas that have an important impact on the project implementation, and for this reason, must be considered in the analysis phase.

One usual suspect here is security. All the considerations about user sessions, what to do with unauthenticated users, and how to manage user logins and such have implications for the feasibility and complexity of the solution, other than having an impact on the user experience. Analog reasoning can be made for performance. As seen in the *Checking for coherence and feasibility* section, small changes in the amount of data and the expected performances of the system may make all the difference. It's not unusual to have non-technical staff neglecting these aspects or expecting unreasonable targets. Agreeing (and negotiating) on the expected result is a good way to prevent issues later in the project.

Other considerations of non-functional requirements and constraints may be particularly relevant in specific use cases. Take into account that this kind of reasoning may also be carried over into the project planning phase, in which constraints in budget or timeframe may drive the roadmap and release plan.

Now, we've gone through the analysis phase in the software requirements life cycle. As expected, we will now approach the specification phase. We will start with a very formal and structured approach and then look at a less structured alternative.

Specifying requirements according to the IEEE standard

The **Institute of Electrical and Electronics Engineers (IEEE)** has driven various efforts in the field of software requirements standardization. As usual, in this kind of industry standard, the documents are pretty complete and extensive, covering a lot of aspects in a very verbose way.

The usage of those standards may be necessary for specific projects in particular environments (for example, the public sector, aviation, medicine). The most famous deliverable by IEEE in this sense is the 830-1998 standard. This standard has been superseded by the ISO/IEEE/IEC 29148 document family.

In this section, we are going to cover both standards, looking at what the documents describe in terms of content, templates, and best practices to define requirements adhering to the standard.

The 830-1998 standard

The **IEEE 830-1998** standard focuses on the **Software Requirement Specification** document (also known as **SRS**), providing templates and suggestions on content to be covered.

Some concepts are pretty similar to the ones discussed in the previous sections. The standard states all the characteristics that a requirement specification must have. Each requirement specification should be the following:

- Correct

- Unambiguous

- Complete

- Consistent

- Ranked for importance and/or stability

- Verifiable

- Modifiable

- Traceable

As you can see, this is similar to the characteristics of requirements. One interesting new concept added here is the ranking of requirements. In particular, the document suggests classifying the requirements by importance, assigning priorities to requirements, such as essential, conditional, optional, and/or stability (stability refers to the number of expected changes to the requirement due to the evolution of the surrounding organization).

Another interesting concept discussed in this standard is **prototyping**. I would say that this is positively futuristic, considering that this standard was defined in 1998. Well before the possibility to cheaply create stubs and mocks, as is normal today, this standard suggests using prototypes to experiment with the possible outcome of the system and use it as a support for requirements gathering and definition.

The last important point I want to highlight about IEEE 830-1998 is the template. The standard provides a couple of samples and a suggested index for software requirements specifications. The agenda includes the following:

- **Introduction**: Covering the overview of the system, and other concepts to set the field, such as the scope of the document, purpose of the project, list of acronyms, and so on.

- **Overall description**: Describing the background and the constructs supporting the requirements. Here, you may define the constraints (including technical constraints), the interfaces to external systems, the intended users of the system (for example, the skill level), and the product functions (intended to give an overview of the product scope, without the details that map to specific requirements).

- **Specific requirements**: This refers to the requirements themselves. Here, everything is expected to be specified with a high amount of detail, focusing on inputs (including validation), expected outputs, internal calculations, and algorithms. The standard offers a lot of suggestions for topics that need to be covered, including database design, object design (as in object-oriented programming), security, and so on.

- **Supporting information**: Containing accessory information such as a table of contents, index, and appendixes.

As you can see, this SRS document may appear a little verbose, but it's a comprehensive and detailed way to express software requirements. As we will see in the next section, IEEE and other organizations have superseded this standard, broadening the scope and including more topics to be covered.

The 29148 standard

As discussed in the previous sections, the 830-1998 standard was superseded by a broader document. The 29148 family of standards represents a superset of 830-1998. The new standard is rich and articulated. It mentions the SRS document, following exactly the same agenda but adding a new section called **verification**. This section refers to specifying a testing strategy for each element of the software, suggesting that you should define a verification for each element specified in the other sections of the SRS.

Other than the SRS document, the 29148 standard suggests four more deliverables. Let's have a quick look at them:

- The **Stakeholder Requirements Specification**: This places the software project into the business perspective, analyzing the business environment around it and the impact it will have by focusing on the point of view of the business stakeholders.

- The **System Requirements Specification**: This focuses on the technical details of the interactions between the software being implemented and the other system composing the overall architecture. It specifies the domain of the application and the inputs/outputs.

- **System Operational Concept**: This describes, from the user's point of view, the system's functionality. It takes a point of view on the operation of the system, policies, and constraints (including supported hardware, software, and performance), user classes (meaning the different kinds of users and how they interact with the system), and operational modes.

- **Concepts of Operations**: This is not a mandatory document. When provided, it addresses the system as a whole and how it fits the overall business strategy of the customer. It includes things such as the investment plan, business continuity, and compliance.

As we have seen, the standards documents are a very polished and complete way to rationalize the requirements and document them in a comprehensive way. However, sometimes it may be unpractical to document the requirements in a such detailed and formalized way. Nevertheless, it's important to take these contents as a reference, and consider providing the same information, even if not using the very same template or level of details.

In the next section, we will have a look at alternative simplified formats for requirements collection and the tools for managing them.

Collecting requirements – formats and tools

In order to manage and document requirements, you can use a tool of your choice. Indeed, many teams use electronic documents to detail requirements and track their progression, that is, in which stage of the requirement life cycle they are. However, when requirements grow in complexity, and the size of the team grows, you may want to start using more tailored tools.

Let's start by having a look at the required data, then we will focus on associated tooling.

Software requirements data to collect

Regardless of the tool of your choice, there is a subset of information you may want to collect:

- **ID**: A unique identifier will be needed since the requirement will be cross-referenced in many different contexts, such as test cases, documentation, and code comments. It can follow a naming convention or simply be an incremental number.

- **Description**: A verbal explanation of the use case to be implemented.

- **Precondition**: (If relevant) the situation that the use case originates from.

- **Essential**: How essential the requirement is, usually classified as *must have, should have, or nice to have*. This may be useful in order to filter requirements to be included in a release.

- **Priority**: A way to order/cluster requirements. Also, a useful way to filter requirements to be included in a release.

- **Source**: The author of the requirement. It may be a department, but it is better if there is also a named owner to contact in case of clarifications being needed.

- **Group**: A way to cluster requirements for functional areas. Also, can be a useful way to collect a set of requirements to implement in a release.

- **Parent**: This is optional, in case you want to implement a hierarchy with a complex/high-level requirement made of a set of sub-requirements.

These are the basic attributes to collect for each software requirement, to enrich with any further column that may be relevant in your context.

You may then want to track the implementation of each requirement. The attributes to do so usually include the following:

- **Status**: A synthetic description of the implementation status, including states such as UNASSIGNED, ASSIGNED, DEVELOPMENT, TESTING, and COMPLETE.

- **Owner**: The team member to whom this requirement is assigned. It may be a developer, a quality engineer, or someone else, depending on the status.

- **Target release**: The software release that is targeted to include this requirement.

- **Blocker**: Whether this requirement is mandatory for this release or not.

- **Depends on**: Whether this requirement depends on other requirements to be completed (and what they are) before it can be worked on.

Also, in this case, this is a common subset of information useful for tracking the requirement status. It may be changed, depending on the tooling and the project management techniques used in your particular context. Let's now have a look at tools to collect and manage this information.

Collecting software requirements in spreadsheets

Looking at the list of attributes described in the previous section, you can imagine that these requirements can be easily collected in spreadsheets. It's a tabular format, with one requirement per row, and columns corresponding to the information we've discussed. Also, you could have the status tracking in the same row or associated by ID in a different sheet. Moreover, you can filter the sheet by attribute (for example, priority, group, status), sort it, and limit/validate the inputs where relevant (for example, restricting values from a specified list). Accessory values may also be added (for example, last modified date).

This is what a requirements spreadsheet might look like:

ID	DESCRIPTION	PRECONDITION	ESSENTIAL	PRIORITY	SOURCE	GROUP	PARENT
1	The application must be awesome	The application was not awesome	Must Have	High	owner@awesome.xyz	Core requirements	-
2	Jedis can log in using the force	The Jedi user is not logged	Should Have	Medium	luke@awesome.xyz	Security	1
3	The user interface must be localized in Wookiee	The user is identified as a Wookie	Nice to Have	Low	chewbe@awesome.xyz	Localization	1

Requirements ▾   Status ▾

Figure 2.4 – A requirements spreadsheet

As mentioned, we can then have a sheet for tracking the progression of each requirement. It may look like the example that follows:

REQ_ID	STATUS	OWNER	TARGET RELEASE	BLOCKER	DEPENDS ON	LAST MODIFIED
1	QE	Obi	0.1	Yes	-	09/07/2020
2	UNASSIGNED	Ian	0.1.1	No	1	14/07/2020
3	DEVELOPMEN	Mace	0.2	No	1	12/07/2020

Requirements ▾   Status ▾

Figure 2.5 – Status tracking sheet

In the next sections, we will have a look at tools that can be used to support requirements gathering and documentation.

Specialized tools for software requirements management

As mentioned in the previous section, with bigger teams and long-term projects, specialized tools for requirements management can be easier to use than a shared document/spreadsheet.

The most useful feature is usually having a centralized repo, avoiding back and forth (and a lack of synchronization), which happens when using documents. Other interesting features to look for are auditing (tracking changes), notifications, reporting, and advanced validation/guided input. Also, integration with source code management (for example, associating features with commits and branches) is pretty common and useful.

The software for requirements management is usually part of a bigger suite of utilities for project management. Here are some common products:

- **Jira** is a pretty widespread project management toolkit. It originated as an issue tracking tool to track defects in software products. It's commonly used for tracking features too. It may also be extended with plugins enriching the functionalities of feature collection, organizing, and reporting.

- **Redmine** is an open source tool and includes many different project management capabilities. The most interesting thing about it is its customizability, enabling you to track features, associate custom fields, reference source code management tools (for example, Git), and define Gantt charts/calendars.

- **IBM Rational DOORS** is commercial software for requirements management, very complete and oriented to mid-large enterprises. It is part of the Rational suite, originally developed by Rational Software (now part of IBM), which is also famous for contributing to the creation of UML notation, which we will discuss in the next chapter.

The selection of a requirements management tool is a complex process, involving cost analysis, feature comparison, and more, which is way beyond the goal of this book.

Spreadsheets versus tools

It is a common debate whether to use specialized tools versus spreadsheets (or documents) for managing lists of requirements. It is a common path to start using a simpler approach (such as spreadsheets) and move to a tool once the project becomes too big or too complex to manage this way. Moreover, managers and non-technical users are more willing to use spreadsheets because they are more comfortable with such technology. Conversely, tech teams find it is often more effective to work with specialized tools. As usual, there is no one size that fits all, but honestly, the benefits of using a dedicated tool are many.

The most immediate is having a centralized repository. Tools for requirement management are made to be used in real time, acting as a central, single source of truth. This allows us to avoid back and forth (and lack of synchronization), which happens when using documents (while you could object here that many Office suites offer real-time sharing and collaborative editing, nowadays).

Other interesting features included with a specialized tool are auditing (tracking changes), notifications, reporting, and advanced validation/guided input.

Also, the integration with the source code management (for example, associating features with commits and branches) is pretty common and appreciated by the development teams. Management can also benefit from planning and insight features, such as charts, aggregated views, and integration with other project management tools.

So, at the end of the day, I strongly advise adopting a full-fledged requirements management tool instead of a simple spreadsheet if that is possible.

In the next section, we will explore requirements validation, as a final step in the software requirements life cycle.

Validating requirements

As we've seen, the final phase of the requirements life cycle involves validating the requirements. In this phase, all the produced documentation is expected to be reviewed and formally agreed by all the stakeholders.

While sometimes neglected and considered optional, this phase is in fact very important. By having a formal agreement, you will ensure that all the iterations on the requirements list, including double-checking and extending partial requirements, still reflect the original intentions of the project.

The business makes sure that all the advantages and benefits will be achieved, while the technical staff will check that the features are correctly mapped in a set of implementable requirements so that the development team will clearly understand what's expected.

This *sign-off* phase could be considered the point at which the project first truly kicks off. At this point, we have a clearer idea of what is going to be implemented. This is not the final word, however; when designing the platform and starting the project plans, you can expect the product to be remodeled. Maybe just a set of features will be implemented, while other functionalities will be put on paper later.

In this section, we took a journey through the requirements life cycle. As already said, most of these phases can be considered iterative, and more than one loop will be needed before completing the process. Let's have a quick recap of the requirements life cycle and the practices we have seen so far:

- **Gathering and vetting**: As we have seen, these two phases are strictly related and involve a cross-team effort to creatively express ideas and define how the final product should look. Here, we have seen techniques for brainstorming such as the lean canvas, Event Storming, and more.

- **Analysis**: This phase includes checking the coherence, testability, and so on.

- **Specification**: This includes the IEEE standard and some less formalized standards and tools.

- **Validation**: This is the formal sign-off and acceptance of a set of requirements. As said, it's not unusual to see a further rework of such a set by going back to the previous phases, in an iterative way.

In the next section, we will continue to look at our mobile payments example, focusing on the requirements analysis phase.

Case studies and examples

Continuing with the case study about our mobile payments solution, we are going to look at the requirements gathering phase. For the sake of simplicity, we will focus only on a small specific scenario: a peer-to-peer payment between two users of the platform.

The mobile payment application example

As we are doing in every chapter, let's have a look at some examples of the concepts discussed in this chapter applied to the mobile payment application that we are using as a case study.

Requirements life cycle

In the real world, the life cycle of requirements will reasonably take weeks (or months), adding up to a lot of requirements and reworking of them, so it is impractical to build a complete example of the requirements life cycle for our mobile payment scenario. However, I think it will be interesting to have a look at how one particular requirement will evolve over the phases we have seen:

1. In the **gathering** phase, it is likely we will end up with a lot of ideas around ease of use and security for each payment transaction. Most of the participants will start to think from an end user perspective, focusing on the user experience, and so it's likely we will have sketches and mockups of the application. Some more ideas will revolve around how to authorize the payment itself along with its options (*how about a secret swipe sequence, a PIN code, a face ID, a* **One-Time Password** (**OTP**)*, or a fingerprint?*).

2. In the **vetting** phase (likely during, or shortly after, the previous phase), we will cluster and clean up what we have collected. The unpractical ideas will be dropped (such as the OTP, which may be cumbersome to implement), while others will be grouped (face ID and fingerprint) under biometric authorization. More concepts will be further explored and detailed: *What does it mean to be fast and easy to use? How many steps should be done to complete the payment? Is entering a PIN code easy enough (in cases where we cannot use biometric authorization)?*

3. It's now time to **analyze** each requirement collected so far. In our case, maybe the payment authorization. It is likely that the user will be presented with a screen asking for biometric authentication. *But what happens if the device doesn't have a supported hardware? Should the customer be asked for other options, such as a PIN code? What should happen if the transaction is not authorized?* And of course, this kind of reasoning may go further and link more than one requirement: *What if a network is not available? What should happen after the transaction is completed successfully?* Maybe the information we have at that moment (where the customer is, what they have bought, the balance of their account) allows for some interesting use cases, such as contextual advertising, offering discounts, and so on.

4. Now that we have clarified our requirements (and discovered new ones), it's time for **specification**. Once we pick a format (IEEE, or something simpler, such as a specialized tool or a spreadsheet), we start inserting our requirements one by one. Now, it's time to go for the maximum level of details. Let's think about bad paths (*what happens when things go wrong?*), corner cases, alternative solutions, and so on.

5. The last phase is the **validation** of what we have collected into our tool of choice. It is likely that only a subset of the team has done the analysis and specification, so it's good to share the result of those phases with everyone (especially with non-technical staff and the project sponsors) to understand whether there is anything missing: maybe the assumptions we have made are not what they were expecting. It's not uncommon that having a look at the full list will trigger discussions about prioritization or brand-new ideas (such as the one about contextual advertising that we mentioned in the analysis phase).

In the next sections, we will see some more examples of the specific phases and techniques.

Lean canvas for the mobile payment application

The lean canvas can be imagined as an elevator pitch for getting sponsorship for this application (such as for getting funds or approval for the development). In this regard, the lean canvas is a kind of conversation starter when it comes to requirements. It could be good to identify and detail the main, most important features, but you will probably need to use other techniques (such as the ones described so far) to identify and track all the requirements with a reasonable level of detail.

With that said, here is how I imagine a lean canvas could look in this particular case. Of course, I am aware that other mobile and contactless solutions exist, so consider this just as an example. For readability purposes, I'm going to represent it as a bullet list. This is a transcribed version, as it happens after collecting all those aspects as sticky notes on a whiteboard:

- **Problem**: The payment procedure is cumbersome and requires cash or card. Payment with card requires a PIN code or a signature. The existing alternatives are credit or debit cards.

- **Customer segment**: Everybody with a not-too-old mobile phone. The early adopters could be people that don't own a credit card or don't have one to hand (maybe runners, who don't bring a wallet but only a mobile phone, or office workers during their lunch/coffee break).

- **Unique value proposition**: Pay with one touch, safely.

- **Solution**: A sleek, fast, and easy-to-use mobile application, allowing users to authorize payment transactions with biometric authentication.

- **Unfair advantage**: Credit/debit cards that don't need biometric authentication. (Of course I am aware, as I said, that contactless payments are available with credit cards, and other NFC options are bundled with mobile phones. So, in the real world, our application doesn't really have an advantage over other existing options.)

- **Revenue streams**: Transaction fees and profiling data over customer spending habits.

- **Cost structure**: App development, hosting, advertising. (In the real world, you may want to have a ballpark figure for it and even have a hypothesis of how many customers/transactions you will need to break even. This will put you in a better position for pitching the project to investors and sponsors.)

- **Key metrics**: Number of active users, transactions per day, average amount per transaction.

- **Channels**: Search engine optimization, affiliation programs, cashback programs.

In the next section, we'll look at Event Storming for peer-to-peer payments.

Event Storming for peer-to-peer payments

As we saw in the *Event Storming* section, in an Event Storming session it's important to have a variety of representations from different departments in order to have meaningful discussions. In this case, let's suppose we have business analysts, chief architects, site reliability engineers, and UX designers. This is what our wall may look like after our brainstorming session:

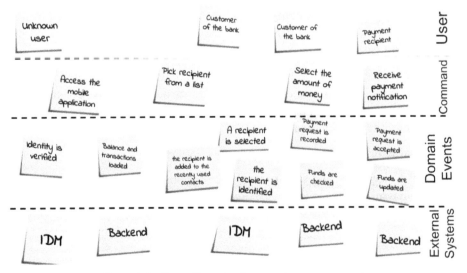

Figure 2.6 – Event Storming for peer-to-peer payment

As you can see from the preceding diagram, even in this simplified example we begin to develop a clear picture of the people involved in this use case and the external systems.

We can see that two systems are identified, **Identity Management** (**IDM**) for dealing with customer profiles and **Backend** for dealing with balances and transactions.

In terms of command and domain events, this is something you may want to reiterate in order to understand whether more interactions are needed, testing unhappy paths and defining aggregates (probably the hardest and most interesting step toward the translation of this model into software objects).

In the next section, we will see what a related spreadsheet of requirements might look like.

Requirements spreadsheet

Now, let's imagine we successfully completed the Event Storming workshop (or even better, a couple of iterations of it). The collected inputs may be directly worked on and translated into software, especially if developers actively participated in the activity. However, for the sake of tracking, double-checking, and completing the requirements list, it's common to translate those views into a document with a different format. While you can complete a standard IEEE requirement document, especially if you can do some further reworking and have access to all the stakeholders, a leaner format is often more suitable.

Now, starting from the features we have identified before, let's start to draft a spreadsheet for collecting and classifying the related requirements:

ID	DESCRIPTION	PRECONDITION	ESSENTIAL	PRIORITY	SOURCE	GROUP	PARENT	NOTES
1	A user is able to access the mobile application	The user is not logged in The user is provisioned in the system The user provides the right credentials	Must Have	High	Ste (Product Management)	Security and Identity	-	-
2	A user sees their balance and transactions list	The user is logged in Transaction list is not empty	Should Have	Medium	Ste (Product Management)	Core Features	-	See mockup
3	The user is greeted with a customized message	The user is logged in A customized message is set	Nice to Have	Low	Alessandra (Marketing)	UX	2	-
4	The user can start a money transfer transaction	The user is logged in The account has payments enabled	Must Have	Medium	Mauro (Product Management)	Core Features		-

Figure 2.7 – Requirements list of a peer-to-peer payment

As you can see, the list is not complete, however, it's already clear that from a concept nice and concisely expressed on a couple of sticky notes, you can potentially derive a lot of rows with requirements and relative preconditions.

Moreover, it is often debated whether you should include all potential paths (for example, including failed logins, error conditions, and other corner cases) in lists like these. The answer is usually common sense; the path is specified if special actions come from it (for example, retries, offering help, and so on). If it's just followed by an error message, this can be specified elsewhere (for example, in the test list and in user acceptance documents).

Another relevant discussion is about supporting information. From time to time, you may have important information to be conveyed in other formats. The most common example is the user interface, commonly specified with graphical mockups. It is up to you whether attaching the mockups somewhere else and referring to them in a field (for example, notes), or directly embedding everything (a list of requirements plus graphic mockups) into the same document is better. This is not very important, however, and it heavily depends on your specific context and what makes your team feel comfortable.

Summary

In this chapter, we have covered a complete overview of software requirements. Knowing the characteristics of a well-defined software requirement, how to collect it, and how to document it is a very good foundation to build software architecture upon. Regardless of the technology and methodologies used in your projects, these ideas will help you to get your project up to speed and to build a collaborative, trusting relationship with your business counterparts.

On the business side, the use of such tools and practices will allow for a structured way to provide input to the technical team and track the progression and coverage of the features implemented.

In the next chapter, we will look at software architecture modeling and what methodologies can be used for representing an architectural design.

Further reading

- Ash Maurya, *The Lean Canvas* (https://leanstack.com/leancanvas)
- Alberto Brandolini, *Introducing Event Storming* (http://ziobrando.blogspot.com/2013/11/introducing-event-storming.html)
- Atlassian, *Jira Software* (https://www.atlassian.com/software/jira)
- Jean-Philippe Lang, *Redmine* (https://www.redmine.org/)
- IBM, *Rational Doors* (https://www.ibm.com/it-it/products/requirements-management)

3
Common Architecture Design Techniques

In the previous chapter, *Chapter 2, Software Requirements – Collecting, Documenting, Managing*, we highlighted techniques to retrieve and analyze the features an application should have. This is done by interacting with the business and other stakeholders and describing what the desired behavior should be. We now have all the ingredients needed to start baking our application. The first—very important—step is to define the architecture.

It is debated as to how much, in terms of resources, you should invest in this phase. Some experts argue that architecture design is the most important phase, while others claim that it's crucial to keep a flexible approach, being able to adapt the architecture while the solution is evolving according to new ideas coming in or shifting external conditions.

For sure, both ideas are interesting and have some strong points. Whatever your point of view on that is, it is really useful to have a clear understanding of what the most common ways of documenting the architectures you will design are.

This is a topic we started to touch on in *Chapter 1*, *Designing Software Architectures in Java – Methods and Styles*. But while in the first chapter the idea was to start sketching some ideas and brainstorm potential solutions, in this chapter, we will cover a detailed design. This means exploring different modeling techniques, walking through notation and diagram types, and creating artifacts that are shareable and clear to understand for other team members. In this chapter, you will learn about the following topics:

- Introducing marchitectures—impactful and purely demonstrative schemas

- Familiarizing ourselves with **Unified Modeling Language** (**UML**) notation

- Exploring **ArchiMate**

- Introducing the **C4 model**

- Other modeling techniques – **Business Process Model and Notation** (**BPMN**), **Decision Model and Notation** (**DMN**), and **arc42**

- Case studies and examples

But first of all, let's start by having a look at a less structured but widely used architectural style, with a funny and a bit of an ugly name: **marchitectures**.

Introducing marchitectures – impactful and purely demonstrative schemas

With its name being a portmanteau of marketing and architecture, as you can imagine, **marchitectures** are a very common tool to pitch your solution and get sponsorship (and often the budget) for your project. You don't need to get into technical details, nor to cover every aspect of the solution; the idea here is to give an idea of what the finished product will look like.

From a content point of view, marchitectures are no more and no less than a polished version of the first whiteboard sketches of a software architecture. This includes the same vague meaning, incomplete vision, and mixed point of view that we discussed in *Chapter 1*, *Designing Software Architectures in Java – Methods and Styles*.

Marchitectures often complement mockups of the **User Interface** (**UI**), marketing research, and industry trends. You want to convince the stakeholders (budget owners, investors, and so on) that your idea is a good one and that the underlying architecture (and implementation) will be rock-solid, yet flexible enough to follow the evolutions that the business will drive.

It is definitely an ambitious goal and is sometimes—inevitably—not fully met. Indeed, the real architecture will often only partially look like what you defined in your marchitecture.

Marchitectures are often used by software vendors and for good reason. If you are pitching a product (or a framework, or a service), you don't want to be too specific on what the finished solution will look like. You just need to give a high-level idea of how your product works. Maybe authentication will be different, and maybe you will need to integrate third-party systems into the final picture, but the important thing is to have a shiny picture of how good your architecture (marchitecture) looks. There is time to get into the nitty-gritty details later. UML notation, which we will look at in the next section, is a very good way to document those details.

Familiarizing ourselves with UML notation

There are things in this book that we need to treat with reverential respect; **UML** is one of them. This modeling language is simply a piece of IT history. You should take into account that UML is a very comprehensive and articulate standard, aimed at modeling and representing a wide number of concepts. For this reason, going through the whole specification is out of the scope of this book.

But by the end of this section, you will have a grasp of the UML philosophy, and we will have covered practical examples of the most widespread UML diagrams.

> **Important Note**
>
> It's worthwhile deepening your knowledge of the UML language. To do this, you will find plenty of resources on the web. I would also suggest you have a look at the official UML website, and at *The Unified Modeling Language User Guide* by Booch, Rumbaugh, and Jacobson (more information is available in the *Further reading* section of this chapter), which is probably the most important UML book, written by the original authors of the language.

Now, let's look at the fascinating UML genesis in the next section, where we will see how UML started as a joint effort by different working groups that were all working to solve a common problem: defining a language to break the barrier between designing and implementing a software solution.

Understanding the background to UML

UML's history began in the 1990s and is strictly related to **object-oriented programming**.

UML was born from an effort to standardize object modeling and the conceptual representation of object-oriented software. A further objective was to create an object that is both human- and machine-readable, supporting the life cycle starting from analysis, and moving toward implementation and testing.

The history of the UML standard starts with a cross-company, meritocratic effort to find a solution to common problems. This looks a lot like the open source development model. Everybody is free to contribute and share ideas with the community, regardless of their role or the company they are working for.

Let's look at an overview of what's contained in the UML framework.

Walking through the UML basics, as we discussed, UML was created to model object-oriented systems, and in theory, diagrams created with UML can be automatically translated into source code.

There are a number of interesting principles in the UML language, making it just as useful and relevant today, more than 20 years after its inception, as when it was created. Let's have a look at some of them here:

- UML is independent of the development methodology, meaning that it can be used even in modern Agile and **DevOps** teams. Some of the diagrams introduced by UML are commonly used in those contexts too. The goal of the language is to visualize, specify, construct, and document OO systems.

- UML is usually associated with diagrams and graphical artifacts. While they are, indeed, a core concept of the language, UML also defines the related semantics. This means that the reasoning for everything is well defined and formalized so that both a trained person and a machine can understand what a UML diagram represents in all its details.

- UML concepts are built upon three different kinds of building blocks—namely, **things**, **relationships**, and **diagrams**. These are further organized into subcategories. For each of those concepts, a graphical representation (symbol) is provided.

These building blocks are covered in detail in the following subsections.

Things

Things are core entities that have the goal of abstracting concepts represented by the system. Things are further grouped into other subtypes, as outlined here:

- **Structural things**: These are the most essential elements in object-oriented programming (such as classes and interfaces)

- **Behavioral things**: These represent interactions (such as messages and actions)

- **Grouping things**: These are used to organize other things (packages are an example)

- **Annotational things**: These support elements to document the models (such as notes)

Relationships

Relationships model the links between things. These are further organized into four main categories, as follows:

- **Dependencies**: Defining a relationship in which the changes made to a thing will influence a dependent thing. Also referred to as a **client-supplier link**, where a change to the supplier requires a change to the client. As an example, think about a `BusinessLogic` component, providing validations, checks, and so on, and a `PaymentService` component, called from the `BusinessLogic` component in order to provide payment functionalities. A change in the methods of the `PaymentService` component will require a change in the BackendAPI that uses it.

- **Associations**: An association, such as a link between classes, is usually modeled as one object holding a reference to one or more instances of the other objects. An example of this is a `PaymentTransaction` component with a user. Each payment must reference at least one user of the platform (that is, the one making the payment).

- **Generalizations**: This represents the parent-child relationship. That's pretty straightforward: a `MobilePayment` interface is a specific type of `PaymentTransaction` component, inheriting from it.

- **Realization**: This helps in modeling the interface-implementation link, representing a contract in terms of methods and signatures, and the realization of it. You can take as an example an `IPaymentService` interface, and its practical implementation—such as `MobilePayment`—implementing one particular way of making a payment (and abstracting the caller from the implementation details).

Diagrams

Diagrams are schemas representing meaningful sets of things. They are technically graphs, which helps them to be easily read and written by machines. Diagrams can be classified as follows:

- **Structural**: Describing the static aspects of a system, such as the structure, grouping, and hierarchy of objects

- **Behavioral**: Describing the interactions between objects

Diagrams are the most widely known concept of UML. It is very likely that you have already seen a class diagram or a sequence one. In my opinion, diagrams are one of the most useful UML concepts. For this reason, we will walk through the main types of diagrams in the following upcoming sections.

In the following diagram, you can see a graphical representation of the UML things we've just seen:

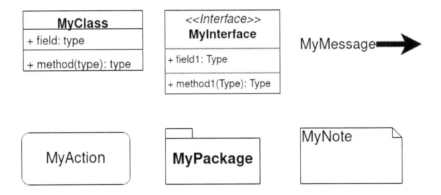

Figure 3.1 – Graphical representation of some UML things

In the next diagram, we represent the graphical symbols of the UML relationships we've described:

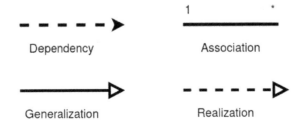

Figure 3.2 – Graphical representation of some UML relationships

With regard to UML diagrams, due to their relevance, we will walk through some of the most common ones in the next sections.

Class diagrams

Since the beginning of this section, we've made it clear that UML is all about **object-oriented** modeling, which is expected since **Java** (probably the most widespread object-oriented language) is one of the pillars of this book, and—of course—modeling classes are one of the most important aspects of object-oriented modeling. I'm pretty sure you've already seen (or used) class diagrams. They're a very common and natural way to represent classes and how they are made, and indeed are used in countless documentation on the internet.

A class diagram is made up of a set of classes (including their fields and methods) and the relationships between them. Interfaces are represented where present, and so is inheritance between classes. As per the other diagrams, a class diagram is conceptually a graph, made up of arcs and vertices.

A class diagram is intended to highlight a specific subset of the whole architecture, so the class represented is part of a given use case or belongs to a specific subdomain.

It's worth noticing that the relationships will represent both the kind of cooperation/responsibility between the classes and the multiplicity of the relationship itself (for example, one-to-many, one-to-one, and similar cardinalities). This is what a basic class diagram looks like:

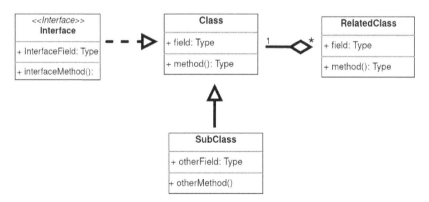

Figure 3.3 – Basic class diagram

As you can see, a class diagram is a great way to model the structure of a logical subdomain of the application (objects and their links). In the next section, we will look at sequence diagrams, which are another very widespread representation, focusing more on the end-to-end interactions needed to implement the functionality.

Sequence diagram

A **sequence diagram** is probably one of the most famous UML diagrams. This diagram is a particular instance of so-called interaction diagrams, which are a representation of a set of objects (such as software components and actors) and how they interact (for example, exchanging messages). In the case of sequence diagrams, the interaction is pictured in a linear way, representing interactions ordered by the temporal dimension.

From a graphical viewpoint, a sequence diagram pictures objects in a row, each one with a line going down vertically (also known as a **lifeline**). Crossing those lifelines, interactions are laid out as horizontal lines, intersecting the involved objects.

A sequence diagram also offers a way to represent conditions and iterations. **Conditions**, **parallelization**, **loops**, and **optional** are represented by drawing a box around the interactions and tagging the box with the right keyword.

Given the nice level of detail that can be expressed by the message flow (including the time ordering) and the expressiveness provided by the structured controls (conditions and such), sequence diagrams are a very nice way to analyze and document functionalities, by breaking them up into smaller operations.

This is what a sequence diagram looks like:

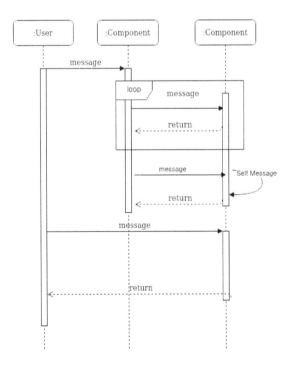

Figure 3.4 – Sequence diagram

As you can see in the first row of the diagram, we have in this case a user and two components, whereby the following applies:

- Each one of them has a lifeline, which is highlighted to represent activation when an interaction is made (for example, a method is called, or a message is sent).

- A self-message is pictured as a curved line, representing the call of a method on the same component.

- A loop is represented as a frame with a tag. In the tag, other than loops, `opt` (**optional**), `par` (**parallel**), and `alt` (**conditional**) are admitted values. `opt` identifies an optional interaction that will happen only if a specified condition is met (such as an `if` block), `par` represents a parallel interaction (such as two methods called in parallel in a multithread fashion), and `alt` matches alternative conditions, such as an `if/else` block.

- The same kind of notation (box with a tag) can be used to represent sub diagrams. In this case, the tag has a `ref` value, while the name of the diagram representing that part is reported in the box. This provides a simple way of breaking down big and complex sequence diagrams into smaller ones.

With this look at sequence diagrams, we have completed our very brief overview of the most common UML ideas.

Wrapping up on UML

As we said at the very beginning of this section, UML is a big and complete framework that is too complex to summarize in just a few pages. However, the essential concepts we have seen so far (including class diagrams and sequence diagrams) are a good way to start getting used to this language and add some useful tools to your toolbox. Of course, my advice is to go deeper and get to know more diagrams and techniques from this awesome language.

In the next section, we are going to explore a technique that shares many similarities with UML: ArchiMate.

Exploring ArchiMate

ArchiMate is an architectural modeling technique aimed at analyzing and documenting enterprise architectures. This means that, while still having roots in technology and software, it's usually adopted in projects with a broader scope, such as documenting the whole enterprise technology landscape (also known as **enterprise architecture**) and modeling the business processes implemented by the underlying technology implementation.

ArchiMate's name is a merging of *architecture* and *animate*, implying that one goal of this framework is to display the enterprise architecture in an intuitive way. ArchiMate was created in the early 2000s in the Netherlands, the result of a concerted effort from players in the government, industry, and academic sectors. Soon after the first drafts of this standard, the governance was transferred to **The Open Group**, an industry consortium regulating many other IT standards, such as **The Open Group Architectural Framework** (**TOGAF**, which is an enterprise architecture standard) and the **Single Unix Specification** (**SUS**, which is a **Portable Operating System Interface** (**POSIX**)-standard superset). The Open Group is also behind other famous standards in the Java world, such as **Service Oriented Architecture** (**SOA**) and **eXtended Architecture** (**XA**).

Let's start with the ArchiMate Core Framework.

The ArchiMate Core and Full Frameworks

The first concept to approach in ArchiMate is the **Core Framework**. The ArchiMate Core Framework is a 3x3 matrix, created by crossing three layers (**Business, Application,** and **Technology**) stacked with three aspects (represented vertically: **Passive Structure, Behavior,** and **Active Structure**).

This is what the Core Framework matrix looks like:

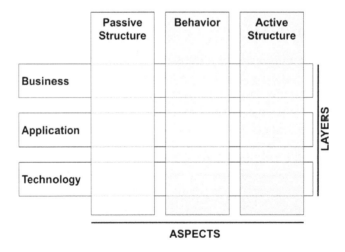

Figure 3.5 – ArchiMate Core Framework

The layers are a way to look at the same concept (or closely related concepts) from three different perspectives. In a way, a concept in one layer makes use of or is linked to concepts in nearby layers. You can see the three layers as a specification from the more abstract (business) to the more concrete (technology), as outlined here:

- The **Business** layer revolves around business capabilities, usually offered to the external world (for example, final customers). This includes business processes, events, and functions related to high-level capabilities.

- The **Application** layer includes the software components offering capabilities to the **Business** layer.

- The **Technology** layer is the technical infrastructure supporting the software components, including hardware and communication.

The aspects are a way to classify objects by their role in an activity, as outlined here:

- **Active Structure** includes the elements starting an action (including actors, devices, and software components).

- **Behavior** includes the action itself being made by something in the **Active Structure** aspect (such as an actor).

- **Passive Structure** includes the objects on which the activity is made (for instance, the recipient of the action itself, such as a data object).

You should take into account the fact that some objects can be part of more than one aspect.

As you will see, the Core Framework provides a simple way to place and categorize objects, and it enables multiple viewpoints. Also, take into account that ArchiMate diagrams do not necessarily follow this matrix layout: this is merely a conceptual way to demonstrate layers and aspects and how they are related.

The ArchiMate standard also provides an extended version of the framework. In this framework, three more layers are added, as outlined here:

- **Strategy**, on top of the **Business** layer, aims to link business functionalities and use cases to the pursuit of strategic objectives.

- **Physical**, technically a subset of the **Technology** layer, is used to represent materials, physical objects, facilities, and so on.

- **Implementation and Migration** is used to model all the temporary components supporting transitory phases during implementation and migration.

A fourth aspect, called **Motivation,** is also included in the extended framework. The goal of this aspect is to map and represent the strategic reasons behind the other architectural choices. In particular, you will see components such as value, goal, and stakeholders used to model the reason behind specific domains or use cases.

This is what the Full Framework looks like:

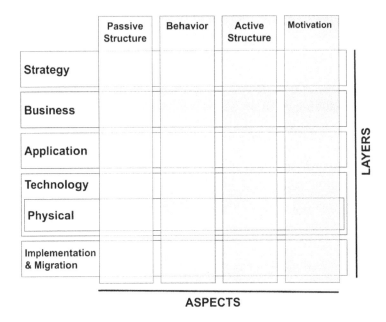

Figure 3.6 – ArchiMate Full Framework

As per the Core Framework, this is just a logical model aimed to highlight areas of overlap between the layers and aspects. ArchiMate compliant schemas will not necessarily come in a matrix format.

In the next section, we will see the components of ArchiMate, which are the objects categorized according to the matrices we've just seen.

Navigating the ArchiMate language tree

The ArchiMate language is conceptually structured as a tree, as follows:

- The top concept is the **model,** defined as a collection of concepts.
- A **concept** is a generic term that can be characterized as an element or a relationship.

- An **element** is a generic item that maps to a definition of the layers—that is, **Behavior**, **Active Structure**, or **Passive Structure**. An element is also allowed as part of the **Motivation** aspect (as per the Full Framework). Composite elements are intended as aggregations of other concepts. **Active Structure** and **Behavior** elements can further be classified as **Internal** or **External**. An event is a further specialization of a **Behavior** element.

- A **relationship** represents the connection between two or more concepts. Relationships are further classified as **Structural** (elements are statically associated to create another element), **Dependency** (elements may be affected by changes in other elements), **Dynamic** (elements have temporal dependencies to other elements), or **Other**.

- **Relationship connectors** are logical junctions (**And, Or**), associating relationships of the same type.

This is what the tree will look like:

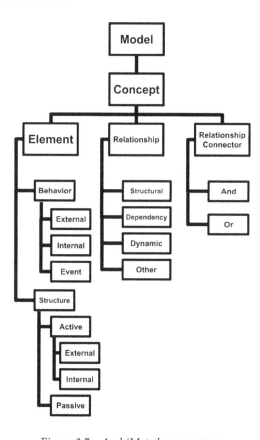

Figure 3.7 – ArchiMate language tree

In this classification, elements are just defined in an abstract way, not dependent on layers. In ArchiMate modeling, concrete implementations of those elements are then instantiated and classified in the relevant layer. As an example, a service is a generic internal **Behavior** element. It will then be used in the form of a business service, application service, or technology service, depending on which layer we are modeling.

Other elements only make sense in a specific layer. For example, a communication network is an element property of the **Technology** layer, classified as an **Active Structure** element that doesn't have a one-to-one correspondence to elements present in other layers.

The combination of elements and relationships can then be organized into custom views, effectively building architectural diagrams, optimized by stakeholders and viewpoints.

In the next section, we are going to compare ArchiMate with UML.

Comparing ArchiMate to UML

As you may have seen, the ArchiMate language shows some similarities to UML. That is not by accident: ArchiMate is indeed inspired by UML, and some concepts of the two frameworks are almost the same.

However, other than specific differences (for instance, concepts present in one framework and not in the other), there are some high-level considerations to take into account when comparing those two frameworks, as outlined here:

- UML is strictly centered around object-oriented modeling, while ArchiMate is not linked to a specific paradigm.

- ArchiMate explicitly defines the **Business** layer and other higher-level concepts (including **Motivation** and **Strategy**) that are usually not contemplated in UML diagrams.

- UML provides a fixed set of diagrams, while ArchiMate is more of a palette of different components and aspects, aimed at building views and viewpoints, explicitly providing ways of customizing the architecture definition.

As we saw when we covered the ArchiMate genesis at the beginning of this section, The Open Group is the organization behind many other standards, including TOGAF. Let's see what the relationship between ArchiMate and TOGAF is.

Comparing ArchiMate to TOGAF

TOGAF is a complete framework, aimed at providing a standardized way of defining, modeling, and implementing architecture projects (for example, classifying the enterprise architecture of an organization). TOGAF is, in a way, complementary to ArchiMate. While TOGAF does not provide a specific architectural notation (as ArchiMate does), ArchiMate does not prescribe a specific process for architecture definition (as TOGAF does).

The core of TOGAF is the **Architecture Development Method** (**ADM**) process. The process is made up of eight steps (plus two special phases: the preliminary phase and requirements collection). A detailed explanation of each step is beyond the scope of this book, but the important takeaway is that each phase of the TOGAF ADM can be mapped as a layer into the ArchiMate framework (for instance, Phase B, which is about the definition of business architecture, of course maps to the **Business** layer, while Phase F, Migration Planning, can be mapped to the **Implementation and Migration** layer).

This concludes the section dedicated to ArchiMate. In the next section, we will go through another very smart architectural modeling technique: the C4 model.

Introducing the C4 model

The **C4 model** is a lightweight methodology for modeling and representing software architecture. It was created in 2006 by Simon Brown, and the official website (under a **Creative Commons (CC) License**) was launched in 2018.

The model is somewhat inspired by UML, but it takes an alternative, leaner approach and, for this reason, is very popular among Agile teams who are looking for a more dynamic and less prescriptive way of designing and documenting software architectures.

Exploring the C4 model

The keyword for understanding the C4 model is *zoom*. This concept means exactly what it does for pictures: the core idea of the C4 model, indeed, is about navigating the architectural representation by widening or narrowing the point of view. The C4 model is built around four main levels, detailed as follows:

- **Context** is a diagram giving the big picture of an application. It shows the whole system represented as a box and depicts interactions with users and other systems.

- **Container** is the view obtained when zooming in one level down. It represents what's inside the system box by modeling the subsystems comprising it. In terms of granularity, a container is something that can be deployed and executed—so, it can represent a backend application, a database, a filesystem, and so on.

- **Component** is another zoom level, looking inside one container. In essence, a component is an abstraction grouping of a set of code instances (for example, a bunch of classes) that implement a functionality.

- **Code** is the maximum level of zoom in this hierarchy and can be omitted. It's used to directly represent source code and configurations. The C4 model does not provide a specific suggestion on how to draw this kind of schema, which is usually represented using UML class diagrams. The reason why it's considered optional is that it is not very easy to keep this view up to date with code changes. A suggestion here is to try to stick to the automatic generation of this diagram if possible (by using plugins for the integrated development environments or other automated procedures).

So, the C4 model, in essence, is made up of three different diagrams (plus an optional one). Each diagram is linked to the others by a different level of zoom, as shown in the following diagram:

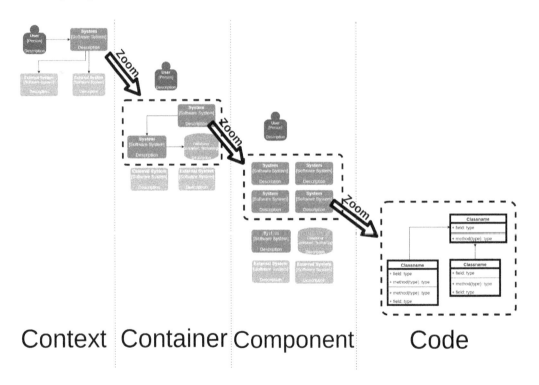

Figure 3.8 – C4 zoom levels

The idea behind this technique is to focus on a different ecosystem based on the level of zoom. Moreover, different views can be aimed at different stakeholders. In the next section, we will see what's inside each level.

Filling in the different levels

The C4 model does not provide any particular notation or symbology. Unlike UML, the kind of shapes, color coding, and so on are not part of the standard. The model simply encourages you to have a consistent representation (for example, once you choose a shape to represent an element, keep that shape in all the diagrams), to add an explicit legend to each diagram, and to comment as much as possible, for better clarity.

C4 is made up of the following elements:

- **Software system**: The top-level element, the center of the *context* representation. This is basically the whole system that we are going to design and implement.

- **Container**: As we mentioned when discussing the level with the same name, a container is roughly something that can be deployed and started/stopped individually. This includes applications, databases, and so on. It's usually completed with a description of the technology and framework used.

- **Component**: As before, this is a concept already introduced when discussing levels. A component is an abstraction aggregated over a subdomain or functionality. It's basically a grouping of code. It may or may not map one-to-one to a Java package.

- **Relationship**: A line (or, more often, an arrow) representing a link between one of the aforementioned elements. It's usually completed with a textual explanation of the kind/scope/goal of the relationship, and technical details where relevant (for instance, the protocol used).

- **Person**: A human interacting with the system.

As you will see, there is no explicit advice for representing the code. It's a common practice to represent it with UML classes but, as we said before, this is something that is only done if strictly necessary.

For the sake of completeness, C4 also includes some additional diagrams, as outlined here:

- **System landscape**: A context diagram showing the whole enterprise, in order to represent the full *neighborhood* of our application.

- **Dynamic**: A diagram representing a use case by numbering the interactions between elements in order to show the temporal progression. It looks quite similar to the UML sequence diagram but is less prescriptive in terms of syntax.

- **Deployment**: This shows the mapping between containers and the underlying infrastructure, which may be a physical server, a virtual machine, a **Linux** container, and so on.

With these diagrams, we have completed our excursus on the C4 model. As you will see, this model is simpler than UML and ArchiMate but still quite complete and expressive, meaning that you can model a lot of architecture types with it.

In the next section, we are going to explore other modeling techniques that are less common and aimed at specific use cases.

Other modeling techniques

The three modeling systems we have seen so far—UML, ArchiMate, and C4—are complete systems with different approaches, aimed at analyzing and representing software architecture end to end.

In this section, we are going to quickly touch on some other techniques that have a more vertical approach, meaning that they are less general-purpose and more detailed when it comes to targeting specific use cases. Those techniques are **Business Process Model and Notation (BPMN)**, **Decision Model and Notation (DMN)**, and **arc42**.

BPMN

BPMN is a standard that was developed and is currently maintained by the **Object Management Group** (**OMG**), the same organization behind UML. BPMN is also a standard that has been recognized by the **International Organization for Standardization (ISO)**.

As may be obvious by its name, this language specializes in representing business processes.

BPMN is usually associated with the activity diagram of UML, as both are flow chart-like diagrams (with a slightly different notation and symbology), aimed at describing a use case in terms of elementary steps and the connections between them (for instance, optional conditions), including a temporal dimension (from-to). But the similarities end there.

UML is wider and aimed at modeling a lot of other things, rather than being fundamentally an object-oriented framework. On the other hand, BPMN focuses just on the modeling of business processes, and its primary goal is to define common ground between technical and business stakeholders. Indeed, the idea behind BPMN is that a businessperson (or better, someone with no technical skills but a good knowledge of processes) can model a diagram that can then be directly imported and executed into a BPMN engine, with little-to-no help from technical staff. This is not something that happens in the real world, as often, BPMN design is still an abstraction, and a number of technical steps are still needed to configure, deploy, and execute a BPMN process.

However, it's true that BPMN is usually at least understandable (if not definable from scratch) by non-technical stakeholders. This is good enough for supporting collaboration between teams and reducing friction when translating business processes into code implementation.

The building blocks of BPMN are categorized as four basic families: **flow objects**, **connecting objects**, **swimlanes**, and **artifacts**. For each of them, a graphical notation is formalized.

Roughly speaking, **flow objects** represent the steps in the diagram and are described in more detail here:

- The most important one is probably the **task**, which is the abstraction of generic activity. This means both non-automatic activities (manually performed outside of the BPMN platform) and automatic activities (such as sending an email or triggering a web service call).

- Other basic flow objects are **start** and **end** events, delimiting the beginning and end of a workflow.

- **Gateways** are another important kind of object, used to model things such as conditional execution or the parallelization of paths.

Connecting objects are used to link flow objects with one another. They can mimic different behaviors, such as sequences, messages, or associations.

Swimlanes are a way to graphically group and organize a business process. With swimlanes, you partition the business process according to the actor (or group of actors) in charge of a specific set of steps.

Finally, **artifacts** are supporting concepts (for example, annotations), aimed at enriching the BPMN flow expressiveness.

This is what all these objects look like graphically:

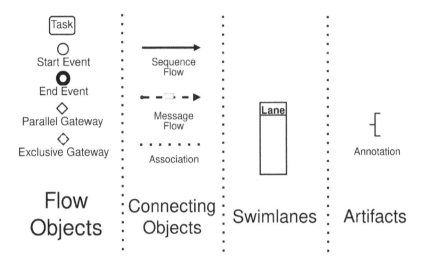

Figure 3.9 – Graphical representation of some BPMN entities

We will talk again about BPMN in *Chapter 7, Exploring Middleware and Frameworks.*

DMN

DMN is a standard published and maintained by OMG, and it's younger than BPMN. DMN is somewhat complementary to BPMN. Instead of being aimed at modeling business processes, the scope of DMN is to model business rules, which are commonly used as one of the tasks in BPMN processes, rather than standalone, outside of BPMN processes.

The goal is exactly the same as BPMN: defining a common language between business and IT personas, allowing for better collaboration.

DMN encompasses elements such as decision tables (tables representing rule outcome based on the combination of a set of inputs) and **Friendly Enough Expression Language** (**FEEL**), an expression language used to formalize the logic behind decisions.

We will talk about DMN again in *Chapter 7, Exploring Middleware and Frameworks.*

arc42

arc42 is not a modeling technique but, instead, a templating model that helps with identifying, in software architecture, what the important concepts to document are and how to document them, by providing a kind of *scaffold*.

arc42 was originally created by Dr. Peter Hruschka and Dr. Gernot Starke and has a completely open source approach (including being free to use in commercial projects). It's an exceptional way to start documenting your system from scratch, from an architectural point of view. From a practical viewpoint, it provides a scaffold (including sections to be fulfilled) on what the documentation should look like.

It is not a substitute for other modeling languages and does not mandate a specific working model or development techniques. Instead, it is expected that you will use concepts and diagrams from other techniques (such as UML or C4) to fill out the sections of arc42-compliant documentation.

Sections include elements such as the introduction, runtime view, cross-cutting concepts, architectural decisions, and more. It is really just a suggestion on the structure of the documentation; it's up to you to choose how deep to dive into each section. If you want to give it a try, you can go to the official website (see the *Further reading* section), download a template, and start to fill out the sections. It really is that easy.

BPMN, DMN, and arc42 cover specific niches and target specific needs. For this reason, they can be a useful complement to the more generic and comprehensive frameworks that we have seen before. With this section, we've completed our overview of architectural modeling techniques. Let's now complete this chapter by looking at some examples.

Case studies and examples

In this chapter, as in previous ones, we will continue our study of the mobile payments application. We will keep exploring this context to see some examples of the diagrams we have discussed so far.

UML class diagrams for mobile payments

As a first example, we will look at UML class modeling. This is a very common diagram in Java projects. It is debated whether it's useful to build and maintain documentation that is so close to code (see also the considerations we discussed in the section on C4), since it may be seen as not adding that much value and being hard to maintain. Moreover, in modern development models (such as cloud-native and microservices), you are supposed to communicate between parts of the application by using established interfaces (such as **REpresentational State Transfer** (**REST**) or **Google Remote Procedure Call** (**gRPC**) and avoid exposing the internal model of your applications for others to tap into.

My personal view is that the truth is in the middle. Unless you are developing something very peculiar (such as a framework, a library, or a plugin extension system), you may not need to document your entire code base as class diagrams. However, in order to analyze impacts and collaborate with other team members on the same code base, it may be worthwhile to at least sketch the critical aspects of your application (this being the core classes and interfaces).

Another useful technique is to rely on the automatic generation of class diagrams. You may find plugins for most commonly used IDEs and also for Maven that can do that for you. Class diagrams can be particularly useful for giving an idea of what the model behind your code looks like (think about a new team member joining the project) and can ease things such as refactoring by giving an idea of what impact a change could have on related classes.

This is what a class diagram for mobile payments will look like (picking just a handful of significant classes):

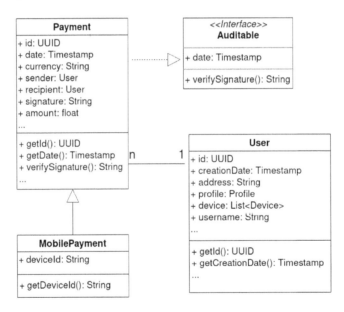

Figure 3.10 – UML class diagram for payment and user objects

As you can see in the preceding diagram, we are representing the Payment and User classes (some methods and fields are omitted for the sake of space).

Some of the notations we used in this diagram are listed as follows:

- MobilePayment is a subclass of Payment (generalization).
- Payment implements the Auditable interface.

- Payment is associated with `User`.

- You can also see the multiplicity (each user can have **n** payments). As we discussed before, this kind of association is very similar to what you can find in an entity relationship diagram representing database tables.

In the next section, we will see some C4 diagrams for mobile payments.

C4 diagrams for mobile payments

In the section dedicated to C4 diagrams, we saw that the C4 technique involves diagramming the system according to four main levels of zoom. As discussed, the last level of zoom (code) is optional, and there are no strict guidelines given on how to represent it. It is common to use class diagrams, as we did in the previous section. Supposing we take that as one of the four representations for our use case, let's see what the path is that takes us to that schematization. Let's start with the context diagram of a module of the mobile payments solution, as follows:

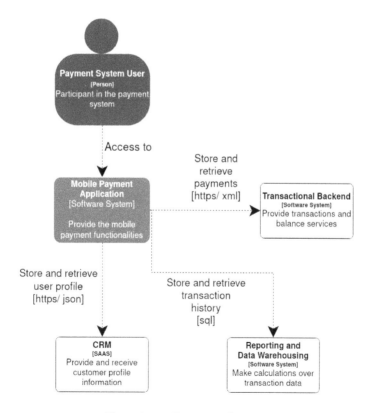

Figure 3.11 – C4 context diagram

As we can see, this is very high-level, aimed at showing the ecosystem of interactions around our system. The mobile payment application is just a big block, and in the diagram, we summarize the external system and the actors interacting with it. There is also a synthetic description of the interactions (including the format/protocol) and the type of each element (software system, person).

Let's now zoom into the container view, as follows:

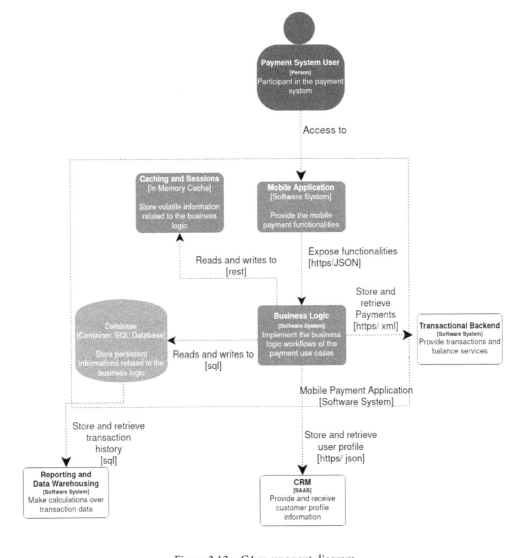

Figure 3.12 – C4 component diagram

Here, we can see a closer representation of the technical pieces comprising our application. Our application is no longer just a box: we can see all the processes (things that can be deployed and started independently from one another) that comprise our system included in the dashed box. External context is still present (for example, the transactional backend). Every interaction has some explanation and protocol. Every container has a generic description of the kind of technology that is implementing it. If you think this diagram is pretty similar to what we saw in *Chapter 1, Designing Software Architectures in Java – Methods and Styles*, you are right.

We are still a bit far from the code/class diagram. The component diagram is the missing link. We can view this here:

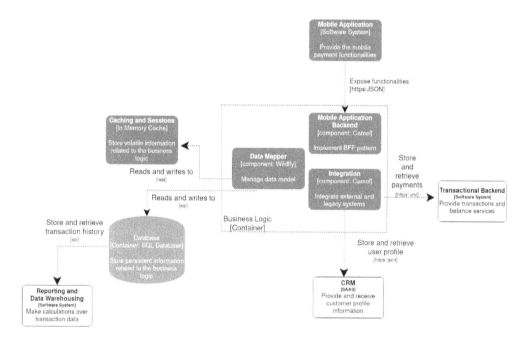

Figure 3.13 – C4 container diagram

As expected, we zoomed in one level deeper, highlighting three components that comprise the business logic container (**Mobile Application Backend**, **Data Mapper**, and **Integration**).

With this container diagram, we are one step above the direct representation of the implementation code (code diagram).

For the sake of space, we are not providing the full code diagram here. However, the classes modeled as UML in the section before can be seen as a partial code diagram of the **Data Mapper** component, somewhat closing the loop.

Those were very basic examples to show some bits of the modeling techniques in practice. Of course, giving detailed examples on every methodology shown in this chapter would have taken a whole book (or more than one) on its own, but I hope to have given you some basics to start from and deep dive into, in case you need to start practicing one of these diagrams for your projects. Let's now recap the main points of this chapter.

Summary

In this chapter, we saw a wide range of techniques for modeling and representing the internal architecture of a software system. We started with UML, which is a consolidated standard that is very widespread and actively used, especially in some of its aspects, such as class diagrams and sequence diagrams.

We then moved on to ArchiMate, which gives an enterprise architecture point of view on the subject and is commonly used in a context that follows the TOGAF approach. We then moved on to the C4 approach, which is a younger standard that is very lightweight and particularly suitable for projects adopting lean methodologies.

We've also seen a handful of specialized languages (BPMN and DMN), which are perfect for modeling specific aspects of our application. Last but not least, we quickly touched on arc42, which is a wonderful template system to start your architecture documentation and ensure that nothing important is missing.

In the next chapter, we will discuss **Domain Driven Design** (**DDD**) and other techniques to flesh out your application, which you can use once you have defined the architecture for it.

Further reading

- The UML official website: `http://uml.org/`

- *The Unified Modeling Language User Guide*, by Grady Booch, James Rumbaugh, and Ivar Jacobson, published by Addison-Wesley, 1999.

- The Open Group, *ArchiMate® 3.1 Specification*: `https://pubs.opengroup.org/architecture/archimate3-doc/`

- InfoQ, *The C4 Model for Software Architecture*: `https://www.infoq.com/articles/C4-architecture-model/`

- The C4 official website: `https://c4model.com/`

- The arc42 official website: `https://arc42.org/`

4
Best Practices for Design and Development

The developers reading this book have probably viewed the previous chapters as appetizers. If that's the case, with this chapter, we are moving on to the main course. While collecting requirements and designing the architecture are crucial steps (I cannot highlight this enough), anyone who comes from a development background will surely want to get their hands dirty with code.

In this chapter, we will focus on how to implement the concepts that we have theorized so far in the source code. Of course, in the real world, the edges are not so smooth, and the architectural design (including **UML** or **C4** schemas) and requirements management will continue during the implementation phase. However, in this chapter, we will focus on some well-known techniques to translate those design ideas into working software.

In this chapter, you will learn about the following topics:

- Understanding **Domain Driven Design (DDD)**
- Introducing **Test Driven Development (TDD)**
- Exploring **Behavior Driven Development (BDD)**
- User story mapping and value slicing
- Case studies and examples

After reading this chapter, you will be able to model complex use cases into elegant software concepts and define domains, objects, and patterns. You will learn how to use TDD and BDD to conduct development activities and implement meaningful use cases with each release. You will understand the concept of **Minimum Viable Products (MVPs)** and the technique of value slicing.

But first, we'll start with DDD, which will provide a solid foundation to build upon.

Understanding Domain Driven Design

DDD takes its name from the book of the same name by Eric Evans (2003). The subtitle beautifully clarifies what the goal is—**Tackling complexity in the heart of software**.

In this section, we will learn about the domain model, ubiquitous language, layered architecture, DDD patterns, and bounded contexts.

DDD is a widely adopted modeling technique to build rich and expressive domains. It is considered to be behind modern approaches such as microservices development.

The idea behind DDD is discovering how to model our software in a way that mirrors the problem we are facing in the real world. It is expected that if properly modeled, our software will be readable, will adhere to requirements, and will work properly.

Of course, there is no magic recipe for that: DDD provides a toolkit of patterns, best practices, and ideas to implement this modeling. This approach works particularly well with complex domains, but it might be overkill for smaller and simpler projects. Additionally, it is true that DDD provides a lot of good ideas, and you might consider adopting it partially if that fits your needs. But first, let's begin with some considerations about the completeness of the domain model.

The anemic domain model

In his seminal paper about this domain model, Martin Fowler defines the **anemic domain model** as an antipattern, which defies any basic purpose of **object-oriented programming**. Of course, I cannot disagree with that at all. Nevertheless, this kind of modeling is far too widespread, as it's a kind of quick and dirty way to design an application.

Essentially, in the anemic domain model, each object maps with its real-world counterpart, including fields and relationships. Those are, in a way, kinds of data objects. What's missing in the anemic domain model's objects is the behavior, meaning the specific actions that are logically associated with that particular concept in the real world. Usually, the objects in an anemic domain model have getter and setter methods, and not much more. All of the behavior is codified as part of specific service objects, operating across all of the other data objects through specific methods.

The issue, here, is that the domain model is simply slipping away from object-oriented programming and toward an overengineered procedural model. This could be good enough in simple scenarios and, indeed, is common in **Create, Read, Update, and Delete (CRUD)** applications over a relational database, where you are, more or less, exposing tables directly as an application, with very limited business logic on top.

If the model is bigger, and it encompasses more complete business logic, this way of modeling starts to show some limits. The data objects become similar, and it's harder to group them and define relationships. The service objects have more and more methods, with growing complexity. You start to gain the cons of both the procedural and object-oriented methods. After all, you have very few (if any) of the pros of object-oriented modeling. DDD aims for the opposite—building rich and expressive object-oriented designs. Let's examine how to start modeling applications on DDD principles.

Understanding ubiquitous language

Indeed, the very first concept of DDD is the principle of good collaboration. To define a good domain model, you have to use both technical language and business language.

This means having a team composed of domain experts besides software developers. *But how will those kinds of people cooperate when they speak different languages?* You will require a lingua franca to ensure they work together.

The concept around the ubiquitous language is simple and brilliant, that is, to define a shared dictionary for a business (for instance, analysts, domain experts, or whoever you want to include) and developers to talk together with fewer misunderstandings. However, it's a kind of abstract concept, and there is no magic recipe to achieve it. You can think about it as a shared culture built into the team. Unfortunately, no one has defined a template document or a kind of diagram that can solve the ubiquitous language challenge for everybody.

Indeed, what's advised in DDD's essential literature is to use UML diagrams (especially class diagrams) and written documents (no particular format is required). However, what's essential is how you get to the shared understanding of ubiquitous language, and there is probably only one way to do this—by working together.

Ubiquitous language is all about how to name the concepts in your model properly. And by concepts, we are not necessarily referring to **Java** classes (as they are an implementation detail), nor to business processes (as, perhaps, they are not mapped one to one in our application). We are referring to something in the middle, that is, a model that is understandable and makes sense for a business and is translatable in meaningful ways into software artifacts by developers.

Of course, the model will comprehend objects, the relationships behind them, and the actions they perform. It is also essential for the team to share the meanings of each operation. Simply defining the name of each interaction might not solve any ambiguities. Once a shared understanding has been reached (it might be a recurrent effort with many cycles), then it must be strictly respected.

This includes using the naming consistently in code and in all of the other artifacts produced (such as analysis documents, test plans, and more), as well as referring to things with the right name in meetings and documenting this shared understanding in some way (as I said, the format is up to you). As we discussed earlier, ubiquitous language is all about creating a shared culture in a working team across different specialties.

The concept might appear abstract; nevertheless, it is essential and can be a useful tool even if you are not fully going with DDD. However, DDD also defines more concrete concepts, such as **layered architecture**, which we will look at in the next section.

Getting familiar with layered architecture

When we start to define the conceptual model around our application, it's natural to wonder where this model practically fits in our implementation and how to keep it pure, regardless of the technology we are using. Think about persistence (the database), the **User Interface** (**UI**), and such. Those technologies probably have constructs that differ from our model. They might not even be object-oriented at all. And for sure, we don't want a change driven by technological reasons (such as the optimization of a query or a change in the UI) to affect our domain model. DDD tackles this concept directly by suggesting a layered architecture approach.

Here, the idea is to partition the application code into different layers, loosely coupled to each other. Then, you implement your domain model into one of those layers, encapsulating the technological details in the other layers, each one with well-defined responsibilities.

A simple and common example of this is with the four layers divided, as follows:

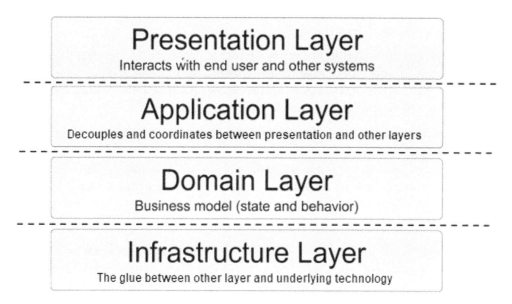

Figure 4.1 – Layered architecture

As you can see, the layered architecture is divided as follows:

- **Presentation Layer**: This layer includes all of the code required to present and collect the data for users. Additionally, this could include machine-to-machine interactions (such as in API calls).

- **Application Layer**: This layer is similar to what's implemented in the **Backends for Frontends** pattern (we'll cover this in *Chapter 6, Exploring Essential Java Architectural Patterns*). Essentially, this layer is a proxy, stateless and without business logic, which simply coordinates the interactions between the presentation layer and the rest of the application.

  The application layer can store session data and perform basic orchestration (such as aggregating or ordering calls to the underlying layers). In my opinion, this layer can be considered optional in some kinds of applications. The risk is that if you avoid it, it will couple the presentation layer tightly to the rest of the architecture. On the other hand, if you decide to adopt it, you should be mindful of the risk of sneaking in too much business logic.

- **Domain Layer**: This layer is, of course, the core of proper DDD. Here lies the whole business model, adherent to what we are representing, in terms of objects, their state, and their behavior. The domain layer exposes the functionalities of the higher levels and uses the underlying layer for technical matters.

- **Infrastructure Layer**: This layer is a supporting layer that deals with all of the other layers. It can be defined as the glue between the layers themselves and the technological layers providing functionalities. Here, a classic feature is persistence—objects in the domain layer use features exposed by the infrastructure layer, which deals with the database (or other persistent technology) using its native protocols and libraries.

This organization might look familiar to you, as it's described in various forms and variants in the software area (you might find some similarities with the **Model-View-Controller** pattern, which we will examine in *Chapter 6, Exploring Essential Java Architectural Patterns*). However, do take into account that this is mostly a way to nicely group responsibilities. It doesn't necessarily mean that each layer should be deployed on its own, as a separate process or artifact.

Having discussed layered architecture, let's focus on the heart of DDD: the **Domain Model** and its parts.

Learning about the domain model

The **Domain Model** is an elegant way to represent reality and implement it in an object-oriented way.

Essentially, you can consider the domain model as the opposite of the anemic model that we looked at earlier. In the anemic model, the objects simply include data and very limited (or even absent) behavior. The domain model of DDD stresses the expressiveness of objects and their behavior.

Put simply, the domain model is simply the concept of comprehending the data and behavior of an application. DDD implements this idea by defining the elements detailed in the following sections, as shown here:

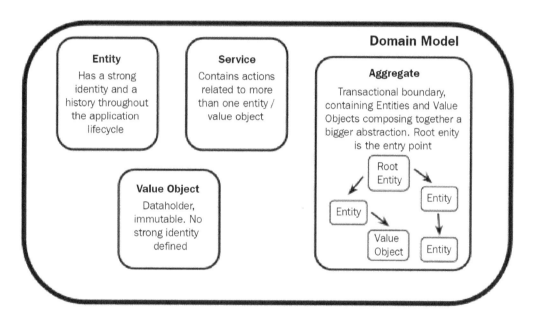

Figure 4.2 – The domain model

We will discuss each of the sections next.

Entities

The **entity** is a core concept of the domain model. Essentially, it is related to modeling objects that have an identity and a history throughout the life cycle of our use case. The keyword for defining an entity is **identity**.

If an object in our use case has a strong identity concept, it will probably map with entities. A classic example is a person: in many use cases (including the mobile payments example that we are carrying throughout this book), a person's identity is strongly defined, regardless of the values contained in its representing objects. In other words, if I have a person object, made up of the classic name, surname, and other details, having two objects with the same name and surname does not necessarily mean I am referring to the same person.

Indeed, I often resort to specific identifying fields (such as a tax code or something specific to my application domain—perhaps an account number) to distinguish a person object from another one. Moreover, the identity concept will still be valid even if the object is persisted (and retrieved). In other words, if I persist (or passivate) an entity object somewhere (such as in a database), it should be clear that it will refer to the same person (in real life) when it's loaded again.

As is clear, defining an entity is a cross-cutting concern between a business and the developers. It is much more than simply identifying a unique field distinguishing objects from one another. Consider bank accounts: they are usually identified by a standard code that is internationally recognized, at the very least, across Europe (IBAN code). However, you might find that a bank account changes the associated IBAN code (such as when a merger between different banks occurs). In this case, *do the two IBAN codes refer to the same account? Will the old account disappear and be replaced by a new one? Should I instead use a third identifier (such as a UUID) to bridge between the two entities and bypass the problem?*

Usually, the answer is that it depends. In this scenario, it depends on the domain around which your use case is modeled. The identity concept can also be different in the same application (in an extended way). Ultimately, an entity object is very much related to the point of view you are considering. However, for sure, it needs to be an object with a very well-defined identity, regardless of the value of its attributes, which links us to a different kind of object—**value objects**.

Value objects

Conceptually, value objects complement entity objects. Simply put, in a value object, the data inside the fields of the object is more important than the object's identity. Value objects simply transport information, and they can be shared, copied, and reused with ease. A typical example of a value object is an address (such as a city, street name, or zip code). It doesn't matter what the identity of each one is; what does matter is the data inside.

Value objects should be immutable. Because they are immutable, they are simpler to use. One common example is multithreading: multiple threads can access the same object instance concurrently, and there is no need for locks, nor any risk of inconsistent value (as the value cannot be changed). It's the same with passing object instances to methods: you can be sure that whatever happens, the value of the object cannot be changed. Essentially, with immutable objects, the life cycle is just easier to manage.

Value objects are usually lighter and safer to manage than entity objects. Additionally, they can be part of an entity, that is, our person entity might have a link to an address value object. However, you should balance the usage of entities and value objects. If you only resort to value objects, you will probably fall into an anemic domain. There is still an important thing to discuss regarding object content, that is, *where can we put the behavior that doesn't belong to either entities or value objects?* The answer is **services**.

Services

As mentioned earlier, entities and value objects are different in terms of identity. Instead, they share the grouping around a logical area, including data and behavior. In other words, both entities and value objects contain data (class attributes), the methods for manipulating it (getters and setters), and more sophisticated behavior (the business logic).

What's missing in this model is the cross-cutting behavior. Indeed, there are some actions that don't feel right when placed in a particular object. That's because those actions involve more than an object type, or they are simply ambiguous. It's important to not force those actions into an unrelated object, as this will impact the expressiveness of the model. Let's think about our mobile payment example again. *Should we put the peer-to-peer payment functionality in the sender or receiver account?*

For all of these scenarios, you can define a **service**. A service explicitly maps actions that are directly linked to the domain as a whole, rather than to a specific object type. In this way, you can nicely group similar actions together without polluting entities or value objects with behavior that doesn't belong there. It's all about keeping the domain model rationally organized, which is also the goal of the next concept: the **aggregate**.

Aggregates

We mentioned the concept of **aggregates** in *Chapter 2, Software Requirements – Collecting, Documenting, Managing*, when discussing event storming. It's worth saying that the whole idea of event storming is strictly related to DDD and one of the ways to put DDD into practice.

Let's return to the concept of aggregates; it's probably one of the most widely known ideas of DDD, and it's also widely used outside of DDD. Put simply, aggregates can be seen as transactional boundaries. The basic idea is to group a set of objects (that is, entities and/or value objects) by data changes. The objects in an aggregate are considered as a whole when it comes to changes to their internal status.

An aggregate has an entry point object, called a **root entity**. Any change to any object part of the aggregate must be carried out through the root entity, which will then perform changes on the linked entities. That's from a technical point of view rather than a domain model point of view. What you are doing is invoking operations (or, even better, actions that are as meaningful in the real world as in the domain model) in the root entity.

This will also mean changing the linked objects under the hood. However, this is an implementation detail. From a logical standpoint, all of the interactions with objects in the aggregate are mediated by the root entity. For this reason, the aggregate is a core concept in DDD. It strictly maps the consistency of the model and can be easily translated into technical concepts such as database transactions. Aggregates can then be seen as a sort of **super object** made by the coordination of different objects. As such, the construction of an aggregate can become complex. For this reason, DDD introduces the **Factory pattern**.

Glancing at DDD patterns

DDD encompasses some patterns to provide support functionalities for the domain model, such as building and managing objects (such as entities and value objects). The factory pattern is the first pattern that we will look at.

Factory

The **factory** pattern is not a new concept. You can refer to the *Design Patterns* book by the Gang of Four, where this has been widely explained. Simply put, if you want to programmatically control the creation of an object (or a set of objects such as an aggregate), and not rely on the logic of a constructor, you can use the factory pattern.

Factory is particularly useful to instantiate an aggregate. By invoking the factory pattern on the root element, you will coordinate the creation of the root itself and all of the other objects linked to the root (entities and value objects). Additionally, you can enforce transactionality on the creation of the objects.

If the creation of one of the objects fails, then you might want to abort the creation of the whole aggregate. The factory pattern can also be used to recreate objects from the database. In this scenario, rather than an instantiation from scratch, it's a retrieval of the existing root entity (and the linked subobjects). That's fine for addressing the retrieval of a known object (given its identity), *but how do you provide different kinds of lookups?* DDD suggests the usage of the **Repository pattern**.

Repository

A **repository**, in the DDD world, is a registry that is used to keep references to objects already instantiated (or persisted on a database). Simply put, a repository can be used to add, remove, and find objects. When used to find objects, typically, a repository acts as a bridge between the domain and the infrastructure layer.

It helps to decouple the features and hide the implementation details of the persistence layer. You can retrieve objects using complex or vendor-specific queries in the infrastructure layer, and this is wrapped by an operation in the repository. It might even be that the infrastructure layer retrieves objects in different ways, such as invoking external web services rather than a database. Nothing will change from a repository point of view. The services exposed by the repository must have an explicit domain meaning, whereas the internal implementation might appear closer to the infrastructure logic.

So far, in all of the concepts that we have examined, we have implicitly assumed that everything falls under one single domain model. Now, we will learn how to make different domains interact with each other by using the concept of a **Bounded Context**.

Bounded Context

It is common to identify a domain model using one application. That's a hard way to delimit the model boundaries. However, that's not always the case. When dealing with large applications, it could be that different models need to coexist. This is because a unified model is impractical (that is, too big or too complex), or because of the model's conflict (that is, an object has different meanings, depending on the point of view and the use cases touching it).

In those scenarios, you might need to define a border around each domain model. A bounded context, then, is the area in which ubiquitous language is valid. If a bounded context can be seen as a country, with defined borders, ubiquitous language is the official (and only) language spoken of that country.

Usually, a bounded context belongs to one team, and it has some well-defined coordinates, such as a portion of the code base and other subsets of related technologies (such as a defined set of tables of the database). Two different bounded contexts cannot share objects, nor call arbitrary methods of each other. The communication must follow well-defined interfaces. In order to support the cooperation between two different bounded contexts, a context map can be used.

A context map is a way to translate, when possible, concepts from one bounded context to another. There are some patterns suggested by DDD to realize context maps. These patterns include the following:

- **Shared kernel**: This is when two bounded contexts share a subset of the domain model. While this technique is easy and intuitive, it can be hard to maintain, since the two teams managing the different bounded contexts must agree on any changes, and in any case, the risk of breaking functionalities in the other context is always present, so every change must be thoroughly tested (automatic is better).

- **Customer supplier**: This is similar, in a way, to the shared kernel approach, but the relationship here is asymmetrical. One of the two bounded contexts (the supplier) will own the interface, developing and maintaining the features, while the customer will simply ask for what is needed. This simplifies the synchronization a bit between the two teams. However, it can still create issues when priorities and milestones start to clash.

- **Conformity**: This shares the customer-supplier type of relationship. The difference here is that the customer domain model completely adopts and imports a subset of the supplier domain model, as it did in the shared kernel approach. However, unlike shared kernel, the relationship stays asymmetric. This means that the customer cannot change (or ask for changes in) the shared model.

- **Anti-corruption layer**: This is a different approach. In this case, there is a translator layer between the two domain objects. This layer acts as a demilitarized zone, preventing objects and behaviors from sneaking from one bounded context to another. This approach is commonly used when dealing with legacy applications, more than when two bounded contexts belong to the same application.

It is worth noting that a proper DDD implementation is not easy to follow. There are several common errors that could slip into a DDD architectural design. The first and most common is the aforementioned anemic domain model, which is the most important reason why you would want to adopt something like DDD. However, it's also common to have some technology considerations slip into the domain model.

That's particularly true when it comes to the **persistent layer**. It is a common practice to design the domain in a way that mimics the database tables and relationships (in this case, we are using a relational database as a persistent backend). Last but not least, one common error is to design the domain model without engaging with domain experts.

We could be tempted to design everything for the IT department, thinking we have a proper understanding of the world we would like to represent. Even if this is partially true, it's still worthwhile engaging with business experts, to better discuss the business jargon (please refer to the *Understanding ubiquitous language* section) and rely on their experience of the specific domain model.

This section concludes our brief overview of DDD. As we have learned, DDD provides elegant ways in which to realize the ideas we have collected in the previous sections (including requirements and architectural designs) and put them into code.

This starts with the concept of ubiquitous language, which we discussed at the beginning of this section and is one of the big ideas of DDD, allowing common ground between all the stakeholders involved in the application development.

Following this, we moved on to the core concepts of DDD, such as the application *shape* (the layered architecture), the definition of objects and methods (the domain model and the encompassed objects), and the recommended practices (patterns) regarding how to address common concerns. A dedicated mention is needed for the concept of bounded contexts, which is a way to structure big applications into more than one *self-contained* model.

As we will learn in *Chapter 9*, *Designing Cloud-Native Architectures*, DDD has some common ideas with microservices architectures.

In the next section, we will look at another common practice to drive the implementation of our design ideas—TDD.

Introducing Test Driven Development

TDD is a development technique based on a simple idea, that is, no code should exist without test coverage.

In order to pursue this goal, TDD inverts our point of view. Instead of developing code, and then writing a unit test to cover its testing, you should start writing a test case. Of course, initially, the test case would intentionally fail while invoking empty or incomplete functions. However, you will have a clear goal, that is, your piece of code is complete when all tests are satisfied.

Starting from the end, you clearly define the boundaries of your software and the extent of its functions. Then, you run the tests, which will all fail. You keep developing the features, piece after piece, until all of the tests are satisfied. Finally, you move to the next piece of code (or class or package)—it's that simple.

Remember that this approach doesn't necessarily guarantee any particular quality or elegance in your code. Having a test pass does not imply that you are using good patterns or efficient solutions. In theory, you might as well simply hardcode the expected results to get a green light.

However, this technique will have a very useful byproduct, that is, you can't forget (or purposefully avoid) to prove/test your code using test cases.

Anyway, there are several factors to take into account. First of all, some features might require external systems to work. You can test the interaction of such systems, simulating them with mocks, but of course, this will mean more code to write, more components (the mocks themselves), and a further degree of approximation (meaning that your test will be less representative of reality). Following this, you might need to test things that are less easy to automate, such as UIs and interactions with devices (for example, mobile devices). Yes, there are a number of solutions for this (such as automating browser navigation), but this will complicate things.

Let me highlight that, even if this will require a significant amount of effort, tests cannot be ditched. Testability is a crucial requirement, and it might also be a drive to rearchitect your code base, increasing modularity and simplifying it, in order to improve testability.

Moreover, you might have dependencies between the features. This means coordinating tests or, worse, having test results depend on the order in which they are running. Of course, this is not easy to maintain and, in general, is not a good idea.

In this specific case, you might want to properly structure your tests, in order to provide adequate setup and teardown phases, making everything simpler and reproducible and greatly increasing the quality of what you are testing. Then, you have to think about the granularity of the tests. It can be tempting to create one generic test and slip in as many hidden features as you can. On the other hand, if your tests are simply unit tests, covering every sub-function, you'll need to aggregate them in a meaningful way, in order to track down the advancements in implementing the features. In other words, shifting your point of view away from testing specific code sections toward testing application behavior.

This is the idea behind BDD.

Exploring Behavior Driven Development

BDD is a technique that extends the TDD approach while also using some DDD concepts. In particular, the workflow is the same as TDD, that is, write a test, run it (initially, it will fail), and implement the feature until the test succeeds.

What changes is how the test is written. Instead of focusing on single functions (or, even better, relying on the developer to pick the right granularity), BDD defines the extent of each test a bit better. In particular, as highlighted in the name of the methodology, BDD tests are focused on the expected behavior, that is, the functionality implemented by each use case.

In this sense, it is an explicit suggestion to keep high-level functionalities, rather than method-by-method unit tests. BDD is also linked to DDD concepts. In particular, it is recommended that you use ubiquitous language as a way to specify each behavior. In this way, you have an explicit mapping between a business use case, expressed with ubiquitous language, translated into an automatic test case.

BDD describes a way to define behaviors. In practice, each behavior is defined as a user story, with a structure given as follows:

- **As a**: This is a person or a role.
- **I want**: This is a specific functionality.
- **So that**: This is when we can get some benefits from using that functionality.

Provided that a number of scenarios are associated with the user story, each scenario is, essentially, an acceptance criterion, which can be easily translated into automated use cases:

- **Given**: This is used for one or more initial conditions.
- **When**: This is used for when something happens.
- **Then**: This is used for when one or more results are expected.

This structure is very self-explanatory. By using a similar template, and sticking to ubiquitous language, you will have a straightforward way in which to define use cases. It is a way that is meaningful for non-technical people and can be easily translated to automated use cases by technical people.

Walking backward, you implement code that will gradually cover the test cases, mapping to a behavior specification that will give direct feedback to the business on which use cases are complete.

This approach offers a structured way to understand what we are implementing and possibly select and prioritize the user story to approach as a development team. This is also the focus of the practice that we will look at in the next section.

Comparing DDD, TDD, and BDD

So far, we have rapidly discussed three different "Something-Driven Development" techniques. It must be clear that such practices should not necessarily be seen as alternatives, but they might have some complementarity.

In particular, DDD relates more to the modeling of the application domain. In this sense, it can be observed from a more architectural point of view, defining how our application is modularized, the different layers, and even how different parts of our broader application (or, if you wish, different teams) should cooperate.

Once we have designed such layers and components, both TDD and BDD can be used as a way to drive our day-to-day development, ensuring we have the right testability and feature coverage requested within our code.

On the other hand, DDD is not a requirement for TDD or BDD, which can be seen as a simple technique that is also applicable to smaller applications, or to software architectures defined with approaches alternative to DDD. As you will often find in this book, those concepts can be viewed as tools, briefly introduced to give you an idea of their potentiality. It's up to you to then take what's needed for your specific project and combine it in a useful way.

Learning about user story mapping

User story mapping is a way to put user stories into context, identify what it takes to implement them, and plan accordingly.

In this section, we will learn what a user story is and how it can be used as a planning method, in order to choose what features to include in each release, following a meaningful pace.

The user story is the same concept that we saw as part of BDD. So, it describes a feature from the point of view of a specific persona (**As a**...), the functionalities required (**I want to**...), and the expected outcome (**So that**...).

As you might observe, all pieces in the puzzle of those seemingly unrelated practices eventually start to match. User story mapping is often described as a **product backlog** on steroids.

We will discuss product backlogs in the next chapter. However, for now, consider them as lists of features to implement. Features are added as long as analysis occurs (or new requirements arise). These are then prioritized and picked by the development team to be implemented.

User story mapping extends this approach by giving more dimensions to the product backlog, enriching the information related to each feature, and linking it to a broader vision of the product. User stories stay on top of the mapping. They describe the high-level features that a system should provide. User stories are organized in a horizontal line and ordered by both importance and the temporal sequence in which they happen, all from the user's point of view.

For each user story, a list of tasks is provided. Essentially, these are the sub-features (also known as **activities**) that each user story encompasses. So, we are detailing each feature, but not yet coming to a level of detail that can be directly mapped into software (at least, not easily). Each task is then attached to a list of subtasks (or task details), which are easier to map to software features. This is what user story mapping looks like:

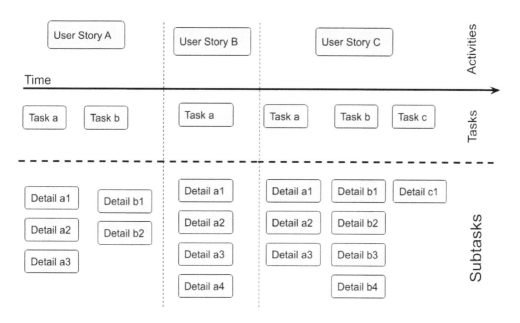

Figure 4.3 – User story mapping

The interesting thing about this model is that you will not have to prioritize the tasks. You just need everything included there (sooner or later); however, you can prioritize the subtasks, gradually improving the completeness of each task, release after release.

This model has a series of positive outcomes. First of all, as with BDD, you have a direct mapping between the subtask the development team is working on and the user story (or activity). Essentially, it gives visibility to the business regarding the finish line toward which we are rowing. Moreover, an interesting practice is applicable to this matrix of tasks and subtasks, namely, **value slicing**. This means picking what to implement for each release.

Given that you will have a finite number of resources (such as programmers, time, and whatever else is required to implement each subtask), you cannot, of course, deliver everything in one release. Well, you could, but it would be risky since you would have to wait a long time before receiving feedback and being able to test the software. We will elaborate more on the **release early, release often approach** (the well-known incremental product releases technique that is widely used in **Agile** and **DevOps**) in the next chapter.

For now, what matters is that it is better to release value incrementally, by picking the subtasks that implement, at least partially, one or more tasks and then the related user stories. Here is what this would look like compared to *Figure 4.3*:

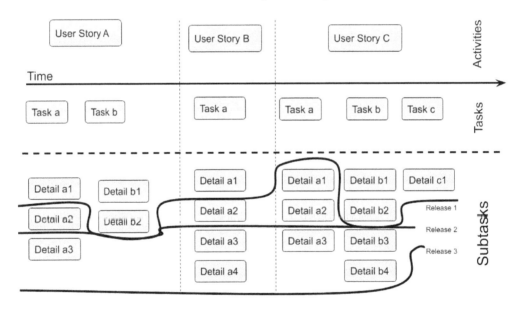

Figure 4.4 – Value slicing

As you can see, the approach here is oriented toward an MVP.

The MVP

The concept of MVP deserves some additional explanation. The term was created before the user story mapping technique and is an independent idea. It is also applicable to products that are different from the software code. The goal of the MVP is to maximize the value of the product (in terms of return of investment or, trivially, how useful, popular, and beneficial your product will be) while minimizing the risks and efforts required to build it. The perfect MVP requires a very low level of effort and risk to build, but it can become greatly popular and appreciated when used (and, optionally, sold).

The purpose of an MVP is to start getting feedback on the product from potential end users (usually, a subset of early adopters). Due to this, the MVP should contain a meaningful subset of features: not too many, to avoid wasting effort in case the product is not well received by customers, but just enough in order to represent what the complete product will be like. Early feedback, in the spirit of Agile development methodologies, could also be beneficial if some steering is required in the product direction, by stressing more on one aspect or another.

In this section, we learned about user story mapping, which is the final technique we will explore in this chapter. In the next section, we will examine some examples of those approaches, as applied to our mobile payments use case.

Case studies and examples

As is easy to imagine, a complete and extensive example of DDD, TDD, BDD, and user story mapping, applied to our mobile payments case study, could easily take more than one book. For this reason, as we mentioned in *Chapter 3*, *Common Architecture Design Techniques*, unfortunately, we can only look at some highlights of those techniques used in our example. However, in this section, I think it is pretty useful to take a look at, even if to just practically visualize some concepts that might appear abstract so far.

The mobile payments domain model

In *Chapter 3*, *Common Architecture Design Techniques*, we looked at the basic modeling of mobile payment objects based on the UML notation. To elaborate more on this, in DDD, you will mostly have the following concepts:

- The user is an entity. This concept is pretty straightforward, that is, the identity is very well defined, and each user has a well-defined life cycle (from registration to deletion).

- Payment is an entity, too. Each user will want to keep track of exactly each transaction, including the time, the amount, the receiver, and more. It is also likely that there will be regulations for you to uniquely identify each payment transaction.

- As we've already mentioned, a peer-to-peer payment is out of place as a method, both in the sender and receiver entity. So, it is probably worth modeling a payment service that can also work as a bridge toward classical CRUD operations in the infrastructure layer.

- On the assumption that our application is operating on a global scale, you will need to manage transactions in different currencies. `ExchangeRate` is a typical example of a value object. It is immutable and composed of currency symbols and a number representing the exchange rate. It is a disposable object and can be easily shared between different payments, as no identity (nor state) is considered.

Once we have defined (a very small subset of) the domain model of mobile payments, we are going to look at the layered architecture of this application.

The layered architecture of mobile payments

If you remember the diagrams designed in *Chapter 1*, *Designing Software Architectures in Java – Methods and Styles*, and the C4 model drafted in *Chapter 3*, *Common Architecture Design Techniques*, you are already familiar with some of the technical components that implement our mobile payment architecture.

There, the mapping between components was pretty much coarse-grained. This is because you would associate the mobile application with the **Presentation Layer**, the business logic with the **Domain Layer**, and so on. However, with DDD, we are progressing further with the analysis of our application. We are going one level down toward something similar to the **C4 Container** diagram (please refer to *Figure 3.8* in *Chapter 3, Common Architecture Design Techniques*) but from a different point of view. My idea of the layered architecture of our application looks similar to the following diagram:

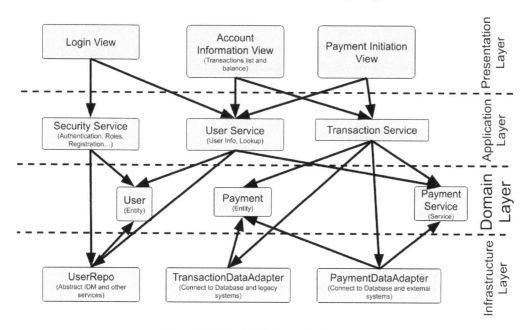

Figure 4.5 – The DDD layered architecture

From the preceding diagram, you might observe the following:

- All of the views in the **Presentation Layer** are a subset of the mobile application functionalities. You will probably have more functionalities in the real world. However, it's interesting to observe how some concepts of the **Domain Layer** (hence, the ubiquitous language) are echoed here. Yet, this is a pretty technical layer, so it does not strictly observe the ubiquitous language.

- The **Application Layer** is a support area, decoupling the needs of the frontend with the services provided by the domain model in the **Domain Layer**. The relationship with the **Presentation Layer** is not one-to-one in this case, but that's up to you to decide according to your context. Additionally, this layer has a dependency on the **Domain Layer**.

- In the **Domain Layer**, we strictly map our domain model. So, the ubiquitous language here is prevalent. Also, this layer should not have a dependency on the neighboring layers in order to stay technologically independent as much as possible (for the sake of clarity, `ExchangeRate` is not represented).

- The **Infrastructure Layer** is the technological glue, providing services to other layers, and abstracting technology-specific details. So, in this case, you can see that **UserRepo** will mediate calls to IDM and other systems (for example, databases or CRMs), while **TransactionDataAdapter** abstracts calls to databases and legacy systems. Consider that in this scenario, there are no direct links between the **Presentation Layer** and the **Infrastructure Layer**, as everything is proxied by the **Application Layer**. However, that's not a strict requirement.

In the next section, I will share my views on how BDD could be applied to mobile payments.

BDD of mobile payments

As we detailed in the *Exploring Behavior Driven Development* section, BDD starts with a user story. A basic user story for mobile payments could be the following:

User Story: Making a payment:

- **As a** registered user.
- **I want** to make a payment to another user.
- **So that** I can transfer money (and benefit from services or goods in exchange for that).

As you might have gathered, this user story implies other user stories (such as **Registration of a user** and **Login**).

The next logical step is to enumerate some scenarios (or acceptance criteria) linked to that story:

- **Given** that I am registered.
- **And** I am logged in.
- **When** I select the payment feature.
- **Then** I am redirected to the payment view.
- **Given** that I am at the payment view.
- **And** I am logged in.

- **And** I enter a valid recipient.

- **And** I enter a valid amount.

- **When** I click on the pay button.

- **Then** a payment transaction is created.

- **And** a notification is sent.

As demonstrated in the preceding examples, each user story usually corresponds to more than one acceptance criteria, which is then codified as a set of (possibly automated) test cases. Following this, you can start to iteratively implement features until each acceptance criteria is met, ultimately fully covering the related user story. Now, let's expand on this user story by means of user story mapping.

User story mapping of mobile payments

In the *Learning about user story mapping* section, we discovered that the top-level element is the user story. So, we will start with the stories that we have just observed in the previous section.

Take into account that while it can be considered as a task attached to each user story, the acceptance criterion is usually considered more like an orthogonal concept, to validate the implementation of each story. Usually, the attached tasks are simply more detailed features composing the story itself. Let's view an example:

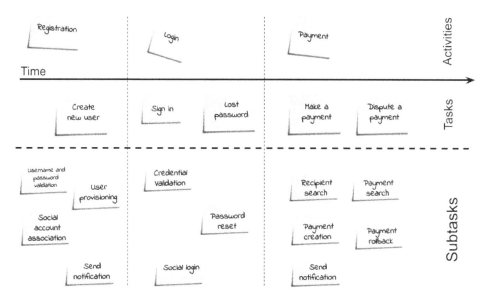

Figure 4.6 – User story mapping example

As you can see in the preceding (simplified) example, for each activity (mapping to a user story, as per the *BDD of mobile payments* section) there are one or more related tasks. Activities and tasks are ordered following a time (and priority) direction. Then, each task is attached to a list of subtasks.

It's a logical next step to plan how to group a set of subtasks as a release, progressively delivering value to the final customer (think about MVPs). We've described this approach as value slicing, which appears as follows:

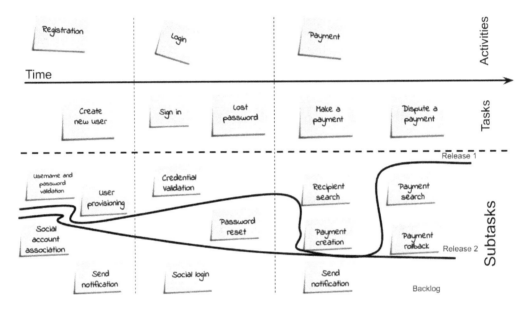

Figure 4.7 – Value slicing example

As you can see in the preceding diagram, we've represented a simple slicing of features as two releases. In the first release (**Release 1**), you will provide the bare minimum functionalities. It will be possible to create users, to **Sign in** (but not using a social account), and to **Make a payment** (but without receiving a notification).

There will be no functionalities regarding lost passwords and payment disputes. In the second release (**Release 2**), no new features will be added to the registration activity, the **Lost password** task will be completed (being made of just one subtask), and the whole **Dispute a payment** task will be completed (in both its two subtasks).

All of the other subtasks are part of the backlog, meaning they are yet to be planned (and more subtasks can then be added). Of course, each line representing a release is drawn together with business/product owners, which will define the priority and helps to aggregate subtasks in a meaningful way. With value slicing, we have completed the objectives of this chapter. Let's look at a quick recap of all the notions we have encountered.

Summary

In this chapter, we looked at a set of techniques to start transforming architecture principles into working software components. DDD is a pretty complete framework that is used to define objects and the way they interact with each other. It puts a number of clever ideas down on paper, such as layered architectures, patterns, and bounded contexts.

Following this, we moved on to Test Driven Design and BDD. You now understand specific ways of structuring the development of new code and mapping it to business features. Finally, we looked at user story mapping as a way to pick functionalities to implement and link them to tasks and activities.

All of these techniques will be better framed in the next chapter, where we will discuss **Agile** methodologies, which include some of the practices that we have just discussed.

Further reading

- *Anemic Domain Model* by Martin Fowler (`https://martinfowler.com/bliki/AnemicDomainModel.html`)

- *Domain-Driven Design: Tackling Complexity in the Heart of Software* by Eric Evans (`https://www.domainlanguage.com/`)

- *Domain-Driven Design Quickly*, by Abel Avram and Floyd Marinescu, published by C4Media (2007)

- *10 Common DDD Mistakes to Avoid* by Jan Stenberg (`https://www.infoq.com/news/2015/07/ddd-mistakes/`)

- *Introducing BDD* by Dan North (`https://dannorth.net/introducing-bdd/`)

- *BDD in Action: Behavior-Driven Development for the Whole Software Lifecycle*, by John Ferguson Smart, published by Manning Publications (2004)

- *User Story Mapping & Value Slicing* by Matt Takane and Ryan DeBeasi (`https://openpracticelibrary.com/practice/user-story-mapping/`)

- *Top 5 Biggest Challenges when Building an MVP and how to Avoid Them* by Ilya Matanov (`https://expertise.jetruby.com/top-5-biggest-challenges-when-building-an-mvp-and-how-to-avoid-them-2969703e5757`)

- *The New User Story Backlog is a Map* by Jeff Patton (`https://www.jpattonassociates.com/the-new-backlog/`)

5
Exploring the Most Common Development Models

In this chapter, we will position some of the notions we have discussed so far into a more complete picture. We are going to elaborate on the most common development models. We've already seen the importance of designing proper architectures, how to collect requirements, and how to translate the architectural ideas into code solutions that answer those requirements.

The software development models that we will see in this chapter revolve around all of those aspects (and some more), arranging them in proper and tested ways, to achieve different results or emphasize certain areas.

In this chapter, we will cover the following topics:

- Learning about Code and Fix
- Glancing at the Waterfall model

- Understanding the Agile methodology

- Introducing Lean software development

- Exploring Scrum

- Learning about other Agile practices

- Understanding DevOps and its siblings

When discussing development models today, everybody goes all-in with **DevOps** and **Agile** techniques. While I do endorse all this enthusiasm for those approaches, my personal experience says that it's not that easy to apply them correctly in all the different contexts. For this reason, it's important to know many different approaches and try to get the crucial lessons from each one. Even if you do not have the ideal conditions for working with DevOps, it doesn't mean that you cannot use some of the good ideas associated with it.

At the end of this chapter, you will have an overview of the most widespread development models, along with their pros and cons. This will help you choose the right model, depending on your project needs.

But first, let's start with the naive development model (or a *non-model*, if you want), that is, **Code and Fix**.

Learning about Code and Fix

Let me get to the point as soon as possible – Code and Fix is not a model. It is something more akin to anarchy. The whole concept here is about diving into coding with no planning at all. For this reason, it is called Code and Fix. In this, you completely skip all the crucial phases highlighted hitherto (requirements collection, architectural design, modeling, and so on) and start coding.

Then, if things go wrong, such as there are bugs or the software does not behave as expected, you start fixing. There is no dedicated time for writing documentation, nor for **automation** and **unit testing**. Versioning of the code is naive, and so is the dependency between modules (or maybe everything is stuck in just one huge module).

As you can imagine, there are few, if any, advantages to adopting this non-model. Let's start with the (obvious) disadvantages:

- You are basically working against whoever will maintain the code (perhaps your future self). All the quick fixes and workarounds that you will stick into your code will come back to bite you when you need to touch it again. This phenomenon is usually known as **technical debt**.

- Since you are not analyzing requirements properly, you risk wasting effort working on a feature that does not provide any value to the customer and the final user.

- Collaboration between developers in the team, and with external teams, is hard, as there is no clear separation of duties (hence, Code and Fix is also known as **cowboy coding**).

- It's hard to estimate the time needed to complete a release.

So, it's easy to say that adopting Code and Fix is not advisable at all. But surprisingly enough, it is still very widespread. These are the main reasons for its widespread application:

- Small teams with no dedicated roles (or with just one developer)

- A lack of skills and experience

- A lack of time (not a good excuse at all, as a bit of structure will probably save time anyway)

However, Code and Fix can be partially justified when working on very small projects that will not require any maintenance or evolution, such as prototypes or projects with a defined, short lifetime.

It's also worth noticing, before diving into more complex and complete techniques, that embracing such methodologies is not a warranty of a successful project, and implementing Scrum, DevOps, or whatever you like is not going to be the perfect way to avoid a technical debt. Indeed, the software development methodologies are suggestions on how to give cadence on a project and what are the meaningful splits of roles and responsibilities, as seen in other projects. But it's ultimately the responsibility of the project team (and yours, as an architect) to ensure that the methodology (if any) is correctly used and that no pieces are left behind, in terms of technical debt, code quality, and project scheduling.

With that said, the natural step after Code and Fix is to provide a bit of structure, sequentially, which is known as the **Waterfall** model.

Glancing at the Waterfall model

As has been said, the Waterfall model is a structured development model based on a sequence of different phases. This means that each phase begins when the previous one has ended.

The Waterfall model probably stems from the application of project management practices coming from other kinds of projects, such as constructing buildings or manufacturing objects. Indeed, while I am no expert on them, it's easy to understand that in order to build a house, you have to precisely follow a sequence of steps, such as calculating the materials and weights, building foundations, and constructing walls.

The Waterfall model originated from a number of different articles and lectures (with the most important coming from Winston Royce) and has also been ratified in an official document by the **US Department of Defense**.

The phases in the Waterfall model are as follows:

- **Requirement management**: You probably have a very good idea of what this phase entails, as per *Chapter 2*, *Software Requirements – Collecting, Documenting, Managing*. In the Waterfall model, the requirement specification must be completed and formally accepted before proceeding with the next phase, while we discussed how, usually, an iterative approach is more natural.

- **Analysis/design**: Sometimes defined as two different phases, the goal is to start from system requirements and then define the solution architecture to satisfy them. As in the previous phases, whoever is in charge of the next phase must formally accept the deliverables coming from this phase (such as system blueprints, diagrams, and pseudocode) in order for the process to continue. This basically means that developers must clearly understand what they are supposed to implement.

- **Implementation**: In this phase, the development team, starting from requirements and from the deliverables produced in the previous phase, must write the code to implement a proper software solution. This phase is, of course, crucial, and the correct completion of this phase basically means the success of the whole project.

- **Testing**: As seen before, the acceptance of the deliverables coming from the preceding step is part of each phase. With testing, the approval is so important (and so complex) that it overlaps with the whole phase. The Waterfall model does not specifically distinguish between different kinds of testing, but this phase is commonly intended as **user acceptance testing**.

- **Operations/maintenance**: This is the final step, facilitated by technical activities ensuring the proper setup of the solution in a production environment, as well as all the planned and unplanned activities to keep it operating properly.

The following diagram demonstrates the phases of the Waterfall model:

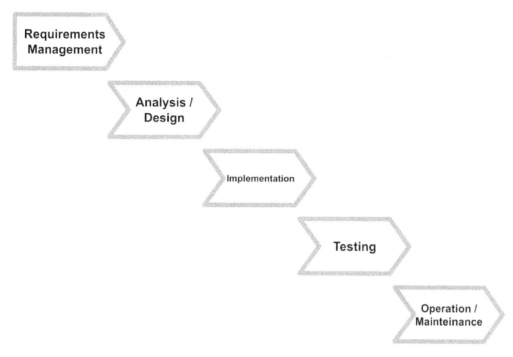

Figure 5.1 – The Waterfall phases

As you can see, the Waterfall model is a big jump when starting from Code and Fix, as we are starting to see a clearer distinction of what should be done in each phase.

Advantages and disadvantages of the Waterfall model

The Waterfall software development model is still widely used. Some of the advantages of this model are as follows:

- There is a clear definition of phases, hence planning is rather easier. Even though the phases should not overlap, it flows sequentially.

- The mechanism encourages a proper handover between teams, including a formal acceptance between one phase and the following, giving greater control over planning and project quality.

However, as you probably already know, there are some disadvantages to this methodology:

- The first and most evident disadvantage is the lack of flexibility. If you implement the Waterfall model entirely, you should not start implementing it before requirements have been collected in their entirety and the design has been carried out in full. In the real world, this is unlikely to happen; as we have seen, the requirement collection keeps flowing and the architecture design evolves while we face (and resolve) implementation issues.

- Moreover, the software that we are building is seen and tested once implementation is complete. This means that you will not receive feedback on your code until very late in the project (maybe too late).

For this reason, the Waterfall model has undergone several transformations, usually shortening the feedback loop, or cycling and jumping between phases (such as adding more requirements while implementing or managing defects identified during the testing phase). But while this model is still used, more flexible methodologies are now widespread, as they offer a less risky approach to development, and it all starts with Agile methodologies.

Understanding the Agile methodology

The **Agile** methodology is a galaxy of best practices and techniques. A lot of still widely used methodologies inspired Agile (such as **Scrum** and **Kanban**), but the official birth of the movement stems from the *Agile Manifesto*, published in 2001. The four very popular key concepts (values) of the *Agile Manifesto* are as follows:

- *Individuals and interactions* over processes and tools
- *Working software* over comprehensive documentation
- *Customer collaboration* over contract negotiation
- *Responding to change* over following a plan

While some of the preceding points can be misinterpreted and result in bad behaviors, such as ditching documentation and planning, it's enlightening to think about such simple but powerful advice. Also, be careful that the manifesto itself advocates against complete anarchy as a result of the following note:

*"While there is value in the items on the right, we value the items on the
left more."*

> **Important Note:**
> The values on the left here are the ones mentioned at the beginning of each value and refer to *freedom* (such as working software), while the ones on the right are the ones at the end, referring to *discipline* (such as comprehensive documentation).

This part is often foreseen by teams looking into Agile methodologies as an excuse to skip the boring parts of the development process. The Agile process appreciates freedom but does not preclude some level of order.

There is moreover a very important observation to make while introducing the topic of Agile. The Agile methodology, and all its implementations described in this chapter, consider it crucial to have the customer (or the business owner, in other words, who is paying for the project) be aware of the methodology and willing to be part of it. Indeed, it's common to see in the following *The Agile principles* section the advice and principles involving the customers, as they're an active part of the software development project by providing inputs and feedbacks in many steps of the process. For such a reason, the adoption of an Agile practice will not be possible if the customer does not agree (implicitly or explicitly) with it.

The *Agile Manifesto* further details the basic values of Agile by providing a list of principles.

The Agile principles

As opposed to Waterfall, Agile bets everything on collaboration (within the team, and with customers too) and releases small chunks of working software often with a view to getting feedback early and adapting planning if necessary. Instead of trying to foresee everything and plan accordingly, Agile teams focus on quickly adapting to changing conditions and acting subsequently. This is well detailed in the Agile principles:

- Our highest priority is to satisfy the customer through early and continuous delivery of valuable software.

- Welcome changing requirements, even late in development. Agile processes harness change for the customer's competitive advantage.

- Deliver working software frequently, from a couple of weeks to a couple of months, with a preference for the shorter timescale.

- Business people and developers must work together daily throughout the project.

- Build projects around motivated individuals. Give them the environment and support they need and trust them to get the job done.

- The most efficient and effective method of conveying information to and within a development team is face-to-face conversation.

- Working software is the primary measure of progress.

- Agile processes promote sustainable development.

- Sponsors, developers, and users should be able to maintain a constant pace indefinitely.

- Continuous attention to technical excellence and good design enhances agility.

- Simplicity – the art of maximizing the amount of work not done – is essential.

- The best architectures, requirements, and designs emerge from self-organizing teams.

- At regular intervals, the team reflects on how to become more effective and then tunes and adjusts its behavior accordingly.

As you may see, other best practices are stressed in these principles, such as focusing on good architecture, privileging simple solutions, and building motivated teams. Of course, those are the general ideas. Before and after the publishing of the manifesto, a number of practices have been built around similar topics. In the next section, we will talk about Lean software development, a practice often associated with Agile development, which has its roots in the manufacturing industry.

Introducing Lean software development

Lean software development is a framework developed after the manufacturing method of the same name, which, in turn, is derived from the **Toyota Production System**. The interesting concept regarding this topic, indeed, is how it translates best practices from industrial production into software production. This is also due to the experience of one of the authors (Mary Poppendieck) in this context. She worked in the manufacturing industry and had the opportunity to learn about the production processes in a factory context directly.

We will quickly cover a selection of the principles of Lean software development in the upcoming sections.

Eliminating waste

Waste is a concept directly mutated from the Toyota Production System. Basically, waste is everything that costs resources without giving any value to the finished product.

Taking it to the extreme, in software development, everything that is not related to analysis or coding could be a waste. This can be seen as another point of view in the **simplicity** Agile principle.

To identify waste in software development, Lean software development suggests looking into its seven main areas:

- **Partially done work**: This area relates to non-completed or non-released features. This means accumulating code, which has to be maintained, without providing any utility to the final customer. Moreover, since incomplete work is never proven in production, you can never be 100% sure that everything works as expected. You can also take into account the fact that releasing the software, which we are building in production, is the only way to understand whether such code is valuable.

  A famous paper by Ron Kohavi states that just one-third of the implemented features provide positive impacts, while the rest are neutral or even negative. The only way to figure it out is to release the code in production and see the feedback of real customers using it.

- **Extra Processes**: This refers to bureaucracy. This means paperwork, approval processes, and similar issues. We all know that there are things that just can't be skipped, such as security checklists and handoff documents for production release. Often, however, those processes are overcomplicated and overengineered. This area should be looked at for simplifications or even automation where relevant. Instead of manually answering security-related questions, maybe you could just run automated tests, as an example.

- **Extra features**: This is a very common pitfall. Perhaps in the requirement analysis, we are just pushing more and more features without any specific thoughts on whether those are useful or not. Or maybe, when implementing a new feature, it's just so easy to add a similar one, which nobody is asking for but *can be useful sooner or later*. This is just wrong. Even if the code is easy to add, it must be maintained, or else it can potentially introduce bugs.

- **Task switching**: Now, it's common sense to know that context switches are time-consuming. That is particularly true in software development, where you have a lot of things to sort out, from setting up your environment (although this can, and should, be automated), to focusing on project structure and code standards, and recalling the team dynamics and latest updates. It is basically as painful as it seems, yet very tempting to juggle multiple projects at a time.

- **Waiting**: This is a very common thing to relate to. We end up waiting for a number of reasons, such as the environment being created and an analysis being completed. While the technical stuff can be mitigated by automation, from a project management standpoint, it is way harder to plan everything to ensure synchronized handoffs between teams. To act against waiting, you may be tempted (or forced) to help out on other tasks and projects, while this can easily transform into other waste (as per the previous point, task switching is not the best idea).

- **Motion**: As introduced previously, we have handoffs between different teams. That's the concept of motion. The longer it takes, the more waste you will have. This includes having a huge amount of back and forth, or simply too many teams cooperating. Handoffs not only include the exchange of artifacts (such as source code) but knowledge in general (such as documents or simply answers).

- **Defects**: Everybody knows what a bug is and how much time it can take to find the causes and solve it. Of course, it's just impossible to write software without any bugs. But there are things you can do to reduce the impact of bugs, such as improving test coverage (including code analysis), which will end up saving time by identifying issues before they move into a snowball effect. Also, as has already been discussed, the sooner you go into production, the sooner you will find bugs (and have the opportunity to enrich your test suite).

To identify waste in your software production cycle, the Lean software development framework provides a very useful tool called **Value Stream Mapping**.

Value Stream Mapping is used to observe the software development process from an external point of view, mapping all the steps necessary (and the waiting time between them) for a requirement to go from inception to production release (usually known as the time to market).

You are supposed to track down this simply with paper and pencil. After tracking down the whole software cycle, you usually end up figuring out that the majority of the time is lost in waiting or in other types of waste, as per the previous list. Now that you have some quantitative data, with good executive sponsorship, you can act by changing the flow to maximize the time spent delivering value and minimizing waste. This will usually include simplifying approval processes and automating manual steps. The efficiency that can be attained here is mind-blowing.

Deciding as late as possible

This section is all about being open to changes. Especially when making expensive choices, it is good to defer the decision as much as possible, as more information may come to light to support the choice. Moreover, making a decision later will reduce the risk of having to get back to redoing part of the work owing to a wrong decision.

However, there are more subtle implications in this principle. What comes to my mind is the mythical quote from Donald Knuth:

"Premature optimization is the root of all evil."

This means that if you make choices (especially hard to undo choices) too soon, you may end up making the wrong choice because of a lack of information, or simply wasting time with a topic that will end up not being that relevant. So, one strong piece of advice from the Lean software development framework is that you shouldn't commit to everything unless you have to, stay open and flexible, and defer from making complex decisions until you have no alternatives.

Translated in the software world, there are a number of different ways to do this, such as using stubs instead of real systems (before deciding which system to use), defining modular options (to facilitate the switching of different implementations), and using feature flags (to elicit specific behaviors directly in production). Just make sure that you find the right trade-off to avoid piling up waste. Implementing tens of different behaviors because you don't know what the final decision will be is, of course, not an option, but there are middle grounds.

A rule of thumb is usually to avoid planning for years or even months in advance. It's better to end up with very detailed planning for the upcoming weeks, which will become less and less detailed going forward in time.

Delivering as fast as possible

This is a concept that I've emphasized a lot, so I will keep it as concise as possible. Organizing the delivery work in small chunks is key. That's what *fast* refers to. You have to plan for releasing often. This will do for having feedback early and perfecting your strategy on the go.

There are several pieces of advice here, such as having a regular rate of release (both in terms of the time window and in terms of the number of features) and moving from a push to a pull approach (there will be more on this when we discuss it in the *Kanban board* section). Personally, I think the most important thing is to avoid keeping the team overloaded. Having some spare capacity will allow the team to work more efficiently.

Optimizing the whole product

As stated previously, optimization is tempting but not necessarily always the answer. The thinking here is about approaching the process (and the system) as a whole. Optimizing just one of the subparts (or the subprocess) may indeed have adverse effects on the final result. Let me explain this with the aid of two practical examples (in the process and system area):

- It may be tempting to reduce the testing phase to improve the time to market. However, if you have a holistic approach, the time spent on fixing bugs will probably be bigger than the saving. And we are not taking into account the impacts of bugs, such as downtimes, bad reputation, and customer churn.

- You may consider optimizing the disk usage of your application in many ways, such as compressing files or using special formats (such as binary). But this may, of course, come at the cost of a slower reading so, overall, it may not be a good idea.

Pros and cons of Lean development

As we have seen, Lean is the first practical implementation of the Agile concepts. For that reason, the advantages over more structured methodologies (such as the Waterfall model, which we have already seen) are evident:

- A greater flexibility, meaning that changes in the planning and requirements are better tolerated

- Enhanced freedom for the teams, where they may choose what works for them locally, that is, both technologically and from an organizational point of view

- A shorter feedback cycle, which means faster time to market and understanding sooner how your software performs (as discussed in *Chapter 4, Best Practices for Design and Development*, when talking about Minimum Viable Products)

The disadvantages of Lean development will definitely vary, based on the team composition and the project complexity. Some common ones are as follows:

- Lean is more of a set of principles (part of the broader set of Agile principles), rather than a structured methodology. This means that the outcome may be less predictable.

- As a further consequence, it doesn't usually work well with less-skilled teams, as it requires high maturity and greatly delegates decisions to each team member.

- In the case of big projects, the modularization for being worked by many small lean teams is accomplished, while the methodology can scale well. It's also hard to keep track of the greater picture and synchronize between each team and subproject.

- Deciding as late as possible means that some architectural decisions are delayed too much. As a consequence, from time to time, some rework may happen (because of wrong choices or simply the lack of any choice).

In this section, we learned about Lean software development, which is a framework full of good ideas, practices, and tools.

We've seen a walk-through of a lot of valuable ideas, such as waste reduction, openness to changes, holistic optimization, and fast feedback loop.

Bear in mind that there is a bit of overlap and mutual influence between the different philosophies in the Agile spectrum. Let's now switch to another well-known one – Scrum.

Exploring Scrum

The **Scrum** methodology was launched by Ken Schwaber and Jeff Sutherland in a paper published in 1995. The authors were also involved in the creation of the *Agile Manifesto* a bit later, so some of those ideas are directly linked.

Scrum differs slightly from Lean software development because, more than principles and high-level advice, it focuses directly on roles, project cadence (via the so-called *events*), and rules. The authors stress the fact that while you can customize the technique a bit, Scrum is intended to be *all or nothing*, meaning that you should accept and practice all the key components before embarking on a Scrum project.

Scrum refers to a phase of rugby and is regarded as an analogy for a cohesive, cross-functional team, pushing together to pursue a common objective.

In this section, we will see the fundamental elements of Scrum: the team composition (roles and responsibilities), the events (meetings and other key appointments of a Scrum project), and artifacts (the tool supporting the Scrum methodology).

Let's start with the team setting.

Understanding the Scrum teams

The Scrum teams are kind of a self-sufficient ecosystem. This means having all the skills needed to deliver tasks (or, in other words, being a cross-functional team), and being self-organized (as long as the team satisfies expectations, it can follow its own rules). The Scrum methodology identifies three main roles: the **Scrum master**, the **product owner**, and the members of the development team.

Development team

The development team, as you can imagine, is the one that will *hands-on* complete the assigned tasks, in the form of implemented and testable features. It is, by design, a flat team (no hierarchy or sub-teams are allowed) and has all the skills needed to complete the tasks (meaning that you can suppose it will not only include developers but also security experts, DBAs, and everyone else that should be needed).

As said, the development team is autonomous in terms of technical choices but is accountable (as a whole) for the outcomes of those choices. One of the main discussions centers on development teams when Scrum is applied to large enterprise environments. Indeed, often, the enterprise has guidelines and policies that have to be respected and, in this sense, are limiting the development team's freedom. Moreover, the need for different kinds of skills may lead to variability in the team's composition (with people temporarily moving between different projects), and that is a mechanism that needs to be sometimes facilitated and monitored, as schedule clashes may occur.

Product owner

The **product owner** is essentially responsible for the development pace. The product owner is the person committed to selecting the working items from a bunch of to-dos (also known as the Product Backlog, as we have briefly seen in *Chapter 4, Best Practices for Design and Development*, when talking about User Story Mapping), and understand which items must be implemented and when.

We will talk more about the Product Backlog soon, in the *Understanding Scrum artifacts* section, but for now, you can imagine how crucial this task is in terms of customer expectations, and how important it is to choose tasks with the right rationales to maximize overall throughput.

Scrum Master

The **Scrum Master** is basically the sponsor and advocate of the Scrum methodology, both internally to the team and externally to the rest of the organization. Their role is to mentor the junior members of the team and, generally, anyone who is not an expert in the methodology.

If the organization is adopting Scrum at scale, all Scrum Masters create community-exchanging best practices on how to achieve results better. Scrum Masters are responsible for facilitating the jobs of the other members of the team by circumventing the blockers that prevent the team from performing at their full potential. The Scrum Master and the product owner are two different roles, and they should be filled by different individuals.

In the next section, we'll be looking at Scrum Events.

Learning about Scrum Events

Scrum Events are the institutionalized project's recurring appointments that set the pace of overall implementations.

Scrum Events are instrumental to a project's success by providing the opportunity for the planning, execution, and reviewing of the work that needs to be done.

The basic unit of measure of this pace in Scrum is the Sprint.

Sprint

A **Sprint** in Scrum is a recurrent iteration, time-boxing a set of development activities. A Sprint is usually considered a mini project, with a fixed timeframe of 2–4 weeks. During the Sprint, there is a fixed set of goals that cannot be changed, and they are picked from the development team in the way they want.

A Sprint is essentially used to implement Agile best practices for working iteratively by releasing working software often and in small batches. This is, of course, very useful in reducing risks. If there is a shift in priorities, or something else goes wrong, your biggest risk in terms of resources is to lose one Sprint's worth of effort.

Sprint planning

Sprint planning is, of course, the meeting at which the whole Scrum team reunites to choose what will be done during a particular Sprint. The product owner clarifies the priorities and the features to be implemented by looking at the Product Backlog. Then, in accordance with the development team (and facilitated by the Scrum Master), the **Sprint Goal** is defined.

The Sprint goal is usually one or more consistent features, representing the objectives for the Sprint. The Sprint goal is then defined as a set of workable items, picked from the Product Backlog. Those items, and the way to achieve them (which is the responsibility of the development team to define), constitute the Sprint Backlog.

Daily Scrum

The **daily Scrum** is a short meeting held every day of the Sprint by the development team. It's usually set up at the beginning of the workday, with a duration of 15 minutes (this is just a rule-of-thumb time slot; it may more or less depend on the team size and project complexity). The Scrum Master and product owner can join, but the meeting is led by the development team.

The goal is to stick to Sprint planning. While there is no fixed agenda, it is usually aimed at reviewing the activities from the day before, planning activities for the current day, and addressing any issue that may put the Sprint goal at risk. Ideally, the daily Scrum should be the only sync meeting for the day, thereby boosting the development team's productivity. However, in the real world, it is not unusual for development teams to have follow-up meetings to address particularly complex issues.

The daily Scrum is also called a *standup meeting*, a naming that is also used in other Agile project methodologies. The reason behind it is that (in theory) it should be done standing up, giving further motivation to the participants to make it quicker (it will be uncomfortable to stand up for an hour during a boring meeting) and to stay active and participate during the meeting.

Sprint review

The **Sprint review** is a recurrent meeting held at the end of each Sprint. The entire Scrum team participates, and relevant business stakeholders are invited by the product owner. The development team has a demo of what was implemented during the Sprint, if possible. There is then a question-and-answer session to address doubts and discuss any issues that arose, if any.

This is also an opportunity to discuss Product Backlog based on current circumstances. This may also include changing priorities. Other *all hands* discussions may occur as well, such as budget, planning, resources, and similar topics. All those interactions usually provide valuable inputs for the next Sprint planning.

Sprint retrospective

The **Sprint retrospective** is a meeting lasting a few hours that takes place after the Sprint review and before the Sprint planning. The meeting involves the entire Scrum team. The goal is to focus on what went well and what needs to improve by looking at the previous Sprint. This meeting is usually focused more on processes, tools, and team interactions. This is also often used as a team-building activity.

It's worth noticing that there is a difference between the review and the retrospective. The Sprint review is focused on what has been implemented (the product); it includes a demo, and the business stakeholders are present and an active part of it. The focus is then on *what* we have done. In the retrospective, the business stakeholders may or may not be invited, and the focus is on *how* we have done whatever we have done. In other words, the spotlight is on the Scrum team, the interactions, and the processes. We may discuss the adopted tools, the choice of frameworks, the architecture, or simply what we liked and didn't like about how we worked in our last Sprint.

The Sprint retrospective meets a common goal of most Agile methodologies, which is continuous improvement. We will come back to this concept later when talking about Kaizen.

Backlog refinement

Backlog refinement is usually a continuous process, more than a fixed appointment. The objective of refinement is reviewing items in the Product Backlog (the project's to-do list; there will be more on this in the *Understanding Scrum artifacts* section). This is done by the product owner and the development team (or part of the development team). They cooperate to detail the items (basically, analyzing technical aspects and revisiting requisites) and refine the estimation (which is the responsibility of the development team).

Priority shifting may happen. Usually, the items with the highest priority (which are likely to happen in the next one or two Sprints) are supposed to be the clearer ones, while the lower-priority items are expected to be reviewed again. In practice, those activities are completed by the team in one or two fixed appointments per Sprint. Scrum suggests using less than 10% of the team's capacity in this sense.

In the next section, we will be learning about Scrum artifacts.

Understanding Scrum artifacts

Scrum artifacts are tools supporting the Scrum activities. This methodology refers to such tools as a way to implement transparency. In this sense, those artifacts should be available to all the teams and the relevant stakeholders.

While digital supports are commonly used, the use of physical items (such as whiteboards and sticky notes) to encourage brainstorming and in-person collaboration is also widespread. The work produced with physical tools should then be digitized for tracking and sharing purposes. Let's now see what those tools are, starting with the Product Backlog, followed by the Spring Backlog.

Product Backlog

We have already referred to the **Product Backlog** a couple of times, so by now you probably already have an idea of what it is, more or less. In simple terms, the Product Backlog is the single source of truth for each thing that should happen in the product, meaning new features, bug fixes, and other developments (improvements, refactoring, and so on).

These are categorized, including a description, unique ID, priority, and the effort required. The effort is constantly evaluated and refined by the development team. Items in the Product Backlog may be attached to test cases and other details, such as mockups and more. The product owner is ultimately accountable for the Product Backlog.

Since the Product Backlog is the funnel ingesting requests to be implemented by the development team, it can be regarded as an *infinite scroll*, meaning that new items will continuously be added to it. As already discussed in *Chapter 4*, *Best Practices for Design and Development*, the **User Story Mapping** technique can be considered a variation or evolution of the Product Backlog, adding more information and dimensions to it.

Sprint Backlog

The **Sprint Backlog** is the chunk of work to be done during each Sprint. It comprises the following:

- The **Sprint goal**, being the feature (sub-feature, or set of features) that we aim to add to the product as a result of the Sprint

- A set of items selected from the Product Backlog that need to be implemented in order to achieve the Sprint goal

- A plan for implementing those items during the Sprint

That's the way to keep work structured at a consistent pace in Scrum.

Advantages and disadvantages of Scrum

It should be evident, at this point, that Scrum is a very well-structured methodology (while still being flexible and adhering to Agile principles). For this reason, it is so widespread, up to the point that there are professional certifications available and plenty of job positions for experienced Scrum professionals.

The main advantages of adopting Scrum could be summarized as follows:

- The roles and responsibilities are very well defined, leaving less room for conflicts and misunderstandings.

- There is a defined timetable and some predictable moments in which updates (and deliverables) are shared with the rest of the team (and made visible to management).

- It's easier to do the planning (even with some expected flexibility and inaccuracy) and have visibility on what's completed and what is left almost constantly (also thanks to the concept of backlog and, in general, to the Scrum artifacts).

The Scrum disadvantages are similar to the ones in the other Agile and Lean methodologies. The following comes to mind:

- The structured process flow and events could be seen as boring and time-intensive, especially when working with highly experienced teams or in long-term projects.

- The coordination of multiple Scrum teams working on different projects may be complex.

- Bigger teams (with more than nine people) usually don't work well in a Scrum setup (hence, they should be modularized into smaller teams, and coordination will be a downside, as per the previous point).

As you have learned in this section, Scrum is a simple but disciplined way to structure the software development process. And due to its simplicity and effectiveness, it has become widespread. So, I hope the information shared in this section has motivated you to learn more and to apply Scrum principles to your projects.

In the next section, we will see some more Agile practices that are not directly linked with Scrum or any other particular framework but are often used complementarily.

Learning about other Agile practices

So far, we have seen the Agile methodologies and had a quick overview of the Lean software development principles and the Scrum framework. All of those ideas are often complemented by a number of practices and tools useful for completing specific phases.

In this section, we'll learn about some of those tools, namely, Kaizen, Planning Poker, Kanban boards, and Burndown charts.

Let's begin with Kaizen.

Kaizen

Kaizen is a principle directly borrowed from the Toyota Production System, which, as we have seen, is a core inspiration for Lean software development. *Kaizen* comes from the Japanese word for *continuous improvement.*

This simple concept is the essence of Kaizen, which articulates it with a comprehensive and elegant philosophy, embodying the concepts of humanization of the workplace, constant change (the opposite of big-bang, huge transformations). It is also responsible for identifying and removing waste (as we discussed in the *Introducing Lean software development* section), encouraging valuable feedback (both internal and external), involving all individuals in the organization (from top managers to lower levels), and so on.

Another core concept of Kaizen (again, very close to some of the Agile principles seen so far) is the shift in the testing process (in a broad sense, as in inspecting the quality of the product) from the end of production to an ongoing process, once again getting feedback early to minimize drift and facilitate constant optimization.

Kaizen is often orchestrated as a loop of five recurring phases:

1. **Observe**: This phase is used to understand what issues should be solved (or which aspect can be improved).
2. **Plan**: This phase is used for setting measurable objectives for achievement.
3. **Do**: This phase is used for putting into practice actions to meet those measurable objectives.
4. **Check**: This phase is used for comparing actual results with expected objectives.
5. **Act**: This phase is used for adjusting (or complementing) the plan to enhance the results and start the loop again.

The following diagram illustrates these phases:

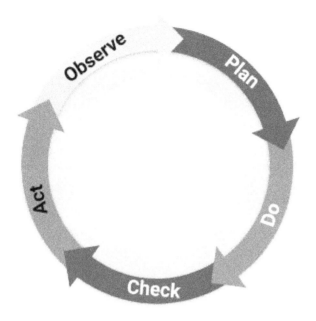

Figure 5.2 – The OPDCA loop

While nicely summarized by the *continuous improvement* concept as seen, Kaizen contains a lot of sage advice and ideas, very close to the whole idea of Lean and Agile.

Planning Poker

Planning Poker is an estimation technique, often used as part of the Scrum framework (but not a mandatory part of the framework itself). When used in Scrum, Planning Poker is done as part of Sprint planning to estimate (or refine the estimation of) the items from the Backlog.

Planning Poker is a way of getting an estimation of the effort of a given item, and it works by inciting the participant to provide a size with no influence from the other team members.

Poker is played by a team of estimators (usually the development team, which will then implement the features), a moderator, and a responsible project participant (which is usually the product owner if the Scrum methodology is used, or otherwise someone with a knowledge of the overall project and roadmap, such as a project manager or other senior staff).

Each estimator team member has a deck of cards (or, commonly, a mobile app) used to represent a difficulty grade. There is no standard here; it is common to use a Fibonacci progression, but your mileage may vary. The Fibonacci sequence has a reasoning behind it: the more the number grows, the more distant they are from each other, and so your choice must be more thoughtful. Another commonly used unit is the t-shirt size (*S*, *M*, *L*, *XL*, and so on).

Also, the expressed value (being a card, a number, or a t-shirt size) may directly map to time (as in days to implement) or not.

When the meeting starts, the moderator acts as a note-taker and master of ceremonies. They read each feature to estimate and start a discussion to clarify the meaning by including estimators and the product owner. Then, the estimators select a unit (by drawing a card, picking a number, or a size) simultaneously (to avoid influencing each other), indicating the estimated difficulty. If there is no consensus, the owner of the highest and lowest estimation has to explain their point of view. Then, everybody again draws a card until a consensus is reached. Consensus rules can be customized, such as having a defined maximum gap from a perfect average or having team members that will own that development to agree on what's an acceptable stop.

Kanban board

A **Kanban board**, in the software development world, is a visual way to represent the flow of items, from the ingestion to the development team to the implementation. It is a subset of the **Value Stream Map** (as seen in the *Introducing Lean software development* section). Kanban is indeed inspired by, and adapted from, the Toyota Production System.

In its simplistic implementation, a Kanban board is a whiteboard (physical or digital), with three vertical swim lanes splitting it into **TO DO**, **DOING**, and **DONE**. Each item is represented as a sticky note moving between those lanes. However, it is common to customize it by adding different columns (such as splitting **DOING** into **Design**, **Code**, and **Test**), or horizontal swim lanes (to represent concepts such as priority by having a kind of fast lane for urgent things such as production issues). The following diagram illustrates this:

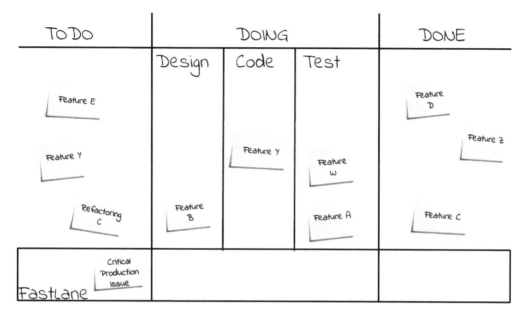

Figure 5.3 – A Kanban board

Kanban boards are just an artifact part of a bigger philosophy (Kanban), which is applied both to software development and industrial production (as Lean).

While describing the entire philosophy is beyond the scope of this book, there are at least a couple of concepts worth mentioning. The first is **Work In Progress** (**WIP**). This is the number of open items that the team is working on. WIP is easily tracked and visualized on the board. As per the Lean methodology, Kanban advises against using context switching; hence, a constraint on WIP should be present at any time.

Another important concept is **pull**. Basically, the Kanban approach puts the working items at disposal of the development team (in the **TO DO** column). As opposed to the push paradigm, the team chooses (pulls) what to do at their own pace. This avoids hogging the team and maximizes throughput.

Burndown chart

A **Burndown chart** is a common artifact (physical or digital) to clearly show a project's progression. It is very useful, regardless of which Agile methodology is used, because it gives real-time insights into planning. As has been mentioned, Agile is against detailed, advanced planning, so having a current snapshot of the project's progression (and maybe some forecasting) is precious for management.

A Burndown chart plots the tasks (usually as a sum of the required effort) as the vertical axis and the timeline as the horizontal axis. Drawing a line from the top left (project start) to the bottom right (project completion) provides an ideal, linear progression. At regular times (such as every day, or at the end of each Scrum Sprint), a dot is plotted that crosses the implemented tasks and the current moment in time. The following diagram is an example of a Burndown chart:

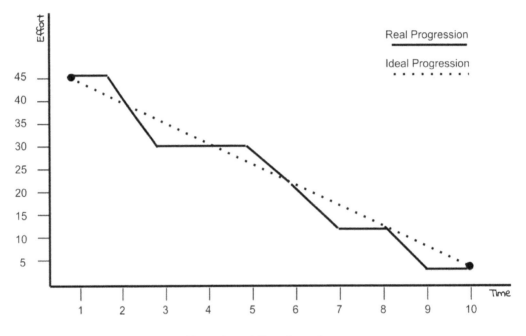

Figure 5.4 – A Burndown chart

As you can see, by drawing a line over those dots, you can compare the ideal project progression versus the actual project's progression. Roughly speaking, if the real project progression is above the ideal one, you are probably late, whereas if it's below, you are ahead. Having minimal deviations from the ideal progression means staying on track, and it's usually a good indicator of a project's health. Also, it gives good hints on when the project (or, at least, the represented list of tasks) will be completed.

In this section, we have seen a nice list of tools that can provide you with support in Agile software development. Regardless of the methodology you are using, if any, such tools can be useful in addressing common use cases, such as optimizing processes and estimating development effort.

In the next section, we will talk about a very hot and debated topic, which seems to be getting all the attention lately – DevOps.

Understanding DevOps and its siblings

At the time of writing, **DevOps** is an overinflated term. It is seen as a silver bullet for every development problem, and a mandatory prerequisite for being considered cool. I'm not going to decrease the hype about DevOps, as I truly believe it's a precious technique useful for ensuring functional and high-performing teams. However, it must be said that DevOps is more a set of best practices, rather than a well-codified, magic recipe. And, as is common in these cases, one size does not fit all.

DevOps can be seen as essentially an extension of Agile methodologies. Indeed, the adoption of Agile practices (not one specifically) can be seen as a prerequisite of DevOps. And, in turn, DevOps is considered to be an essential condition for the adoption of cutting-edge approaches such as **microservices** (more on this in *Chapter 9, Designing Cloud-Native Architectures*).

The essential characteristic of DevOps is cooperation between different roles. This commonly means, in practical terms, a small team, encompassing all the different skills needed to build and maintain a software product in production.

In this section, we will cover some core aspects of the DevOps movement, such as team composition, roles and responsibilities, and variants of DevOps, which are about including more functions in this collaboration method. But let's start with a common consideration covered in DevOps regarding team size.

DevOps team size

When it comes to team size, the Scrum guide says that a team should be small enough to stay lean but large enough to develop a reasonable number of features in each Sprint. A common rule of thumb is to have a team of around 10 or fewer people.

This rule of thumb is commonly accepted and has echoes in other stories, such as the famous two pizzas team, which states that it should be possible to feed the team with two large pizzas (so, again, roughly fewer than 10 people).

This depends on the logic of links. In a functional team, each team member should have a link with the others. This means that in a team of 10 people, you will have 90 links. That's the reason why the team should not grow much above 10, or else you will have too many internal interactions to manage, which quickly impacts productivity.

But what about the internal team's responsibility?

Roles and responsibilities in a DevOps team

As is obvious from the name, DevOps aims at blurring the responsibilities between developers and operations. This does not mean that everybody should be capable of doing everything; it is more about having a shared goal.

One of the most hateful dynamics in IT teams is the lack of accountability in case of issues.

The most commonly involved teams are Ops, who are the operations and system engineers responsible for the infrastructure (and for the uptime of production systems), and Devs, which are, well, the developers, of course.

Ops will always blame Devs' buggy code when something goes wrong in production, and Devs will throw code at Ops for releasing without caring about the release outcome, to the battle cry of *works on my machine.*

While these dynamics are purposefully exaggerated, you can agree that the relationship between Devs and Ops is not always the best. DevOps starts here. Everybody is accountable for production – *you build it, you run it.*

This means that the team (and the individuals) must shift from a skill perspective (I'm a specialist only accountable for my limited piece) to a product perspective (my first responsibility is to have a fully functional product in production, and I will use my skills for this goal). The goal of this is to build high-quality products (everybody is committed to a fully functional production service) in less time (you eliminate handovers between different departments).

Taking apart the philosophy and motivations behind DevOps, there are some direct technological impacts, which can be seen both as a prerequisite and fundamental benefit of adopting DevOps:

- **Pervasive automation**, also known as **infrastructure as code**: Everything, including environment definitions, should be declarative, versioned (usually in a code versioning system such as **Git**), and repeatable. This avoids drifting (environments strictly adhere to the expected configurations) and reduces the time for recovering from failures (it's easy to spin up new copies of the environment). This is something usually driven by the team members with prevalent Ops skills. It is common, in this regard, to see a shift toward **Site reliability engineering** practices, meaning that Ops will intentionally use an increasing part of their time to develop automation and other production support tools, instead of doing exclusively production-related tasks (even manually).

- **Shifting quality into software development**: This means embedding all the feedback coming from production exposure into software development. This often means increasing observability (to support troubleshooting and performance tuning in production), improving code testing (to reduce the defects found in production), and everything that's necessary for safer, high-quality production releases (such as automated rollbacks in case of failures, supporting auto-scaling, and modularizing releases).

It is now safe to try to extend this philosophy beyond Devs and Ops.

Devs, Ops, and more

It is natural to try to extend such good practices, such as borderless collaboration (breaking silos) and tooling support (automating everything) beyond development and operations.

DevSecOps is a clear example of that. This is all about shifting security concerns into all phases of product development. This means, of course, integrating security specialists in the DevOps team. Very often, the approach to security is to run specific tests against the finished product soon before (or shortly after) the production release. The result is that, often, it's too late and maybe you don't have the time (or it's costly) to fix the security findings.

At the opposite end, DevSecOps impacts the production process in several ways. The first is to embed best practices in the development of code, then to automate testing against security principles and rules, and lastly, continuously check compliance with those principles as part of production operation practices. This extension is particularly well accepted in highly regulated environments (such as banks, government institutions, and healthcare), and it has a positive impact in terms of the time to market and overall security.

BizDevOps is another variant, breaking another wall and making business owners (analysts, budget owners, and even marketing) part of the team. The collaboration model used here is less structured than with Devs and Ops (and security, if you want), since some of the activities are not perfectly overlapped, nor comprehensible between technicians and business people.

However, if you think about it, Agile methodologies (and DevOps, by extension) inherently encourage cooperation with business by emphasizing short and frequent feedback loops, and openness to changes in the product life cycle. What's probably a distinct characteristic of BizDevOps is the crossed visibility on KPIs.

This includes the technical team having insights into business KPIs (things such as budget, the number of users, sales trends, and more) in order to try to figure out how technical choices (new releases, changes in the infrastructure, and resource efficiency) impact on it. And it's also true the other way around; that is, the business team could have a look at the technical teams' *tuning wheels* (the size of the team, resources, and the number of changes) and how they impact the end-to-end process, in terms of development speed, costs, and so on.

Lastly, **NoOps** is a trending topic, gaining visibility as a result of the assonance with DevOps. As it's easy to imagine, the idea here is to get rid of the Operations team completely. While it is theoretically possible, as a result of using heavily automated environments such as **Platform as a Service** and **Cloud** (there is more on this in *Chapter 9, Designing Cloud-Native Architectures*), to have developers capable of basic Ops tasks, such as the provisioning of new environments and deployments, *I strongly believe NoOps is a dead end (at least for the foreseeable future).* It can be applied when reducing Ops resources in small contexts (such as serverless applications; this topic will be discussed in *Chapter 9, Designing Cloud-Native Architectures*), but this seems more like outsourcing. You basically do not need to care about the infrastructure because someone else is taking care of it for you (a cloud provider, or maybe another department).

Personally, I feel such an approach is completely the opposite of DevOps. You will end up having a huge gap between platform users (Devs) and the team running the infrastructure (Ops, which are indeed not even part of the project).

DevOps and the bigger organization

A model that is commonly seen as a large-scale implementation of DevOps is the **Spotify** development model, which is famous because it has been created and used in the homonym company building the music streaming app.

Even though, in their seminal work, theorized in a publicly available paper entitled *Scaling Agile @ Spotify*, there is no mention of the word *DevOps*, you can recognize some common principles.

You will find the link to the full paper in the *Further reading* section. For now, it's enough to consider that DevOps must solve the conflicting needs of having a multidisciplinary team focused on delivery (and production quality) with knowledge and best practice sharing. In the Spotify model, this is resolved with a matrix organization, in which individuals belong to one team (so-called *squads* and *tribes*) with product-delivery purposes but share interests with people of the same skills (such as DBAs or frontend developers) for knowledge sharing and personal growth purposes (in the so-called *chapters* and *guilds*).

The Spotify model suggests a number of other mechanisms for boosting collaboration. It's an interesting point of view and gives some practical advice. However, considering that every organization is different, and has different challenges and strengths, so the first piece of advice is flexibility. No model will simply work out of the box; you have to look at the company's objectives and people skills and keep adapting to changing conditions.

Pros and cons of DevOps

We anticipated some impacts of DevOps, both in positive and negative ways, in the previous sections. However, to summarize, here are some advantages of adopting a DevOps model:

- It's a high-performance methodology, meaning that, when working properly, it enables us to deliver high-quality software frequently. Hence, it's rapidly responding to changing conditions, such as new requirements or production issues.
- It copes well (and often is seen as a requirement) with modern architectures, such as cloud-based and microservices applications.
- It's challenging and rewarding for team members, meaning that there is a lot of room for learning, as each team member can easily enrich his/her skills and responsibilities.

The disadvantages can be summarized as follows:

- It's a huge paradigm shift and can be hard to accept for more traditional organizations, as it requires many people to get out of their comfort zone and start thinking about their role in a different way (stretching everybody's responsibilities).
- It may be difficult to map from an organizational point of view, as it will require breaking the traditional silos and setting up cross-department, product-oriented teams.
- It requires highly skilled and motivated team members. It may be stressful in the long term.

With this section, we have completed our overview of DevOps.

We have seen what the founding principles of such a methodology are and why it claims to boost efficiency, along with some of the variants, such as DevSecOps.

In the next section, we will have a look at some examples and case studies.

Case studies and examples

In this section, we will model an ideal Product Backlog in the Scrum way, applied to our mobile payments example.

The official Scrum guide does not provide any example of a Product Backlog, and there are no standards as regards the fields that should be included. Based on my personal experience, a Product Backlog should look like this:

ID	DESCRIPTION	STORY	CATEGORY	PRIORITY	DEPENDENCIES	EFFORT	NOTE
57	Credential Validation	Account Management	Feature	Medium	36,17	10	See doc http://intranet/xyz
73	Payment rollback	Payment	Feature	Medium	22	10	-
74	Recipient selection performances	Payment	Fix	High		15	Check data collected from production
49	Social Login	Account Management	Enhancement	Low	24	TBD	-

Figure 5.5 – Mobile payments Product Backlog

This is, of course, just a small subset, but several considerations can be made:

- **Items are identified by ID and DESCRIPTION**: Most likely, **ID** will link to a detailed requirements document or at least a more detailed description. Also, every item is likely categorized as part of a bigger user **STORY**. As discussed previously, User Story Mapping is a different way to visualize this kind of relationship.

- **Items are categorized**: Usually, at least features and fixes are categorized, while more types, such as enhancements and technical terms (for things such as refactoring and other internal tasks), may be used.

- **Dependencies**: This is a way to help choose items through the links to other items.

- **Effort**: This is something that may be roughly evaluated when adding items to the backlog. However, this is likely to change over time when more details will be known.

You can see some similarities with the requirements template seen in *Chapter 2*, *Software Requirements – Collecting, Documenting, Managing*, and indeed the goals are similar. However, a different level of detail is evident, as those two artifacts have different goals in the project cycle.

With this simple example, we have covered all the topics relevant to this chapter.

Summary

In this chapter, we have seen a complete overview of the development models. Starting with the more traditional approaches, such as Code and Fix and Waterfall, we then moved to the core of the chapter, focusing on Agile.

As we have seen, Agile is a broad term, including more structured frameworks (such as Scrum) and other tools and best practices (such as Lean and some other techniques, such as Kanban), which can be mixed and matched to better suit the needs of other projects. As a last big topic, we discussed DevOps (and some extensions of it). While not being a well-codified practice, the huge potential of this approach is clear, which is now seeing widespread adoption in many innovative projects. DevOps, indeed, is the prerequisite for some advanced architectures that we will see in the forthcoming chapters, such as microservices.

In the next chapter, we will focus on Java architectural patterns. We will cover some essential topics, including multi-tier architectures, encapsulation, and practical tips regarding performance and scalability.

Further reading

- *The pros and cons of Waterfall Software Development* (https://www.dcsl.com/pros-cons-waterfall-software-development/), DCSL GuideSmiths

- *The Waterfall Model: Advantages, disadvantages, and when you should use it* (https://developer.ibm.com/articles/waterfall-model-advantages-disadvantages/), by Aiden Gallagher, Jack Dunleavy, and Peter Reeves

- *The Waterfall model: Advantages and disadvantages* (https://www.blocshop.io/blog/waterfall-advantages-disadvantages/), Blocshop

- The Agile Manifesto (https://agilemanifesto.org), by Kent Beck, Mike Beedle, Arie van Bennekum, Alistair Cockburn, Ward Cunningham, Martin Fowler, James Grenning, Jim Highsmith, Andrew Hunt, Ron Jeffries, Jon Kern, Brian Marick, Robert C. Martin, Steve Mellor, Ken Schwaber, Jeff Sutherland, and Dave Thomas

- *Lean Software Development: An Agile Toolkit, Mary Poppendieck and Tom Poppendieck, Pearson Education* (2003)

- *Implementing Lean Software Development: From Concept to Cash, Mary and Tom Poppendieck, Pearson Education* (2006)

- *Lean Software Development in Action, Andrea Janes and Giancarlo Succi, Springer Berlin Heidelberg* (2014)

- *Agile Metrics in Action: How to measure and improve team performance, Christopher Davis, Manning Publications* (2015)

- *The Surprising Power of Online Experiments* (`https://hbr.org/2017/09/the-surprising-power-of-online-experiments`), by Ron Kohavi and Stefan Thomke

- *The Art of Lean Software Development, Curt Hibbs, Steve Jewett, and Mike Sullivan, O'Reilly Media* (2009).

- The Scrum guide (`https://www.scrumguides.org`), by Jeff Sutherland and Ken Schwaber

- *Scrum: The Art of Doing Twice the Work in Half the Time, Jeff Sutherland, Random House* (2014)

- *9 retrospective techniques that won't bore your team to death* (`https://www.atlassian.com/blog/teamwork/revitalize-retrospectives-fresh-techniques`), by Sarah Goff-Dupont

- *6 Effective Sprint Retrospective Techniques* (`https://www.parabol.co/resources/agile-sprint-retrospective-ideas`), Parabol

- *DevOpsCulture* (`https://martinfowler.com/bliki/DevOpsCulture.html`), by Rouan Wilsenach

- *Scaling Agile @ Spotify with Tribes, Squads, Chapters & Guilds* (`https://blog.crisp.se/wp-content/uploads/2012/11/SpotifyScaling.pdf`), by Henrik Kniberg and Anders Ivarsson

- *Create Your Successful Agile Project: Collaborate, Measure, Estimate, Deliver, Johanna Rothman, Pragmatic Bookshelf* (2017)

- *Operations Anti-Patterns, DevOps Solutions, Jeffery D. Smith, Manning Publications* (2020)

Section 2: Software Architecture Patterns

Are there any recognizable architecture patterns or reusable best practices? In this section of the book, we will get an overview of different kinds of architectural patterns and their most widespread implementation in Java.

We will discuss some basic Java architectural patterns, such as encapsulation, MVC, and event-driven. Another big topic will be middleware and frameworks, including JEE application servers, as well as frameworks for microservices implementation. We will also deal with integration and business automation, which are two other common middleware concepts. This section will be completed with elements of cloud-native architectures, concepts of user interfaces, and an overview of data storage and retrieval.

This section comprises the following chapters:

- *Chapter 6, Exploring Essential Java Architectural Patterns*
- *Chapter 7, Exploring Middleware and Frameworks*
- *Chapter 8, Designing Application Integration and Business Automation*
- *Chapter 9, Designing Cloud-Native Architectures*
- *Chapter 10, Implementing User Interaction*
- *Chapter 11, Dealing with Data*

6
Exploring Essential Java Architectural Patterns

In the last chapter, you had an overview of the most common development models, from the older (but still used) **Waterfall model** to the widely used and appreciated **DevOps** and **Agile**.

In this chapter, you will have a look at some very common architectural patterns. These architectural definitions are often considered basic building blocks that are useful to know about in order to solve common architectural problems.

You will learn about the following topics in this chapter:

- Encapsulation and hexagonal architectures
- Learning about multi-tier architectures
- Exploring Model View Controller
- Diving into event-driven and reactive approaches
- Designing for large-scale adoption
- Case studies and examples

After reading this chapter, you'll know about some useful tools that can be used to translate requirements into well-designed software components that are easy to develop and maintain. All the patterns described in this chapter are, of course, orthogonal to the development models that we have seen in the previous chapters; in other words, you can use all of them regardless of the model used.

Let's start with one of the most natural architectural considerations: encapsulation and hexagonal architectures.

Encapsulation and hexagonal architectures

Encapsulation is a concept taken for granted by programmers who are used to working with object-oriented programming and, indeed, it is quite a basic idea. When talking about encapsulation, your mind goes to the getters and setters methods. To put it simply, you can hide fields in your class, and control how the other objects interact with them. This is a basic way to protect the status of your object (internal data) from the outside world. In this way, you decouple the state from the behavior, and you are free to switch the data type, validate the input, change formats, and so on. In short, it's easy to understand the advantages of this approach.

However, encapsulation is a concept that goes beyond simple getters and setters. I personally find some echoes of this concept in other modern approaches, such as APIs and microservices (more on this in *Chapter 9, Designing Cloud-Native Architectures*). In my opinion, encapsulation (also known as **information hiding**) is all about modularization, in that it's about having objects talk to each other by using defined contracts.

If those contracts (in this case, normal method signatures) are stable and generic enough, objects can change their internal implementation or can be swapped with other objects without breaking the overall functionality. That is, of course, a concept that fits nicely with interfaces. An interface can be seen as a *super contract* (a set of methods) and a way to easily identify compatible objects.

In my personal view, the concept of encapsulation is extended with the idea of hexagonal architectures. Hexagonal architectures, theorized by Alistair Cockburn in 2005, visualize an application component as a hexagon. The following diagram illustrates this:

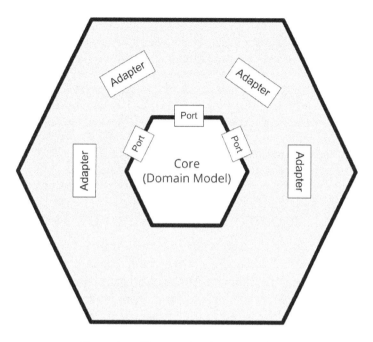

Figure 6.1 – Hexagonal architecture schema

As you can see in the preceding diagram, the business logic stays at the core of this representation:

- **Core**: The core can be intended to be the domain model, as seen in *Chapter 4, Best Practices for Design and Development*. It's the real distinctive part of your application component – the one solving the business problem.

- **Port**: Around the core, the ports are represented. The domain model uses the ports as a way to communicate with other application components, being other modules or systems (such as databases and other infrastructures). The ports are usually mapped to use cases of the module itself (such as sending payments). However, more technical interpretations of ports are not unusual (such as persisting to a database).

- **Adapter**: The layer outside the ports represents the adapters. The Adapter is a well-known pattern in which a piece of software acts as an interpreter between two different sides. In this case, it translates from the domain model to the outside world, and vice versa, according to what is defined in each port. While the diagram is in the shape of a hexagon, that's not indicative of being limited to six ports or adapters. That's just a graphical representation, probably related to the idea of representing the ports as discrete elements (which is hard to do if you represent the layers as concentric circles). The hexagonal architecture is also known as **Ports and Adapters**.

> **Important Note:**
> There is another architectural model implementing encapsulation that is often compared to hexagonal architectures: **Onion architectures**. Whether the hexagonal architecture defines the roles mentioned earlier, such as core, ports, and adapters, the Onion architecture focuses the modeling on the concept of layers. There is an inner core (the Domain layer) and then a number of layers around it, usually including a repository (to access the data of the Domain layer), services (to implement business logic and other interactions), and a presentation layer (for interacting with the end user or other systems). Each layer is supposed to communicate only with the layer above itself.

Hexagonal architectures and Domain Driven Design

Encapsulation is a cross-cutting concern, applicable to many aspects of a software architecture, and hexagonal architectures are a way to implement this concept. As we have seen, encapsulation has many touchpoints with the concept of **Domain-Driven Design** (**DDD**). The core, as mentioned, can be seen as the domain model in DDD. The Adapter pattern is also very similar to the concept of the Infrastructure layer, which in DDD is the layer mapping the domain model with the underlying technology (and abstracting such technology details).

It's then worth noticing that DDD is a way more complete approach, as seen in *Chapter 4, Best Practices for Design and Development*, tackling things such as defining a language for creating domain model concepts and implementing some peculiar use cases (such as where to store data, where to store implementations, how to make different models talk to each other). Conversely, hexagonal architectures are a more practical, immediate approach that may directly address a concern (such as implementing encapsulation in a structured way), but do not touch other aspects (such as how to define the objects in the core).

Encapsulation and microservices

While we are going to talk about microservices in *Chapter 9, Designing Cloud-Native Architectures*, I'm sure you are familiar with, or at least have heard about, the concept of microservices. In this section, it's relevant to mention that the topic of encapsulation is one of the core reasonings behind microservices. Indeed, a microservice is considered to be a disposable piece of software, easy to scale and to interoperate with other similar components through a well-defined API.

Moreover, each microservice composing an application is (in theory) a product, with a dedicated team behind it and using a set of technologies (including the programming language itself) different from the other microservices around it. For all those reasons, encapsulation is the basis of the microservices applications, and the concepts behind it (as the ones that we have seen in the context of hexagonal architectures) are intrinsic in microservices.

So, as you now know, the concept of modularization is in some way orthogonal to software entities. This need to define clear responsibilities and specific contracts is a common way to address complexity, and it has a lot of advantages, such as testability, scaling, extensibility, and more. Another common way to define roles in a software system is the multi-tier architecture.

Learning about multi-tier architectures

Multi-tier architectures, also known as **n-tier architectures**, are a way to categorize software architectures based on the number and kind of tiers (or layers) encompassing the components of such a system. A tier is a logical grouping of the software components, and it's usually also reflected in the physical deployment of the components. One way of designing applications is to define the number of tiers composing them and how they communicate with each other. Then, you can define which component belongs to which tier. The most common types of multi-tier applications are defined in the following list:

- The simplest (and most useless) examples are **single-tier applications**, where every component falls into the same layer. So, you have what is called a monolithic application.

- Things get slightly more interesting in the next iteration, that is, **two-tier applications**. These are commonly implemented as client-server systems. You will have a layer including the components provided to end users, usually through some kind of graphical or textual user interfaces, and a layer including the backend systems, which normally implement the business rules and the transactional functionalities.

- **Three-tier applications** are a very common architectural setup. In this kind of design, you have a presentation layer taking care of interaction with end users. We also have a business logic layer implementing the business logic and exposing APIs consumable by the presentation layer, and a data layer, which is responsible for storing data in a persistent way (such as in a database or on a disk).

- More than three layers can be contemplated, but that is less conventional, meaning that the naming and roles may vary. Usually, the additional tiers are specializations of the business logic tier, which was seen in the previous point. An example of a four-tier application was detailed in *Chapter 4, Best Practices for Design and Development*, when talking about the layered architecture of DDD.

The following diagram illustrates the various types of multi-tier architectures:

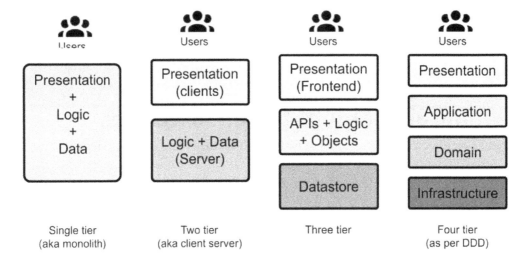

Figure 6.2 – Multi-tier architectures

The advantages of a multi-tier approach are similar to those that you can achieve with the modularization of your application components (more on this in *Chapter 9, Designing Cloud-Native Architectures*). Some of the advantages are as follows:

- The most relevant advantage is probably **scalability**. This kind of architecture allows each layer to scale independently from each other. So, if you have more load on the business (or frontend, or database) layer, you can scale it (vertically, by adding more computational resources, or horizontally, by adding more instances of the same component) without having a huge impact on the other components. And that is also linked to increased stability overall: an issue on one of the layers is not so likely to influence the other layers.

- Another positive impact is improved **testability**. Since you are forced to define clearly how the layers communicate with each other (such as by defining some APIs), it becomes easier to test each layer individually by using the same communication channel.

- **Modularity** is also an interesting aspect. Having layers talking to each other will enforce a well-defined API to decouple each other. For this reason, it is possible (and is very common) to have different actors on the same layer, interacting with the other layer. The most widespread example here is related to the frontend. Many applications have different versions of the frontend (such as a web GUI and a mobile app) interacting with the same underlying layer.

- Last but not least, by **layering** your application, you will end up having more parallelization in the development process. Sub teams can work on a layer without interfering with each other. The layers, in most cases, can be released individually, reducing the risks associated with a big bang release.

There are, of course, drawbacks to the multi-tier approach, and they are similar to the ones you can observe when adopting other modular approaches, such as microservices. The main disadvantage is to do with **tracing**.

It may become hard to understand the end-to-end path of each transaction, especially (as is common) if one call in a layer is mapped to many calls in other layers. To mitigate this, you will have to adopt specific monitoring to trace the path of each call; this is usually done by injecting unique IDs to correlate the calls to each other to help when troubleshooting is needed (such as when you want to spot where the transactions slow down) and in general to give better visibility into system behavior. We will study this approach (often referred to as tracing or observability) in more detail in *Chapter 9, Designing Cloud-Native Architectures.*

In the next section, we will have a look at a widespread pattern: Model View Controller.

Exploring Model View Controller

At first glance, **Model View Controller** (MVC) may show some similarities with the classical three-tier architecture. You have the classification of your logical objects into three kinds and a clear separation between presentation and data layers. However, MVC and the three-tier architecture are two different concepts that often coexist.

The three-tier architecture is an architectural style where the elements (presentation, business, and data) are split into different deployable artifacts (possibly even using different languages and technologies). These elements are often executed on different servers in order to achieve the already discussed goals of scalability, testability, and so on.

On the other hand, MVC is not an architectural style, but a design pattern. For this reason, it does not suggest any particular deployment model regarding its components, and indeed, very often the Model, View, and Controller coexist in the same application layer.

Taking apart the *philosophical* similarity and differences, from a practical point of view, MVC is a common pattern for designing and implementing the presentation layer in a multi-tier architecture.

In MVC, the three essential components are listed as follows:

- **Model**: This component takes care of abstracting access to the data used by the application. There is no logic to the data presented here.

- **View**: This component takes care of the interaction with the users (or other external systems), including the visual representation of data (if expected).

- **Controller**: This component receives the commands (often mediated by the view) from the users (or other external systems) and updates the other two components accordingly. The **Controller** is commonly seen as a facilitator (or glue) between the **Model** and **View** components.

The following diagram shows you the essential components of MVC:

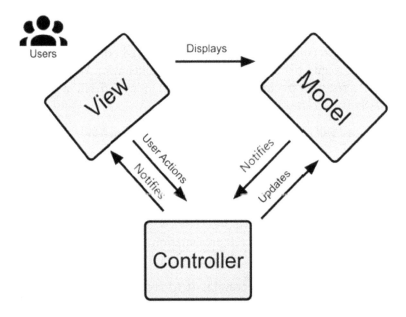

Figure 6.3 – MVC components

Another difference between MVC and the three-tier architecture is clear from the interaction of the three components described previously: in a three-tier architecture, the interaction is usually linear; that is, the presentation layer does not interact directly with the data layer. MVC classifies the kind and goal of each interaction but also allows all three components to interact with each other, forming a triangular model.

MVC is commonly implemented by a framework or middleware and is used by the developer, specific interfaces, hooks, conventions, and more.

In the real world, this pattern is commonly implemented either at the server side or the client side.

Server-side MVC

The **Java Enterprise Edition** (**JEE**) implementation is a widely used example (even if not really a modern one) of an MVC server-side implementation. In this section, we are going to mention some *classical* Java implementations of web technologies (such as JSPs and servlets) that are going to be detailed further in *Chapter 10, Implementing User Interaction*.

In terms of relevance to this chapter, it's worthwhile knowing that in the JEE world, the MVC model is implemented using Java beans, the view is in the form of JSP files, and the controller takes the form of servlets, as shown in the following diagram:

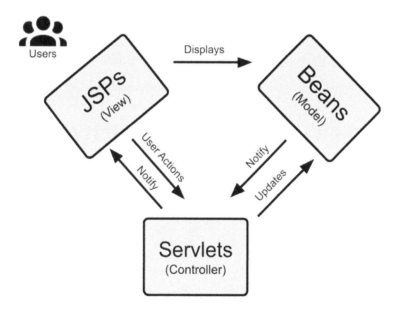

Figure 6.4 – MVC with JEE

As you can see, in this way, the end user interacts with the web pages generated by the **JSPs** (the **View**), which are bound to the Java **Beans** (the **Model**) keeping the values displayed and collected. The overall flow is guaranteed by the **Servlets** (the **Controller**), which take care of things such as the binding of the Model and View, session handling, page routing, and other aspects that *glue* the application together. Other widespread Java MVC frameworks, such as **Spring MVC**, adopt a similar approach.

Client-side MVC

MVC can also be completely implemented on the client side, which usually means that all three roles are played by a web browser. The de facto standard language for client-side MVC is **JavaScript**.

Client-side MVC is almost identical to **single-page applications**. We will see more about single-page applications in *Chapter 10, Implementing User Interaction*, but basically, the idea is to minimize page changes and full-page reloads in order to provide a near-native experience to users while keeping the advantages of a web application (such as simplified distribution and centralized management).

The single-page applications approach is not so different from server-side MVC. This technology commonly uses a templating language for views (similar to what we have seen with JSPs on the server side), a model implementation for keeping data and storing it in local browser storage or remotely calling the remaining APIs exposed from the backend, and controllers for navigation, session handling, and more support code.

In this section, you learned about MVC and related patterns, which are considered a classical implementation for applications and have been useful for nicely setting up all the components and interactions, separating the user interface from the implementation.

In the next section, we will have a look at the event-driven and reactive approaches.

Diving into event-driven and reactive approaches

Event-driven architecture isn't a new concept. My first experiences with it were related to GUI development (with **Java Swing**) a long time ago. But, of course, the concept is older than that. And the reason is that events, meaning *things that happen*, are a pretty natural phenomenon in the real world.

There is also a technological reason for the event-driven approach. This way of programming is deeply related to (or in other words, is most advantageous when used together with) asynchronous and non-blocking approaches, and these paradigms are inherently efficient in terms of the use of resources.

Here is a diagram representing the event-driven approach:

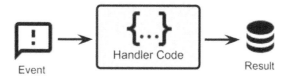

Figure 6.5 – Event-driven approach

As shown in the previous diagram, the whole concept of the event-driven approach is to have our application architecture react to external events. When it comes to GUIs, such events are mostly user inputs (such as clicking a button, entering data in text fields, and so on), but events can be many other things, such as changes in the price of a stock option, a payment transaction coming in, data being collected from sensors, and so on.

Another pattern worth mentioning is the **actor model** pattern, which is another way to use messaging to maximize the concurrency and throughput of a software system.

I like to think that **reactive programming** is an evolution of all this. Actually, it is probably an evolution of many different techniques.

It is a bit harder to define reactive, probably because this approach is still relatively new and less widespread. Reactive has its roots in functional programming, and it's a complete paradigm shift from the way you think about and write your code right now. While it's out of the scope of this book to introduce functional programming, we will try to understand some principles of reactive programming with the usual goal of giving you some more tools you can use in your day-to-day architect life and that you can develop further elsewhere if you find them useful for solving your current issues.

But first, let's start with a cornerstone concept: events.

Defining events, commands, and messages

From a technological point of view, an event can be defined as something that changes the status of something. In an event-driven architecture, such a change is then propagated (notified) as a message that can be picked up by components *interested* in that kind of event.

For this reason, the terms **event-driven** and **message-driven** are commonly used interchangeably (even if the meaning may be slightly different).

So, an **event** can be seen as a more abstract concept to do with new information, while a message can be seen as how this information is propagated throughout our system. Another core concept is the **command**. Roughly speaking, a command is the expression of an action, while an event is an expression of something happening (such as a change in the status of something).

So, an event reflects a change in data (and somebody downstream may need to be notified of the change and need to do something accordingly), while a command explicitly asks for a specific action to be done by somebody downstream.

Again, generally speaking, an event may have a broader audience (many consumers might be interested in it), while a command is usually targeted at a specific system. Both types of messages are a nice way to implement loose coupling, meaning it's possible to switch at any moment between producer and consumer implementations, given that the contract (the message format) is respected. It could be even done live with zero impact on system uptime. That's why the usage of messaging techniques is so important in application design.

Since these concepts are so important and there are many different variations on brokers, messages, and how they are propagated and managed, we will look at more on messaging in *Chapter 8, Designing Application Integration and Business Automation*. Now, let's talk about the event-driven approach in detail.

Introducing the event-driven pattern and event-driven architecture

The **event-driven pattern** is a pattern and architectural style focused on reacting to things happening around (or inside of) our application, where notifications of actions to be taken appear in the form of events.

In its simplest form, expressed in imperative languages (as is widespread in embedded systems), event-driven architecture is managed via infinite loops in code that continuously poll against event sources (queues), and actions are performed when messages are received.

However, **event-driven architecture** is orthogonal to the programming style, meaning that it can be adopted both in imperative models and other models, such as object-oriented programming.

With regard to **Object-Oriented Programming** (**OOP**), there are plenty of Java-based examples when it comes to user interface development, with a widely known one being the Swing framework. Here you have objects (such as buttons, windows, and other controls) that provide handlers for user events. You can register one or more handlers (consumers) with those events, which are then executed.

From the point of view of the application flow, you are not defining the order in which the methods are executing. You are just defining the possibilities, which are then executed and composed according to the user inputs.

But if you abstract a bit, many other aspects of Java programming are event-driven. Servlets inherently react to events (such as an incoming HTTP request), and even error handling, with try-catch, defines the ways to react if an unplanned event occurs. In those examples, however, the events are handled internally by the framework, and you don't have a centralized middleware operating them (such as a messaging broker or queue manager). Events are simply a way to define the behavior of an application.

Event-driven architecture can be extended as an architectural style. Simply put, an event-driven architecture prescribes that all interactions between the components of your software system are done via events (or commands). Such events, in this case, are mediated by a central messaging system (a broker, or bus).

In this way, you can extend the advantages of the event-driven pattern, such as loose coupling, better scalability, and a more natural way to represent the use case, beyond a single software component. Moreover, you will achieve the advantage of greater visibility (as you can inspect the content and number of messages exchanged between the pieces of your architecture). You will also have better manageability and uptime (because you can start, stop, and change every component without directly impacting the others, as a consequence of loose coupling).

Challenges of the event-driven approach

So far, we have seen the advantages of the event-driven approach. In my personal opinion, they greatly outweigh the challenges that it poses, so I strongly recommend using this kind of architecture wherever possible. As always, take into account that the techniques and advice provided in this book are seldom entirely prescriptive, so in the real world I bet you will use some bits of the event-driven pattern even if you are using other patterns and techniques as your main choice.

However, for the sake of completeness, I think it is worth mentioning the challenges I have faced while building event-driven architectures in the past:

- **Message content**: It's always challenging to define what should be inside a message. In theory, you should keep the message as simple and as light as possible to avoid hogging the messaging channels and achieve better performance. So, you usually have only a message type and references to data stored elsewhere.

However, this means that downstream systems may not have all the data needed for the computation in the message, and so they would complete the data from external systems (typically, a database). Moreover, most of the messaging frameworks and APIs (such as **JMS**) allow you to complete your message with metadata, such as headers and attachments. I've seen endless discussions about what should go into a message and what the metadata is. Of course, I don't have an answer here. My advice, as always, is to keep it as simple as possible.

- **Message format**: Related to the previous point, the message format is also very relevant. Hence, after you establish what information type should be contained in each message, the next step is to decide the shape this information should have. You will have to define a message schema, and this should be understandable by each actor. Also, message validation could be needed (to understand whether each message is a formally valid one), and a schema repository could be useful, in order to have a centralized infrastructure that each actor can access to extract metadata about how each message should be formatted.

- **Transactional behavior**: The write or read of a message, in abstract, constitutes access to external storage (not so different from accessing a database). For this reason, if you are building a traditional enterprise application, when you are using messaging, you will need to extend your transactional behavior.

It's a very common situation that if your consumer needs to update the database as a consequence of receiving a message, you will have a transaction encompassing the read of the message and the write to the database. If the write fails, you will roll back the read of the message. In the Java world, you will implement this with a two-phase commit. While it's a well-known problem and many frameworks offer some facilities to do this, it's still not a simple solution; it can be hard to troubleshoot (and recover from) and can have a non-negligible performance hit.

- **Tracing**: If the system starts dispatching many messages between many systems, including intermediate steps such as message transformations and filtering, it may become difficult to reconstruct a user transaction end to end. This could lead to a lack of visibility (from a logical/use case point of view) and make troubleshooting harder. However, you can easily solve this aspect with the propagation of transaction identifiers in messages and appropriate logging.

- **Security**: You will need to apply security practices at many points. In particular, you may want to authenticate the connections to the messaging system (both for producing and consuming messages), define access control for authorization (you can read and write only to authorized destinations), and even sign messages to ensure the identity of the sender. This is not a big deal, honestly, but is one more thing to take into account.

As you can see, the challenges are not impossible to face, and the advantages will probably outweigh them for you. Also, as we will see in *Chapter 9, Designing Cloud-Native Architectures*, many of these challenges are not exclusive to event-driven architecture, as they are also common in distributed architectures such as microservices.

Event-driven and domain model

We have already discussed many times the importance of correctly modeling a business domain, and how this domain is very specific to the application boundaries. Indeed, in *Chapter 4, Best Practices for Design and Development*, we introduced the idea of bounded context. Event-driven architectures are dealing almost every time with the exchange of information between different bounded contexts.

As already discussed, there are a number of techniques for dealing with such kinds of interactions between different bounded contexts, including the shared kernel, customer suppliers, conformity, and anti-corruption layer. As already mentioned, unfortunately, a perfect approach does not exist for ensuring that different bounded contexts can share meaningful information but stay correctly decoupled.

My personal experience is that the often-used approach here is the shared kernel. In other words, a new object is defined and used as an event format. Such an object contains the minimum amount of information needed for the different bounded contexts to communicate. This does not necessarily mean that the communication will work in every case and no side effects will occur, but it's a solution good enough in most cases.

In the next section, we are going to touch on a common implementation of the event-driven pattern, known as the actor model.

Building on the event-driven architecture – the actor model

The **actor model** is a stricter implementation of the event-driven pattern. In the actor model, the actor is the most elementary unit of computation, encapsulating the state and behavior. An actor can communicate with other actors only through messages.

An actor can create other actors. Each actor encapsulates its internal status (no actor can directly manipulate the status of another actor). This is usually a nice and elegant way to take advantage of multithreading and parallel processing, thereby maintaining integrity and avoiding explicit locks and synchronizations.

In my personal experience, the actor model is a bit too prescriptive when it comes to describing bigger use cases. Moreover, some requirements, such as session handling and access to relational databases, are not an immediate match with the actor model's logic (though they are still implementable within it). You will probably end up implementing some components (maybe core ones) with the actor model while having others that use a less rigorous approach, for the sake of simplicity. The most famous actor model implementation with Java is probably **Akka**, with some other frameworks, such as **Vert.x**, taking some principles from it.

So far, we have elaborated on generic messaging with both the event-driven approach and the actor model.

It is now important, for the purpose of this chapter, to introduce the concept of **streaming**.

Introducing streaming

Streaming has grown more popular with the rise of Apache Kafka even if other popular alternatives, such as Apache Pulsar, are available. Streaming shares some similarities with messaging (there are still producers, consumers, and messages flowing, after all), but it also has some slight differences.

From a purely technical point of view, streaming has one important difference compared with messaging. In a streaming system, messages persist for a certain amount of time (or, if you want, a specified number of messages can be maintained), regardless of whether they have been consumed or not.

This creates a kind of *sliding window*, meaning that consumers of a streaming system can rewind messages, following the flow from a previous point to the current point. This means that some of the information is moved from the messaging system (the broker, or bus) to the consumers (which have to maintain a cursor to keep track of the messages read and can move back in time).

This behavior also enables some advanced use cases. Since consumers can see a consolidated list of messages (the stream, if you like), complex logic can be applied to such messages. Different messages can be combined for computation purposes, different streams can be merged, and advanced filtering logic can be implemented. Moreover, the offloading of part of the logic from the server to the consumers is one factor that enables the management of high volumes of messages with low latencies, allowing for near real-time scenarios.

Given those technical differences, streaming also offers some conceptual differences that lead to use cases that are ideal for modeling with this kind of technology.

With streams, the events (which are then propagated as messages) are seen as a whole information flow as they usually have a constant rate. And moreover, a single event is normally less important than the sequence of events. Last but not least, the ability to rewind the event stream leads to better consistency in distributed environments.

Imagine adding more instances of your application (scaling). Each instance can reconstruct the status of the data by looking at the sequence of messages collected until that moment, in an approach commonly defined as **Event Sourcing**. This is also a commonly used pattern to improve resiliency and return to normal operations following a malfunction or disaster event. This characteristic is one of the reasons for the rising popularity of streaming systems in microservice architectures.

Touching on reactive programming

I like to think of **reactive programming** as event-driven architecture being applied to data streaming. However, I'm aware that that's an oversimplification, as reactive programming is a complex concept, both from a theoretical and technological point of view.

To fully embrace the benefits of reactive programming, you have to both master the tools for implementing it (such as **RxJava**, **Vert.x**, or even **BaconJS**) and switch your reasoning to the reactive point of view. We can do this by modeling all our data as streams (including changes in variables content) and writing our code on the basis of a declarative approach.

Reactive programming considers data streams as the primary construct. This makes the programming style an elegant and efficient way to write asynchronous code, by observing streams and reacting to signals. I understand that this is not easy at all to grasp at first glance.

It's also worth noting that the term *reactive* is also used in the context of reactive systems, as per the **Reactive Manifesto**, produced in 2014 by the community to implement responsive and distributed systems. The Reactive Manifesto focuses on building systems that are as follows:

- **Responsive**: This means replying with minimal and predictable delays to inputs (in order to maximize the user experience).

- **Resilient**: This means that a failure in one of the components is handled gracefully and impacts the whole system's availability and responsiveness as little as possible.

- **Elastic**: This means that the system can adapt to variable workloads, keeping constant response times.

- **Message-driven**: This means that systems that adhere to the manifesto use a message-driven communication model (hence achieving the same goals as described in the *Introducing the event-driven pattern and event-driven architecture* section).

While some of the goals and techniques of the Reactive Manifesto resonate with the concepts we have explored so far, reactive systems and reactive programming are different things.

The Reactive Manifesto does not prescribe any particular approach to achieve the preceding four goals, while reactive programming does not guarantee, per se, all the benefits pursued by the Reactive Manifesto.

A bit confusing, I know. So, now that we've understood the differences between a reactive system (as per the Reactive Manifesto) and reactive programming, let's shift our focus back to reactive programming.

As we have said, the concept of data streaming is central to reactive programming. Another fundamental ingredient is the **declarative approach** (something similar to functional programming). In this approach, you express what you want to achieve instead of focusing on all the steps needed to get there. You declare the final result (leveraging standard constructs such as filter, map, and join) and attach it to a stream of data to which it will be applied.

The final result will be compact and elegant, even if it may not be immediate in terms of readability. One last concept that is crucial in reactive programming is **backpressure**. This is basically a mechanism for standardizing communication between producers and consumers in a reactive programming model in order to regulate flow control.

This means that if a consumer can't keep up with the pace of messages received from the producer (typically because of a lack of resources), it can send a notification about the problem upstream so that it can be managed by the producer or any other intermediate entity in the stream chain (in reactive programming, an event stream can be manipulated by intermediate functions). In theory, backpressure can bubble up to the first producer, which can also be a human user in the case of interactive systems.

When a producer is notified of backpressure, it can manage the issue in different ways. The most simple is to slow down the speed and just send less data, if possible. A more elaborate technique is to buffer the data, waiting for the consumer to get up to speed (for example, by scaling its resources). A more destructive approach (but one that is effective nevertheless) is to drop some messages. However, this may not be the best solution in every case.

With that, we have finished our quick look at reactive programming. I understand that some concepts have been merely mentioned, and things such as the functional and declarative approaches may require at least a whole chapter on their own. However, a full deep dive into the topic is beyond the scope of this book. I hope I gave you some hints to orient yourself toward the best architectural approach when it comes to message- and event-centric use cases.

In this section, you learned about the basic concepts and terms to do with reactive and event-driven programming, which, if well understood and implemented, can be used to create high-performance applications.

In the next section, we will start discussing how to optimize our architecture for performance and scalability purposes.

Designing for large-scale adoption

So far, in this chapter, we have discussed some widespread patterns and architectural styles that are well used in the world of enterprise Java applications.

One common idea around the techniques that we have discussed is to organize the code and the software components not only for better readability, but also for performance and scalability.

As you can see (and will continue to see) in this book, in current *web-scale* applications, it is crucial to think ahead in terms of planning to absorb traffic spikes, minimize resource usage, and ultimately have good performance. Let's have a quick look at what this all means in our context.

Defining performance goals

Performance is a very broad term. It can mean many different things, and often you will want to achieve all performance goals at once, which is of course not realistic.

In my personal experience, there are some main performance indicators to look after, as they usually have a direct impact on the business outcome:

- **Throughput**: This is measured as the number of transactions that can be managed per time unit (usually in seconds). The tricky part here is to define exactly what a transaction is in each particular context, as probably your system will manage different transaction types (with different resources being needed for each kind of transaction). Business people understand this metric instantaneously, knowing that having a higher throughput means that you will spend less on hardware (or cloud) resources.

- **Response time**: This term means many different things. It usually refers to the time it takes to load your web pages or the time it takes to complete a transaction. This has to do with customer satisfaction (the quicker, the better). You may also have a contractual **Service Level Agreement** (**SLA**); for example, your system must complete a transaction in no more than x milliseconds. Also, you may want to focus on an average time or set a maximum time threshold.

- **Elapsed time**: This basically means the amount of time needed to complete a defined chunk of work. This is common for batch computations (such as in big data or other calculations). This is kind of a mix of the previous two metrics. If you are able to do more work in parallel, you will spend less on your infrastructure. You may have a fixed deadline that you have to honor (such as finishing all your computations before a branch opens to the public).

Performance tuning is definitely a broad topic, and there is no magic formula to easily achieve the best performance. You will need to get real-world experience by experimenting with different configurations and get a lot of production traffic, as each case is different. However, here are some general considerations for each performance goal that we have seen:

- To enhance throughput, your best bet is to parallelize. This basically means leveraging threading where possible. It's unbelievable how often we tend to chain our calls in a sequential way. Unless it is strictly necessary (because of data), we should parallelize as much as we can and then merge the results.

  This entails, basically, splitting each call wherever possible (by delegating it to another thread), waiting for all the subcalls to complete in order to join the results in the main thread, and returning the main thread to the caller. This is particularly relevant where the subcalls involve calling to external systems (such as via web services). When parallelizing, the total elapsed time to answer will be equal to the longest subcall, instead of being the sum of the time of each subcall.

  In the next diagram, you can see how parallelizing calls can help in reducing the total elapsed time needed to complete the execution of an application feature:

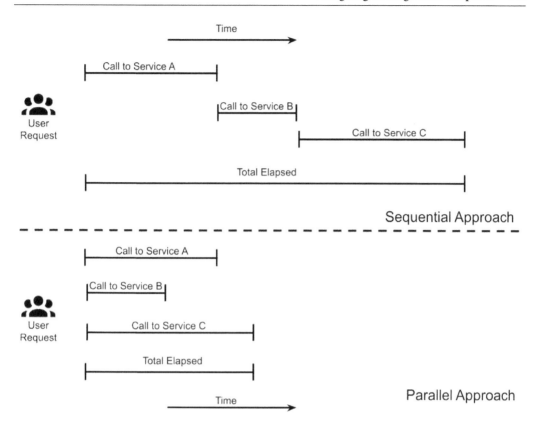

Figure 6.6 – Sequential versus parallel approach

- There should be a physical separation of our service based on the load and the performance expectations (something greatly facilitated by containers and microservices architecture). Instead of mixing all your APIs, you may want to dedicate more resources to the more critical ones (perhaps even dynamically, following the variation of traffic) by isolating them from the other services.

- For better response times, async is the way to go. After reviewing the previous sections for advice, I suggest working with your business and functional analysts and fighting to have everything be as asynchronous as possible from a use case perspective.

It is very uncommon to have really strict requirements in terms of checking everything on every backend system before giving feedback to your users. Your best bet is to do a quick validation and reply with an acknowledgment to the customer. You will, of course, need an asynchronous channel (such as an email, a notification, or a webhook) to notify regarding progression of the transaction. There are countless examples in real life; for example, when you buy something online, often, your card funds won't even be checked in the first interaction. You are then notified by email that the payment has been completed (or has failed). Then, the package is shipped, and so on. Moreover, optimizing access to data is crucial; caching, pre-calculating, and de-duplicating are all viable strategies.

- When optimizing for elapsed time, you may want to follow the advice previously given: parallelizing and optimizing access to data is key. Also, here, you may want to rely on specialized infrastructure, such as scaling to have a lot of hardware (maybe in the cloud) and powering it off when it is not needed, or using infrastructures optimized for input/output. But the best advice is to work on the use case to maximize the amount of parallelizable work, possibly duplicating part of the information.

We will learn more about performance in *Chapter 12, Cross-Cutting Concerns*. Let's now review some key concepts linked to scalability.

Stateless

Stateless is a very recurrent concept (we will see it again in *Chapter 9, Designing Cloud-Native Architectures*). It is difficult to define with simple words, however.

Let's take the example of an ATM versus a workstation.

Your workstation is something that is usually difficult to replace. Yes, you have backups and you probably store some of your data online (in your email inbox, on the intranet, or on shared online drives). But still, when you have to change your laptop for a new one, you lose some time ensuring that you have copied any local data. Then, you have to export and reimport your settings, and so on. In other words, your laptop is very much stateful. It has a lot of local data that you don't want to lose.

Now, let's think about an ATM. Before you insert your card, it is a perfectly empty machine. It then loads your data, allows cash withdrawal (or whatever you need), and then it goes back to the previous (empty) state, ready for the next client to serve. It is stateless from this point of view. It is also engineered to minimize the impact if something happens while you are using it. It's usually enough to end your current session and restart from scratch.

But back to our software architecture: *how do we design an architecture to be stateless?*

The most common ways are as follows:

- **Push the state to clients**: This can mean having a cookie in the customer browser or having your APIs carry a token (such as a **JWT**). Every time you get a request, you may get to choose the best instance for your software (be it a container, a new JVM instance, or simply a thread) to handle it – *which will it be: the closest to the customer, the closest to the data, or simply the one with the least amount of load at that moment?*

- **Push the state to an external system**: You can offload the state to a dedicated system, such as a distributed cache. Your API (and business logic) only need to identify the user. All the session data is then loaded from a dedicated system. Any new instance can simply ask for the session data. Of course, your problem is then how to scale and maximize the uptime of such a caching system.

Whatever your approach is, think always about the *phoenix*; that is, you should be able to reconstruct the data from the ashes (and quickly). In this way, you can maximize scaling, and as a positive side effect, you will boost availability and disaster recovery capabilities. As highlighted in the *Introducing streaming* section, events (and the event sourcing technique) are a good way to implement similar approaches. Indeed, provided that you have persisted all the changes in your data into a streaming system, such changes could be replayed in case of a disaster, and you can reconstruct the data from scratch.

Beware of the concept of **stickiness** (pointing your clients to the same instance whenever possible). It's a quick win at the beginning, but it may lead you to unbalanced infrastructure and a lack of scalability. The next foundational aspect of performance is data.

Data

Data is very often a crucial aspect of performance management. Slow access times to the data you need will frustrate all other optimizations in terms of parallelizing or keeping interactions asynchronous. Of course, each type of data has different optimization paths: indexing for relational databases, proximity for in-memory caching, and low-level tuning for filesystems.

However, here are my considerations as regards the low-hanging fruit when optimizing access to data:

- **Sharding**: This is a foundational concept. If you can split your data into smaller chunks (such as by segmenting your users by geographical areas, sorting using alphabetical order, or using any other criteria compliant with your data model), you can dedicate a subset of the system (such as a database schema or a file) to each data shard.

This will boost your resource usage by minimizing the interference between different data segments. A common strategy to properly cluster data in shards is **hashing**. If you can define a proper hashing function, you will have a quick and reliable way to identify where your data is located by mapping the result of the hashing operation to a specific system (containing the realm that is needed). If you still need to access data across different shards (such as for performing computations or for different representations of data), you may consider a different sharding strategy or even duplicating your data (but this path is always complex and risky, so be careful with that).

- **Consistency point**: This is another concept to take care of. It may seem like a lower-level detail, but it's worthwhile exploring. To put it simply: *how often do you need your data to persist?* Persistence particularly common in long transactions (such as ones involving a lot of submethods). Maybe you just don't need to persist your data every time; you can keep it in the memory and batch all the persistence operations (this often includes writing to files or other intensive steps) together.

  For sure, if the system crashes, you might lose your data (and whether to take this risk is up to you), but are you sure that incongruent data (which is what you'd have after saving only a part of the operations) is better than no data at all? Moreover, maybe you can afford a crash because your data has persisted elsewhere and can be recovered (think about streaming, which we learned about previously). Last but not least, *is it okay if your use case requires persistence at every step?* Just be aware of that. Very often, we simply don't care about this aspect, and we pay a penalty without even knowing it.

- **Caching**: This is the most common technique. Memory is cheap, after all, and almost always has better access times than disk storage. So, you may just want to have a caching layer in front of your persistent storage (database, filesystem, or whatever). Of course, you will end up dealing with stale data and the propagation of changes, but it's still a simple and powerful concept, so it's worth a try.

  Caching may be implemented in different ways. Common implementations include caching data in the working memory of each microservice (in other words, in the heap, in the case of Java applications), or relying on external caching systems (such as client-server, centralized caching systems such as Infinispan or Redis). Another implementation makes use of external tools (such as Nginx or Varnish) sitting in front of the API of each microservice and caching everything at that level.

We will see more about data in *Chapter 11, Dealing with Data*, but for now, let me give you a spoiler about my favorite takeaway here: you must have multiple ways of storing and retrieving data and using it according to the constraints of your use case. Your mobile application has a very different data access pattern from a batch computation system. Now, let's go to the next section and have a quick overview of scaling techniques.

Scaling

Scaling has been the main mantra so far for reaching performance goals and is one of the key reasons why you would want to architect your software in a certain way (such as in a multi-tier or async fashion). And honestly, I'm almost certain that you already know what scaling is and why it matters. However, let's quickly review the main things to consider when we talk about scaling:

- **Vertical scaling** is, somewhat, the most traditional way of scaling. To achieve better performance, you need to add more resources to your infrastructure. While it is still common and advisable in some scenarios (such as when trying to squeeze more performance from databases, caches, or other stateful systems), it is seldom a long-term solution.

  You will hit a blocking limit sooner or later. Moreover, vertical scaling is not very dynamic, as you may need to purchase new hardware or resize your virtual machine, and maybe downtime will be needed to make effective changes. It is not something you can do in a few seconds to absorb a traffic spike.

- **Horizontal scaling** is way more popular nowadays as it copes well with cloud and PaaS architectures. It is also the basis of stateless, sharding, and the other concepts discussed previously. You can simply create another instance of a component, and that's it. In this sense, the slimmer, the better. If your service is very small and efficient and takes a very short time to start (*microservices, anyone?*), it will nicely absorb traffic spikes.

  You can take this concept to the extreme and shut down everything (thereby saving money) when you have no traffic. As we will see in *Chapter 9, Designing Cloud-Native Architectures*, scaling to zero (so that no instance is running if there are no requests to work with) is the concept behind serverless.

- We are naturally led to think about scaling in a **reactive way**. You can get more traffic and react by scaling your components. The key here is identifying which metric to look after. It is usually the number of requests, but memory and CPU consumption are the other key metrics to look after. The advantage of this approach is that you will consume the resources needed for scaling *just in time*, hence you will mostly use it in an efficient way. The disadvantage is that you may end up suffering a bit if traffic increases suddenly, especially if the new instances take some time to get up and running.

- The opposite of reactive scaling is, of course, **proactive scaling**. You may know in advance that a traffic spike is expected, such as in the case of Black Friday or during the tax payment season. If you manage to automate your infrastructure in the right way, you can schedule the proper growth of the infrastructure in advance. This may be even more important if scaling takes some time, as in vertical scaling. The obvious advantage of this approach is that you will be ready in no time in case of a traffic increase, as all the instances needed are already up and running. The disadvantage is that you may end up wasting resources, especially if you overestimate the expected traffic.

With this section, we achieved the goals of this chapter. There was quite a lot of interesting content. We started with hexagonal architectures (an interesting example of encapsulation), before moving on to multi-tier architectures (a very common way to organize application components). Then, you learned about MVC (a widely used pattern for user interfaces), event-driven (an alternative way to design highly performant applications), and finally, we looked at some common-sense suggestions about building highly scalable and performant application architectures.

It is not possible to get into all the details of all the topics discussed in this chapter. However, I hope to have given you the foundation you need to start experimenting and learning more about the topics that are relevant to you.

And now, let's have a look at some practical examples.

Case studies and examples

As with other chapters in this book, let's end this chapter with some practical considerations about how to apply the concepts we've looked at to our recurrent example involving a mobile payment solution. Let's start with encapsulation.

Encapsulating with a hexagonal architecture

A common way to map hexagonal concepts in Java is to encompass the following concept representations:

- The core maps into the domain model. So, here you have the usual entities (**Payment**, in our example), services (**PaymentService**, in this case), value objects, and so on. Basically, all the elements in the core are **Plain Old Java Objects** (**POJOs**) and some business logic implementations.

- Here, the ports are the interfaces. They are somewhere in the middle, between a logical concept in the domain realm (enquire, notify, and store, in our example) and the respective technical concepts. This will promote the decoupling of the business logic (in the core) and the specific technology (which may change and evolve).

- The adapters are implementations of such interfaces. So, an enquire interface will be implemented by **SoapAdapter**, **RestAdapter**, and **GraphQLAdapter**, in this particular case.

- Outside of the hexagon, the external actors (such as the mobile app, databases, queues, or even external applications) interact with our application domain via the adapters provided.

The following diagram illustrates the preceding points:

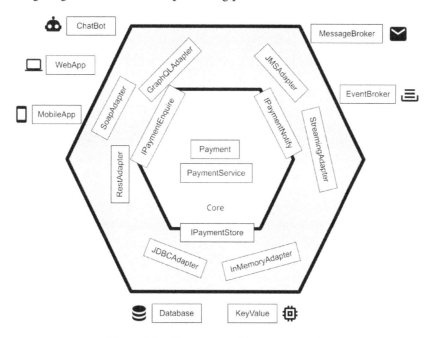

Figure 6.7 – Hexagonal architecture example

Here are some key considerations:

- The cardinality is completely arbitrary. You are not limited to six ports or adapters. Each port can map to one or more adapters. Each external system can be bound to one or more adapters. Each adapter can be consumed by more than one external system (they are not exclusive unless you want them to be).

  This logical grouping can be seen at the level that you want. This could be an application, meaning that everything inside the hexagon is deployed on a single artifact – an **Enterprise Application Archive** (**EAR**) or a **Java Application Archive** (**JAR**) – in a machine, or it could be a collection of different artifacts and machines (as in a microservices setup). In this case, most probably you will decouple your interfaces with REST or something similar, to avoid sharing dependencies across your modules.

- The advantage in terms of test coverage is obvious. You can switch each adapter into a mock system, to test in environments that don't have the complete infrastructure. So, you can test your notifications without the need for a queue, or test persistence without the need for a database. This, of course, will not replace end-to-end testing, in which you have to broaden your test and attach it to real adapters (such as in automating tests that call REST or SOAP APIs) or even external systems (such as in testing the mobile app or the web app itself).

As usual, I think that considering hexagonal modeling as a tool can be useful when implementing software architecture. Let's now have a quick look at multi-tier architecture.

Componentizing with multi-tier architecture

Multi-tier architecture gives us occasion to think about componentization and, ultimately, the evolution of software architectures. If we think about our mobile payment application, a three-tier approach may be considered a good fit. And honestly, it is. Historically, you probably wouldn't have had many other options than a pure, centralized, client-server application. Even with a modern perspective, starting with a less complex approach, such as the three-tier one, it can be a good choice for two reasons:

- It can be considered a *prototypization* phase, with the goal of building a **Minimum Viable Product** (**MVP**). You will have something to showcase and test soon, which means you can check whether you have correctly understood the requirements or whether users like (and use) your product. Moreover, if you designed your application correctly (using well-designed APIs), maybe you can evolve your backend (toward multi-tier or microservices) with minimal impact on the clients.

- It can be a good benchmark for your domain definition. As per the famous Martin Fowler article (*Monolith First*), you may want to start with a simpler, all-in-one architecture in order to understand the boundaries of your business logic, and then correctly decomponentize it in the following phase (maybe going toward a cloud-native approach).

In the next diagram, you can see a simple representation of an application's evolution from three-tier to microservices:

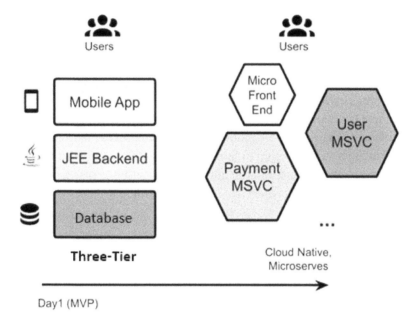

Figure 6.8 – Tier segmentation evolution

As you can see in the previous diagram, each component change has a role and name. There are some key considerations to make about this kind of evolution:

- We will see more about microservices in *Chapter 9, Designing Cloud-Native Architectures*. For now, consider the fact that this example will only represent architectural evolution over time and how your tier segmentation can evolve. Microservices is probably not similar to multi-tier architecture, as some concepts (such as responsibilities in terms of data representation in views) are orthogonal to it (in other words, you can still have concepts from three-tier on top of microservices).

- We are starting with three tiers because it is simply an antipattern to have business logic mixed together with your data in terms of being deployed to the database (with stored procedures and such). However, in my opinion, having an external database does not constitute a data layer *per se*. So, in this example, the three-tier architecture can also be seen as a two-tier/client-server architecture, with the external database simply being a technological detail.

- In the real world, there is no defined boundary between one architectural view (such as three-tier) and another alternative (such as microservices). It's not as if one day you will transition from client-server (or three-tier) to microservices. You will probably start adding more layers, and then reorganize some capabilities into a complete microservice from the ground up and offload some capabilities to it.

  In general, it is possible to have a few differing architectural choices coexisting in the same application, perhaps just for a defined (limited) time, leading to a transition to a completely different architecture. In other words, your three-tier architecture can start with some modularized microservices alongside tiers (making it a hybrid architecture, bringing different styles together), and then the tiered part can be progressively reduced and the microservices part increased, before a final and complete move to a microservices implementation.

Once again, this is designed to give you some food for thought as to how to use some key concepts seen in this chapter in the real world. It's important to understand that it's rare (and maybe wrong) to completely and religiously embrace just one model, for instance, starting with a pure three-tier model and staying with it even if the external conditions change (if you start using a cloud-like environment, for example).

Planning for performance and scalability

As seen in the previous sections, performance is a broad term. In our example, it is likely that we will want to optimize for both throughput and response time. It is, of course, a target that is not easy to reach, but it is a common request in this kind of project:

- **Throughput** means a more sustainable business, with a lower cost for each transaction (considering hardware, power, software licenses, and so on).

- **Response time** means having a happier customer base and, ultimately, the success of the project. Being an online product, it is expected today that access to this kind of service (whether it is for making a payment or accessing a list of transactions) happens with zero delay; every hiccup could lead a customer to switch to alternative platforms.

Also, you may want to have a hard limit. It is common to have a timeout, meaning that if your payment takes more than 10 seconds, it is considered to have failed and is forcefully dropped. That's for limiting customer dissatisfaction and avoiding the overloading of the infrastructure.

But how do you design your software architecture to meet such objectives? As usual, there is no magic recipe for this. Performance tuning is a continuous process in which you have to monitor every single component for performance and load, experiment to find the most efficient solution, and then switch to the next bottleneck. However, there are a number of considerations that can be made upfront:

- First of all, there is **transactional behavior**. We will see in *Chapter 9, Designing Cloud-Native Architectures*, how heavily decentralized architectures, such as microservices, do not cope well with long and distributed transactions. Even if you are not yet in such a situation and you are starting with a simpler, three-tier architecture, having long transaction boundaries will cause serialization in your code, penalizing your performance.

  To avoid this, you have to restrict the transaction as much as possible and handle consistency in different ways wherever possible. You may want to have your transaction encompass the payment request and the check of monetary funds (as in the classic examples about transactions), but you can take most of the other operations elsewhere. So, notifications and updates of non-critical systems (such as CRMs or data sources only used for inquiries) can be done outside of the transactions and retried in the case of failures.

- As a follow-up from the previous point, it should be taken into account that you don't have to penalize the most common cases to avoid very remote cases unless they have dramatic consequences. So, it is okay to check funds before making the payments in a strict way (as in the same transaction), because a malfunction there can cause bad advertising and a loss of trust in your platform, with potentially devastating consequences.

  But you can probably afford to have a notification lost or sent twice from time to time if this means that 99% of the other transaction are performing better. And the rules can also be adapted to your specific context. Maybe the business can accept skipping some online checks (such as anti-fraud checks) in payment transactions of small amounts. The damage of some fraudulent transactions slipping through (or only being identified after the fact) may be lower than the benefit in terms of performance for the vast majority of licit traffic.

- In terms of **asynchronous behavior**, as has been seen, it is expected that you only do synchronously what's essential to do synchronously. So, apart from the obvious things such as notifications, every other step should be made asynchronous if possible – for example, updating downstream systems.

  So, in our use case, if we have a transactional database (or a legacy system) storing the user position that is used to authorize payments, it should be checked and updated synchronously to keep consistency. But if we have other systems, such as a CRM that stores the customer position, perhaps it's okay to place an update request in a queue and update that system after a few seconds, when the message is consumed and acted upon.

- Last but not least, in terms of **scaling**, the more your component will be stateless, the better. So, if we have each step of the payment process carrying over all the data needed (such as the customer identifier and transaction identifier), maybe we can minimize the lookups and checks on the external systems.

  In the case of more load, we can (in advance, if it is planned, or reactively if it is an unexpected peak) create more instances of our components. Then, they will be immediately able to take over for the incoming requests, even if they originated from existing instances.

  So, if you imagine a payment transaction being completed in more than one step (as in first checking for the existence of the recipient, then making a payment request, then sending a confirmation), then it may be possible that each of those steps is worked on by different instances of the same component. Think about what would happen if you had to manage all those steps on the same instance that started the process because the component stored the data in an internal session. In cases of high traffic, new instances would not be able to help with the existing transactions, which would have to be completed where they originated. And the failure of one instance would likely create issues for users.

This completes the content of this chapter. Let's quickly recap the key concepts that you have seen.

Summary

In this chapter, you have seen a lot of the cornerstone concepts when it comes to architectural patterns and best practices in Java. In particular, you started with the concept of encapsulation; one practical way to achieve it is the hexagonal architecture. You then moved to multi-tier architectures, which is a core concept in Java and JEE (especially the three-tier architecture, which is commonly implemented with beans, servlets, and JSPs).

There was a quick look at MVC, which is more a design pattern than an architectural guideline but is crucial to highlight some concepts such as the importance of separating presentation from business logic. You then covered the asynchronous and event-driven architecture concepts, which apply to a huge portion of different approaches that are popular right now in the world of Java. These concepts are known for their positive impacts on performance and scalability, which were also the final topics of this chapter.

While being covered further in other chapters, such as *Chapter 9, Designing Cloud-Native Architectures*, and *Chapter 12, Cross-Cutting Concerns*, here you have seen some general considerations about architecture that will link some of the concepts that you've seen so far, such as tiering and asynchronous interactions, to specific performance goals.

In the next chapter, we will look in more detail at what middleware is and how it's evolving.

Further reading

- *Hexagonal architecture*, by Alistair Cockburn (`https://alistair.cockburn.us/hexagonal-architecture/`)

- *Java Performance: The Definitive Guide: Getting the Most Out of Your Code*, by Scott Oaks, published by O'Reilly Media (2014)

- *Kafka Streams in Action*, by William P. Bejeck Jr., published by Manning

- *Scalability Rules 50 Principles for Scaling Web Sites*, by Martin L. Abbott and Michael T. Fisher, published by Pearson Education (2011)

- *Monolith First*, by Martin Fowler (`https://www.martinfowler.com/bliki/MonolithFirst.html`)

7
Exploring Middleware and Frameworks

In this chapter, we will start talking about the concept of middleware and how it has evolved over time. In particular, we will focus on the **Java Enterprise Edition** (**JEE**) standard, including the **Jakarta EE** transition. We will see a notable open source implementation, which is **WildFly** (formerly known as **JBoss Application Server**), and we will start exploring how the concept of middleware is evolving into cloud-native frameworks – in our case, **Quarkus**.

You will learn the following topics in this chapter:

- The JEE standard
- The WildFly application server
- The most common JEE APIs
- Beyond JEE
- Quarkus

Our picture of middleware will be completed in the next chapter, in which we will see the approach to integration, which is another cornerstone of what's traditionally called middleware.

After reading this chapter, you will know the differences and similarities between the JEE standard and its cloud-native alternative, MicroProfile. Moreover, we will have seen the most common and useful APIs provided by both standards.

But first of all, let's start with the most popular middleware standard for Java developers, which is, of course, JEE.

Technical requirements

Please make sure that you have a supported **Java Virtual Machine** (**JVM**) installed on your machine.

You can find the source code used in this chapter on GitHub: `https://github.com/PacktPublishing/Hands-On-Software-Architecture-with-Java/tree/master/Chapter7`.

Introducing the JEE standard

We (as programmers, who are well versed with the digital world) know that Java is a powerful and expressive language. It is a widely used tool for building applications, both in a traditional way (as it is already done in a majority of enterprise contexts) and more and more in a cloud-native way too (as we will see in this chapter).

According to the JVM Ecosystem Report 2021 by Snyk, roughly 37% of production applications use JEE (with Java EE, referring to the older version, still being used by a majority compared to newer JakartaEE implementations). Spring Boot counts for 57%, while Quarkus, which we are going to see in this chapter, is growing and is currently at 10%.

So, Java doesn't need an introduction per se. Everybody (at least, everybody who is reading this book) knows that it's a powerful and expressive language that aims to be available across platforms (write once, run everywhere – I love it!) and that it is based on the compilation of bytecode, which can then be executed by the virtual machine.

It's a technology platform that includes a programming language, specifications, documentation, and a set of supporting tools, including runtimes (the JVM), a compiler, and so on. The tools are provided by different vendors (with the major ones being Oracle, IBM, and Red Hat) and comply with the standards. The language is currently owned by Oracle. So far, so good.

Then, we have the **Enterprise Edition**. There are a number of standards that are not really needed in the *plain* version of the Java technology. Features such as transactions and messaging are specifically targeted at server-side enterprise scenarios, such as banking applications, CRMs, and ERPs. For this reason, such features are standardized as an extension of the Java platform, namely the Enterprise Edition.

However, in 2017, Oracle decided to donate the rights of the Enterprise Edition to the Eclipse open source community while holding the rights to the Java language (and brand). For this reason, the Enterprise Edition has been renamed Jakarta EE after a community vote.

This transition caused some slight changes in the specification process, basically making it more open to cooperation and less linked to just one vendor. The old process was named the **Java Community Process** (**JCP**), while the new one is called the **Eclipse Foundation Specification Process** (**EFSP**). The most important concepts stay the same, such as the **Java Specification Request** (**JSR**), which is a way of specifying the new features, and the **Technology Compatibility Kits** (**TCKs**), which are used to certify adherence to the standard. Jakarta starts from version **8**, based on **Java EE 8**. At the time of writing, **Jakarta EE 9** is available. The examples in this chapter are tested against **JEE 8** (because it's the most widely used version right now) but should work properly in **JEE 9** too.

It's worth noting that in this section, we will install the WildFly application server in order to start playing with JEE (and later on, we will start working with Quarkus to learn about MicroProfile). In both cases, the only requirement on your machine is a compatible version of the JVM. If you are in doubt, you can download the version you need for free from the OpenJDK website (`http://openjdk.java.net/`).

Diving into JEE implementations

As we said, the JEE specification (before and after the transition to Jakarta) provides TCKs. TCKs are suites of tests to certify compliance with the JEE standards. JEE currently provides a full profile and a web profile. The web profile is basically a subset of the specifications included in the full profile, aiming at a lighter implementation for some scenarios.

There are a number of application servers that are JEE compliant. In my personal experience, the most widely adopted servers are as follows:

- **WildFly** is a fully open source JEE application server, and it has a commercially supported version named JBoss Enterprise Application Platform (by Red Hat).

- **WebSphere Application Server**, developed by IBM, is distributed in many different versions, including the open source Open Liberty.

- **Oracle WebLogic Server** is developed and distributed by Oracle (the full profile only).

Among the other servers fully implementing JEE specifications, Payara and GlassFish are worth mentioning. There are also a number of other interesting projects (such as Tomcat and Jetty) that are not fully JEE certified, but they implement most of the APIs and can plug into some of the others via external dependencies. In this chapter, we will work with WildFly, but thanks to the JEE standard, if we change some dependencies, everything should work in the other servers too.

Introducing the WildFly application server

WildFly is by far the application server that I've come across most often in my daily job. It's probably the most widespread Java application server. It was renamed from JBoss, as a contraction of **Enterprise Java Beans (EJB)** and **Open-Source Software (OSS)**, EJBoss then becoming JBoss for copyright reasons relating to the EJB trademark. Since 2014, after a community vote, JBoss was renamed WildFly in its upstream distribution. This was to reduce the confusion in names between the project (WildFly), the community (JBoss. org), and the product family commercially supported by Red Hat (including **JBoss EAP**).

It is worth mentioning that JBoss EAP is made of the same components as WildFly. There are no hidden features available in the commercial distribution. JBoss EAP is simply a frozen distribution of the WildFly components at a certain version, which is used to provide stability, certifications, and commercial support for enterprise environments. WildFly is developed in Java.

Exploring the WildFly architecture

If you have had a decent experience with WildFly, you may remember that JBoss used to be huge and sometimes slow. For this reason, a long time ago (around 2011), the server, then named **JBoss AS 7**, was rearchitected from the ground up. The resulting version was a modular and fast application server, which provided the basis for all the later releases (including the current one).

Major changes were about class loading (made more granular), core feature implementation (moved to a modular, lazy-loading system), management, and configuration (unified into one single file). The result of this rearchitecting was then used for the **JBoss EAP 6** (and the following versions) commercial distribution. The latest version of WildFly (**22.0**) starts on my laptop in around 1 second. WildFly can be started in standalone mode (everything running in one Java process) or in domain mode, which is a way to centrally manage a fleet of instances from a single point.

> **Important Note:**
>
> A very common misconception is the overlapping of the concept of domain with the concepts of clustering and high availability. They are actually orthogonal concepts.

We can have an arbitrary number of standalone servers, individually managed and configured to be clustered in a highly available fashion, or a domain managing a fleet of non-clustered instances. Also, it's worth noting that we can have multiple server instances on a single machine (whether a physical or virtual host) by operating different port offsets (to avoid TCP port clashing) and different subdirectories.

The server is distributed as a `.zip` file and the most significant folders are as follows:

- `bin` directory: This contains the executable scripts (both for **Linux** and **Windows**, so `.sh` and `.bat`) for starting the server, along with some other utilities, such as for adding users and configuring vaults.

- `modules` directory: This contains the system dependencies of the application server, which implement the core JEE features (and other supporting subsystems).

- `standalone` directory: This is used as a root directory when the server is started in standalone mode. It includes subdirectories such as `configuration` (used for storing configuration files), `data` (where the persistent data from the deployed applications is stored), `tmp` (used to store temporary files used by applications), `log` (the default location for the server and applications logfiles), and `deployments` (which can be used to deploy applications by dropping deployable files and is used for development purposes).

- `domain` directory: This is similar to `standalone`, but it doesn't contain the `deployments` folder (which is used for drop-in deployment, which is when we deploy new applications by copying the artifact in the directory and expect the application server to pick it and deploy it. This is not supported in domain mode). It contains a `content` directory (supporting some system functionalities, specific to the domain operating mode) and a `server` directory, which contains a subdirectory for each server instance hosted in the current machine, in turn containing `tmp`, `data`, and `log` folders used by that particular server.

So far, we've been introduced to the WildFly architecture; now let's see how to run a WildFly server.

Running the WildFly server

For the sake of this chapter, we will be running the WildFly server in standalone mode. Before we get started, please make sure that you have a supported JVM installed on your machine. We will need to download the latest server distribution from `https://www.wildfly.org/downloads/`.

We'll use the following steps to install the WildFly server runtime:

1. After downloading the required suitable files, we'll unzip them and run the following command on the terminal:

    `/bin/standalone.sh`

    We can also use the `.bat` script (if you are on Windows) to run the server. We get the following output:

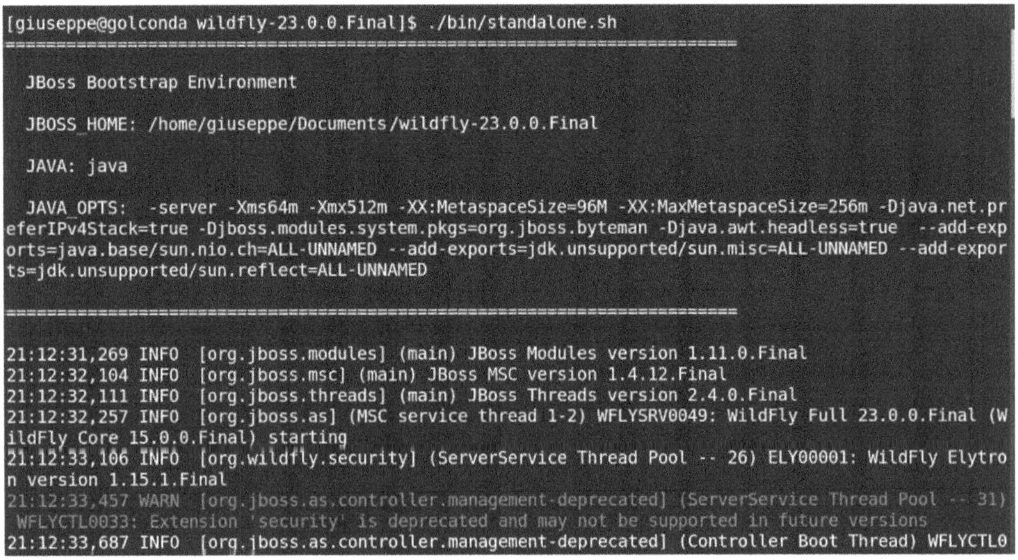

```
[giuseppe@golconda wildfly-23.0.0.Final]$ ./bin/standalone.sh
=====================================================================

  JBoss Bootstrap Environment

  JBOSS_HOME: /home/giuseppe/Documents/wildfly-23.0.0.Final

  JAVA: java

  JAVA_OPTS:  -server -Xms64m -Xmx512m -XX:MetaspaceSize=96M -XX:MaxMetaspaceSize=256m -Djava.net.pr
eferIPv4Stack=true -Djboss.modules.system.pkgs=org.jboss.byteman -Djava.awt.headless=true  --add-exp
orts=java.base/sun.nio.ch=ALL-UNNAMED --add-exports=jdk.unsupported/sun.misc=ALL-UNNAMED --add-expor
ts=jdk.unsupported/sun.reflect=ALL-UNNAMED

=====================================================================

21:12:31,269 INFO  [org.jboss.modules] (main) JBoss Modules version 1.11.0.Final
21:12:32,104 INFO  [org.jboss.msc] (main) JBoss MSC version 1.4.12.Final
21:12:32,111 INFO  [org.jboss.threads] (main) JBoss Threads version 2.4.0.Final
21:12:32,257 INFO  [org.jboss.as] (MSC service thread 1-2) WFLYSRV0049: WildFly Full 23.0.0.Final (W
ildFly Core 15.0.0.Final) starting
21:12:33,106 INFO  [org.wildfly.security] (ServerService Thread Pool -- 26) ELY00001: WildFly Elytro
n version 1.15.1.Final
21:12:33,457 WARN  [org.jboss.as.controller.management-deprecated] (ServerService Thread Pool -- 31)
WFLYCTL0033: Extension 'security' is deprecated and may not be supported in future versions
21:12:33,687 INFO  [org.jboss.as.controller.management-deprecated] (Controller Boot Thread) WFLYCTL0
```

Figure 7.1 – Initializing WildFly

2. Once the server is started, we can deploy our application by dropping our artifact (`.jar` / `.war` / `.ear`) into the `deployments` folder (which is not advised for production purposes) or, better yet, we can deploy by using the JBoss **Command-Line Interface** (**CLI**).

3. The JBoss CLI can be executed by running the `jboss-cli.sh` (or `.bat`) script from the `bin` directory. The CLI can be used to connect, configure, and manage WildFly setups (both locally and over the network). In order to use it to connect to a local standalone WildFly instance, we can simply use this command:

```
./jboss-cli.sh --connect
```

4. We will then enter the interactive WildFly CLI. To deploy our application, we can use this command:

```
deploy /pathToArtifact/myArtifact.war
```

5. We can then exit the WildFly CLI with the `exit` command:

```
exit
```

Now that we know the basics of configuring and operating the WildFly server, we can start playing with simple JEE examples.

Understanding the most common JEE APIs

Now that we have seen an overview of the JEE technology and implemented it with application servers, we will learn about the most common JEE APIs that are used in enterprise projects. We will have a look at some examples of those APIs at the end of this chapter, in the *Case studies and examples* section.

Dependency injection

I remember the times when dependency injection was simply not available in JEE, and we had to rely exclusively on **EJB version 2** (unfortunately) to wire our dependencies. This was probably one of the reasons behind the growth in popularity of the Spring Framework, which became widespread by offering a lightweight alternative to wiring, based on dependency injection, and avoiding verbose and error-prone configuration files. But that's another story that is out of the scope of this book.

Dependency Injection (DI) or **Contexts and Dependency Injection (CDI)** is a concept that extends and implements the **Inversion of Control (IoC)** principle. The idea here is that instead of letting each class instantiate the required classes, we can let an external entity (sometimes referred to as an IoC container) do that. This allows us to just use Java interfaces at design time and lets the container pick the right implementations, thus boosting flexibility and decoupling. Moreover, the CDI concept rationalizes the application structure by making the wiring points of one class with the others explicit.

In the current implementations of the CDI standard (**version 3.0** in **JEE 9**), CDI is so easy to use that we can start developing with it without knowing much about it. The implementation is designed around annotations that decorate the classes identifying the contact points between each other. The most popular one is the `@Inject` annotation.

By marking a field (or a setter method, or a constructor) of our class with this annotation, we are basically telling the framework that we want that field instantiated and provided for us. The container tries to identify a class in the application that may satisfy that dependency. The objects that can be injected are almost any kind of Java class, including special things that provide access to JEE services, such as persistence context, data sources, and messaging.

But how does the container identify the class to use? Skipping the trivial case in which just one possible implementation is provided, of course, there are ways to define which compatible class to inject. One way is to use qualifiers.

Qualifiers are custom annotations that can be created to specify which class to use from a list of compatible ones. Another widely used technique is to use the `@Named` annotation. With this annotation, we can provide each compatible class with a name and then specify which one to use in the injection.

Last, but not least, it's possible to mark a class with `@Default` and the other implementations with `@Alternatives` to identify which one we want to be selected. `@Alternatives` can then be given an order of priority.

CDI also provides the management of the life cycle of the objects, which means when the objects should be created and when they should be destroyed. The CDI scopes are configured by using annotations, as per the injection that we have just seen. The most commonly used scopes are as follows:

- **@ApplicationScoped**: This binds the creation and destruction of the objects with the life cycle of the whole application. This means that one instance is created at application startup and destroyed at shutdown. Only one instance will be managed by the container and shared by all the clients. In this sense, this annotation is an implementation of the singleton pattern.

- **@Dependent**: This is the default scope that creates a class, which is linked to the life cycle of the object using it, and so it's created and destroyed concurrently with the object in which it is injected.

- **@SessionScoped**: This links the life cycle of the object with the HTTP session in which it is referenced (and so makes it a good tool for storing user and session information).

- **@RequestScoped**: This binds the object life cycle to the life cycle of the HTTP request where it is referenced.

- **@TransactionScoped**: This associates the life cycle of the object with the duration of the transactional boundary in which it is utilized.

Here is the diagram for CDI scopes:

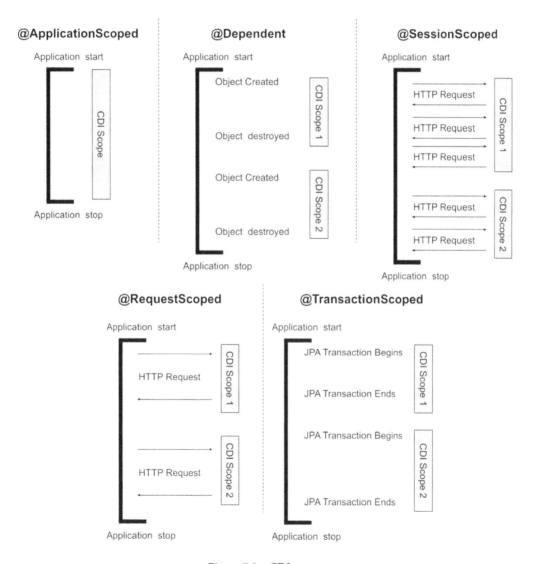

Figure 7.2 – CDI scopes

CDI specifications also provide hooks to specific life cycle events. The most commonly used are @PostConstruct and @PreDestroy, which are called immediately after object creation and before destruction, respectively.

Jakarta RESTful Web Services

Another essential piece of Java applications nowadays is RESTful Web Services. The Jakarta EE standard for RESTful Web Services (**JAX-RS**) provides a nice and declarative way to implement the classic JSON over HTTP web service communication. Once we enable our application to use this specification (a common way is to add a class that extends the JAXRSApplication class to the class path), all we have to do is create a bean for mapping the resource that we want to expose and annotate it accordingly.

In the most common use case, we will have to map the whole class to the path we want to expose by annotating the class with the @Path annotation. We may then want to specify the media types that the class produces and consumes (usually JSON) by using the @Produces and @Consumes annotations.

Each method of the class can be mapped to HTTP methods by using annotations such as @Get, @Post, @Delete, @Put, and @Head, and we can bind these methods to sub-paths by using the same @Path annotation. As an example, we can have the whole class bound to /myPath (with GET, POST, and other HTTP methods referring to that path) and then have the specific methods annotated to refer to /myPath/mySubPath.

Another very common scenario is the binding of method parameters with HTTP path parameters (@PathParam), parameters on the query string (@QueryParam), and HTTP headers (@HeaderParam). We should make a special mention of the **Jakarta JSON Binding (JSON-B)** specification, which, acting behind the scenes, can provide the JSON to **Plain Old Java Objects (POJOs)** (and vice versa) mapping for our beans, provided that they have a simple structure (and without needing any complex configuration). This is true for the most common use cases, meaning the Java classes with simple type fields with getters and setters. But of course, it's possible to provide customizations and implement special cases, if we need to.

As part of the JAX-RS specification, we can create REST clients too (to query REST services). In order to do that, a common way is to use the ClientBuilder class, which provides a fluent API to specify the usual parameters of an HTTP client (such as timeouts, filters, and similar settings). We can then create a so-called WebTarget object, which is an object that allows us to specify the path to invoke using the client. Acting on WebTarget, it is possible to send requests by passing parameters and getting results (usually in form of JSON objects).

An interesting twist of the JAX-RS specification is the possibility to manage **Server Sent Events (SSEs)**. SSEs were introduced with the **HTML5** standardization and are a way to provide data from a server to a client in the form of events by using an open connection.

What happens is that the client initiates the request to the server, but instead of getting all the data in one shot and closing the connection, it will keep the connection open and fetch the data as it comes from the client (eventually closing it at some point, or being disconnected by the server). The advantage here is that we can reuse the same connection to reduce the overhead, and we can get (and visualize) the data in real time without needing to poll the server for updates. The client could be a Java client or a web page in a browser.

In order to implement this behavior, JAX-RS provides `Sse` and `SseEventSink` resources that can be injected into our method with the `@Context` annotation, as follows:

```
@GET
@Path("/serverSentExample ")
@Produces(MediaType.SERVER_SENT_EVENTS)
public void serverSentExample(@Context SseEventSink
   sseEventSink, @Context Sse sse)
```

Once we have those two resources, we can use `sse` to build new events and `sseEventSink` to send such events. Once we've completed our interactions with the client, we can use `sseEventSink` to close the connection:

```
OutboundSseEvent event = sse.newEventBuilder()
.mediaType(MediaType.APPLICATION_JSON_TYPE)
.data(MyEventData.class, myEventData)
.build();
eventSink.send(event);
...
eventSink.close();
```

So, let's summarize:

- We create an `event` object by invoking `newEventBuilder` on the `sse` object injected in our class.

- We set `mediaType` to JSON.

- We add the data we want to send, specifying the class type and the object instance containing the data.

- We call the `build` method to create the `event` instance.

- We invoke the `send` method on the `eventSink` object, passing the event instance we just created.

- Eventually, we can close the connection by calling `close` on the `eventSink` object. Of course, in a real-world scenario, we may want to send a number of events (such as a consequence of something happening) before closing the connection. It doesn't make much sense to have `sse` just to send one event.

One interesting scenario generated by SSE is the possibility to implement broadcast scenarios. In such scenarios, instead of having each client connected to a different thread (and receiving different messages), we can have clients all receiving the same message. In this way, we will have clients subscribing (usually calling a specific REST service) and then getting the data (calling another one). Here is a code example (simplified):

```java
    @GET
  @Path("broadcast")
  public Response broadcast(@Context Sse sse) {
     SseBroadcaster sseb = sse.newBroadcaster();
     OutboundSseEvent event = sse.newEventBuilder()
     .mediaType(MediaType.APPLICATION_JSON_TYPE)
     .data(MyEventData.class, myEventData)
     .build();
     sseb.broadcast(event);
     ...
}
@GET
    @Path("subscribe")
    @Produces(MediaType.SERVER_SENT_EVENTS)
    public void subscribe(@Context SseEventSink
      sseEventSink){
          broadcaster = sse.newBroadcaster();
     broadcaster.register(sseEventSink);
     }
```

Let's summarize what this code does:

- We create a `broadcast` method and annotate it to indicate that it will be associated with an HTTP GET method that is exposed on the `broadcast` path.

- This `broadcast` method will be injected with an `sse` object instance present in the context.

- We create a `broadcaster` object by invoking the `newBroadcaster` method on the `sse` object.

- We create an `OutboundSseEvent` object by invoking the `newEventBuilder` method on the `sse` object.

- We set `mediaType` to JSON.

- We add the data we want to send, specifying the class type and the object instance containing the data.

- We call the `build` method to create the `event` instance. We invoke the `broadcast` method on the `broadcaster` object, passing the `event` instance we just created.

- We create a `subscribe` method and annotate it to indicate that it will be associated with an HTTP `GET` method that is exposed on the `subscribe` path and that will produce answers with the `SERVER_SENT_EVENTS` media type.

- The `subscribe` method will be injected with an `SseEventSink` object instance present in the context.

- We create an instance of a `broadcaster` object by invoking the `newBroadcaster` method on the `sse` instance.

- We register `sseEventSink` by passing it to the `register` method on the `broadcaster` object.

On the client side, we will most likely interact with SSE by using a framework such as Vue or Angular. But in any case, under the hood it will use the JavaScript `EventSource` object:

```
var source = new EventSource('mySSEEndpoint');
source.onmessage = function(e) { ... do something...}
```

As mentioned, we can also interact with SSE by using a Java client. Similar to the JavaScript version, the SSE implementation in Java provides an `EventSource` object too:

```
Client client = ClientBuilder.newBuilder().build();
WebTarget target = client.target("mySSEBroadcastEndpoint");
SseEventSource source =
  SseEventSource.target(target).build();
```

```
source.register(event -> { ... do something ... });
sseEventSource.open();
```

A personal consideration here is to thoroughly test this kind of implementation in real production scenarios and manage and monitor exceptions. We must also consider some alternatives in case of unexpected disconnects that may be due to clients with unstable connections (such as mobile clients) or network devices misbehaving in the overall infrastructure. Frameworks usually also provide some resiliency features, such as connection retries (in case the backend is momentarily unavailable). Resiliency must be also considered from a backend perspective, hence if a failure occurs while sending a message (and an exception is thrown), you should consider handling (including retries). But since this is basically non-transactional (because of network connections being potentially unreliable), you should consider edge cases including duplicate events or message loss.

WebSocket

Jakarta EE includes support for WebSocket technology. Using this technology, we can implement full-duplex communication between client and server, supporting the development of a rich user experience in web applications. WebSocket sits directly on top of TCP, so it doesn't rely on HTTP. However, it is compatible with HTTP, meaning that, from a connection point of view, it uses a compatible handshake and may be transported over HTTP and HTTPS standard ports (80 and 443), so it is compatible with most network infrastructures.

In order to implement the WebSocket capabilities on the backend, you need to annotate a class with @ServerEndpoint, specifying the path on which the capabilities will be published. With this class, we can then annotate methods with @OnMessage, @OnOpen, @OnClose, and @OnError to intercept the message received, the client connected, the client disconnected, and error events, respectively. After the connection of a client, in the method annotated with @OnOpen, it's possible to retrieve and store a session object. This object can then be used to send messages to the clients, hence implementing full-duplex communication, as shown here:

```
@ServerEndpoint("/myWebSocket")
public class WebSocketEndpoint {
    @OnMessage
    public String onMessage(String message) {
        System.out.println("Message received: "+ message);
        return message;
    }
}
```

```
    @OnOpen
  public void onOpen(Session session) {
    System.out.println("Client connected");
//Session object can be stored and used to send messages
  back
  }


  @OnClose
  public void onClose() {
    System.out.println("Connection closed");
  }
    @OnError
    public void onError(Session session, Throwable
      throwable)
    {
        System.out.println("Error in session " +
          session.getId() + " " + throwable.getMessage());
    }
}
```

As we saw when discussing server-sent events, WebSocket's applications are usually implemented on the client side using frameworks. However, JavaScript exposes a WebSocket object that can be used to mirror the server-side life cycle (OnOpen, OnMessage, and OnError) and the message-sending capabilities. As with SSE, my suggestion is to test this kind of interaction on an infrastructure that's comparable to the production one and be ready with alternatives in case something goes wrong with network connectivity, such as having graceful fallbacks. A nice implementation of this could be the **circuit breaker** pattern, as we are going to see in *Chapter 9, Designing Cloud-Native Architectures*.

Messaging

Messaging is another key component in modern applications. In the cloud-native microservices world, the **Java Message Service** (**JMS**) is considered to be an *enterprise-y*, complex manner of communication, often used together with other technologies, such as Kafka and AMQP. However, for many years, from **version 2.0** onward (**3.0** was just released at the time of writing), JMS has become very easy to use (at least in basic use cases).

The idea behind the messaging standard in JEE (which is one of the things I like about the application servers in general) is that we can keep the code simple and compact and offload the configurations to the application server. This also has the advantage of separating the code from the configuration. This also has advantages in terms of clarity, portability, and testability.

In order to send messages, we can use the injection of JEE resources – in this case, JMSContext. With the same approach, we can inject an object representing our target queue. The API then allows us to create a producer from the JMSContext object and use it to send a message against the queue, such as in the following code snippet:

```
@Resource (mappedName = "java:jboss/jms/queue/testQueue")
private Queue testQueue;
@Inject
JMSContext context;
...
context.createProducer().send(testQueue,msg);
...
```

With a similar kind of API, we can consume messages by creating a consumer and invoking the receive method against it. But this is not how it is done commonly.

The most widely used way is to use a **Message Driven Bean** (**MDB**), which is natively designed to be triggered asynchronously when a message is received. The code to use an MDB involves the implementation of the MessageListener interface and the use of some annotations to configure the queue to attach to. The code is quite self-explanatory:

```
@MessageDriven(name = "TestMDB", activationConfig = {
    @ActivationConfigProperty(propertyName =
    "destinationLookup", propertyValue = "queue/TestQueue"),
    @ActivationConfigProperty(propertyName =
    "destinationType", propertyValue = "javax.jms.Queue")})
public class TestMDB implements MessageListener {
    public void onMessage(Message msg) {
        TextMessage myMsg =(TextMessage) rcvMessage;
        LOGGER.info("Received Message " + myMsg
        .getText());
```

In both the consumer (MDB) and the producer example, the code looks for the default JMS connection factory, which is supposed to be bound to `java:/ConnectionFactory`. It is possible to explicitly state an alternative connection factory if we want to (such as when our application server must be connected to different brokers).

In order to set the properties to connect to a broker, such as a host, port, username, and password (and associate it with a **Java Naming and Directory Interface** (**JNDI**) name, such as the default `java:/ConnectionFactory`), we will have to configure the application server. This is, of course, specific to the server we choose. In WildFly, we commonly do that by using a CLI (as we have seen when deploying applications) or by directly editing the configuration file.

Persistence

Persistence is often one of the must-have properties for Java EE applications. While other persistence alternatives are now widely used, such as NoSQL stores and **InMemory** caches, database persistence is unlikely to disappear anytime soon.

Persistence in JEE is regulated by the **Java Persistence API** (**JPA**) specification. In the earlier versions, JPA was clumsy and painful to use (as was the EJB specification). This is not true anymore, and JPA is now very easy to use. As you may probably know, JPA is built around the **Object-Relational Mapping** (**ORM**) idea, which aims for relational database tables to be mapped to objects (in our case, Java objects).

So, the first thing to do to use JPA is to define our objects and how they map to database tables. As you can imagine, this is easily done by using annotations. The relevant annotations here are `@Entity` to identify the class and map it to the database, `@ID` to mark the field linked to the primary key, `@GeneratedValue` to define the strategy for the key generation, `@Table` to configure the table name (which defaults to the class name), and `@Column` to configure the column name for each class field (also, in this case, it defaults to the field name). This is what the code looks like:

```
@Entity
@Table(name="MyTableName")
public class MyPojo {
@Id
@GeneratedValue(strategy = GenerationType.IDENTITY)
private int id;

@Column(name="myColumn")
```

```
private String myField;
...
```

After we have our classes linked to our database tables, it's time to interact with the database itself. You can easily do that by injecting the so-called `EntityManager` where it's needed. The entity manager is associated with a persistence context, which is essentially the set of configurations that you set into the application and the application server to make it aware of where the database should connect to, such as the **Java Database Connectivity** (**JDBC**) string and other properties.

You can use the entity manager to retrieve objects from the database by using the JPA query language (which is similar to SQL) to create new objects, delete them, and so on. Here is a code example:

```
@PersistenceContext(unitName="userDatabase")
private EntityManager em;
Query query = em.createQuery("Select p from MyPojo p");
(List<MyPojo>) query.getResultList();
em.getTransaction().begin();
        MyPojo pojo = new MyPojo();
        pojo.setMyField ("This is a test");
        em.persist(pojo);
em.getTransaction().commit();
em.close();
...
```

As you can see, consistent with the other APIs that we have seen so far, JPA is pretty easy to use. It will nicely decouple business logic (in your Java code) from configuration (in the application server) and standardize the implementation of common aspects such as table-to-POJO mapping and transaction usage.

What's missing in Java EE

One of the reasons why some developers are moving away from the JEE specification is that the evolution of the standard is a bit slow. One goal of the platform is to include a big list of vendors providing reference implementations and to give long-term stability to the standard users, so it will take time to evolve JEE. At the time of writing, a number of things were missing from JEE that need to be overcome by using third-party libraries.

We will try to summarize the most common criticisms in this area:

- **Observability**: Since the beginning, some advanced monitoring capabilities have been missing from the JEE specification. **Java Management Extension (JMX)** was provided in the Java platform as a first attempt to provide some metrics and monitoring, and JDK Mission Control was donated to open source communities, providing some more advanced capabilities in terms of profiling.

    However, enterprises commonly complement such technologies with third-party software, sometimes proprietary software. As of today, more advanced monitoring capabilities, such as tracing, are commonly required for fully controlling the application behavior in production. Moreover, metric collections and display technologies based on stacks such as Prometheus and Grafana have become a de facto standard. Observability also includes things such as health and readiness probes, which are special services exposed by the application that can be useful for checking for application availability (and send an alert or implement some kind of workaround if the application is not available).

- **Security**: While JEE and Java, in general, are pretty rich in terms of security, including role-based access control at different architectural levels, support for encryption, multi-factor authentication, and authorization facilities is missing. There are some other features, such as OpenID Connect and JSON Web Token, that are still missing from the core specification.

- **Fault tolerance**: In heavily decentralized environments, such as microservices and cloud-native, it's crucial to defend the application from issues in external components, such as endpoints failing or responding slowly. In JEE, there is no standardized way to manage those events (other than normal exception handling).

- **OpenAPI**: REST services are widespread in the JEE world. However, JEE does not specify a way to define API contracts for REST services, as it's done by the OpenAPI standard.

Other features less likely to be standardized, such as alternative datastores (think about NoSQL databases) and alternative messaging (such as streaming platforms or AMQP), are also missing. All those functionalities are normally added by third-party libraries and connectors. As we will see in the upcoming sections, **MicroProfile** provides a way to overcome those limitations in a standard way.

What's great about Java EE

While some useful and modern technology is missing in the vanilla specification of JEE, as we have just said (but can be easily added via third-party libraries most of the time, such as what's provided by the `smallrye.io` project), I still think that JEE technology is just great and is here to stay. Some reasons are as follows:

- **Vendor ecosystem**: As we saw at the beginning of this chapter, there are a number of alternative implementations, both paid and free, providing JEE compatibility. This will ensure long-term stability and (where needed) commercial support, which can be crucial in some environments.

- **Operations**: While there is no fixed standard, as each vendor implements it in their own way, JEE enforces some configurability points on an application. This means that a JEE application can be easily fine-tuned for things such as thread pool size, timeouts, and authentication providers. While this is, of course, possible even while using other approaches, JEE tends to be more operation-friendly. Once the system administrators know about the specifics of the application server in use, they can easily change those aspects, regardless of the kind of application deployed.

- **Battle-tested for enterprise needs**: JEE still provides things that are very useful (sometimes essential) in the enterprise world. We are talking about distributed transactions, connectors for legacy or enterprise systems, robust deployment standards, and so on. You are likely to find some of those features in alternative stacks, but they will often be fragmentary and less robust.

This completes our quick overview of JEE's pros and cons. As you may know, a detailed explanation of JEE may take a whole (huge) book. However, in these sections, we have seen a simple selection of some basic APIs that are useful for building modern applications, including RESTful Web Services, JPA persistence, and messaging. We have also seen the pros and cons of the JEE framework.

In the next section, we will start talking about alternatives to application servers.

Going beyond Java Enterprise Edition

In *Chapter 1, Designing Software Architectures in Java – Methods and Styles*, we had a very quick look at containerizing Java applications.

We will now look into alternatives and extensions to Java Enterprise, including lightweight Java servers and **fat JAR** applications. Here, we will see a quick overview of why and how to implement fat JAR applications.

Packaging microservices applications

A **fat JAR** (also known as an **Uber JAR**) is likely to be one of the starting points in the inception of application service alternatives (and microservices runtimes). Frameworks such as Dropwizard, Spring Boot, and, more recently, Quarkus have been using this approach.

The idea of fat JAR is that you package all you need into a single `.jar` file so that we have a self-contained and immutable way to deploy your applications.

The advantages are easy to imagine:

- **Deployment is simplified**: Just copy the `.jar` file.

- **Behavior is consistent between different environments**: You can test the application on a laptop without needing a full-fledged app server.

- **Full control of the dependencies**: Versions and implementation of the supporting libraries are fixed at build time, so you will have fewer variables in production (and you are not forced to stick with what the app server provides).

Of course, all of this comes at a cost. Here are some not-so-obvious disadvantages of this approach:

- It's less standard (think about configurations). There are some de facto standards, such as `.yaml`, application properties files, or system properties. But this usually varies from app to app, even when using the same technology stack. Conversely, app servers tend to be more prescriptive in terms of what can be configured and where to put such configurations.

- While you can pick and choose the dependencies you need, you have to carry over such dependencies with each deployment (or scale). And if you use many dependencies, this will be impactful in terms of network usage and time lost (and compiling time too). With the application servers, you take for granted that such dependencies are already waiting for you in the application server.

- When it comes to supportability, either you get support services from a vendor or simply adhere to internal standards. You are normally bound to a fixed set of libraries and versions that have probably been tested to be compatible with your environment and to adhere to security and performances standards. With a fat JAR, you have less control over this at runtime and deployment time. You will have to move such controls at build time, and maybe double-check that the content of the fat JAR adheres to standards before putting it into your production environment.

As we discussed in *Chapter 1, Designing Software Architectures in Java – Methods and Styles*, containers changed the rules of the game a bit.

With container technology, you can create a full portable environment, including a base operating system (sort of) with a filesystem in which you can place dependencies and resources together with your application. This means that you don't need a self-consistent application to deploy, as these features are provided by container technology. And, as already discussed, this may also be harmful when used together with containers, as they are designed to work in a layered way. So, you can use this feature to package and deploy only the upper level (containing your application code) instead of carrying over the whole dependency set.

So, while still convenient in some cases (such as local testing), fat JAR is not necessary right now.

But as we have seen, other than a different packaging approach, there are some features that may be very useful in the cloud-native and microservices world. These features are missing in JEE, such as observability and support for alternative technologies. It used to be common for microservices runtimes to define custom solutions to fill those gaps.

But as previously mentioned, lack of standards is a known issue with microservices. This used to be a minor issue because early adopters were usually deeply technically skilled teams relying on self-support and that didn't need support from a third-party vendor.

However, nowadays, the adoption of microservices, cloud-native, and general extensions to JEE is growing a lot. And factors such as long-term stability, enterprise support, and an open ecosystem are becoming more and more essential. That's one of the reasons behind MicroProfile.

Introducing MicroProfile

MicroProfile started with a focus on extending the JEE specification with features offered by microservices. The development is backed by a consortium of industry players, such as IBM, Red Hat, Oracle, and Microsoft. The specification lives in parallel to Jakarta EE, sharing some functionality, evolving some others, and adding some more that are not part of JEE.

This works because the MicroProfile consortium, part of the Eclipse Foundation, has chosen a less bureaucratic and more frequent release model.

This means that modern Java development can now basically take two parallel roads:

- **Jakarta EE**: We can choose this if long-term stability and enterprise features are more important (or if you want to maintain and modernize existing code bases).

- **MicroProfile**: We can choose this if cloud-native features and a frequent release cycle are priorities.

But what are the features added by MicroProfile?

MicroProfile specifications

It's really challenging to print a snapshot of something (that is changing very frequently) on paper. At the time of writing, MicroProfile releases a new version every 3 to 6 months. The most important features to highlight are the following:

- **Configuration**: This is a practical approach to separate the configuration repository (such as .xml files, system environments, and properties files) from the application itself. This provides the facilities for accessing the configuration values and checking for changes without needing to restart the application (in supported implementations).

- **Fault tolerance**: This is a way to choreograph the reaction to failures (such as failing to call an external service) by using patterns such as circuit breaker, retry, and fallback.

- **OpenAPI**: This provides support for the OpenAPI standard, which is a way to define contracts for REST services, similar to what a WSDL schema provides to SOAP web services.

- **OpenTracing**: This is a modern approach to monitoring and managing chains of calls in a distributed environment by passing an ID and introducing concepts such as spans and traces.

- **Health**: This is a standardized way to create liveness and readiness probes in order to instrument an application for checking the correct behavior of an application (when it's live, that is, to verify whether it is up or down) and its readiness (when it's ready to take requests).

- **Metrics**: This is an API for providing facilities for exporting monitorable values from your applications. This is usually used for things such as capacity planning and overall understanding of the application performances (such as the number of current transactions).

As you may have noticed, most of the preceding features exactly match what we highlighted in the *What's missing in Java EE* section.

We will explore some of those techniques in more detail in *Chapter 9, Designing Cloud-Native Architectures*.

It's also important to highlight that MicroProfile encompasses specifications included in JEE (such as JAX-RS, JSON-B, and CDI, as we saw in the *Introducing the JEE standard* section). While MicroProfile tends to align the version of such shared libraries with one target JEE version, it may be that some of those versions are out of sync (being probably more up to date in the MicroProfile edition).

It's also worth noticing that MicroProfile does not imply any specific packaging model for applications. Some implementations, such as **Helidon** (backed by Oracle) and Quarkus (backed by Red Hat) tend to use fat JARs and similar, while others, such as **OpenLiberty** (provided by IBM) and WildFly (provided by Red Hat) run in a more traditional way (deployed into a lightweight running server).

For the upcoming sections, we will start seeing more about Quarkus, which is an implementation of the MicroProfile standard and is becoming more and more popular and widely used.

Exploring Quarkus

Quarkus is an open source Java framework that aims to be optimized for cloud-native and microservices. It was born in the container and **Kubernetes** world, and for this reason, it's been optimized by design for container and Kubernetes-based cloud-native applications.

Quarkus comes from an engineering team with experience in many interesting projects, such as **Hibernate**, **Vert.X**, and **RESTEasy**, and so reuses a lot of good ideas and best practices from these famous communities.

This is what a Quarkus application looks like when started from a terminal console:

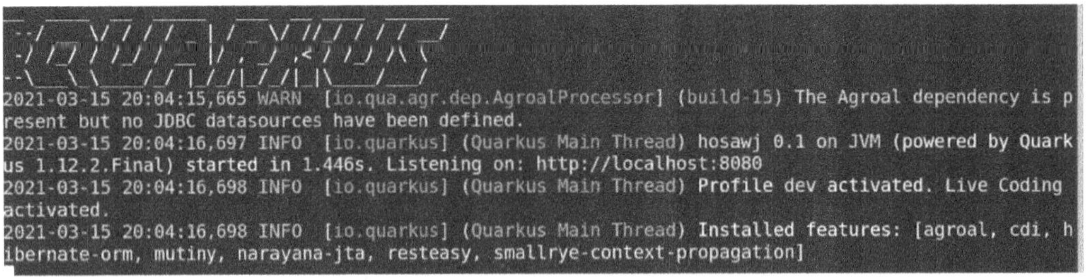

```
2021-03-15 20:04:15,665 WARN  [io.qua.agr.dep.AgroalProcessor] (build-15) The Agroal dependency is p
resent but no JDBC datasources have been defined.
2021-03-15 20:04:16,697 INFO  [io.quarkus] (Quarkus Main Thread) hosawj 0.1 on JVM (powered by Quark
us 1.12.2.Final) started in 1.446s. Listening on: http://localhost:8080
2021-03-15 20:04:16,698 INFO  [io.quarkus] (Quarkus Main Thread) Profile dev activated. Live Coding
activated.
2021-03-15 20:04:16,698 INFO  [io.quarkus] (Quarkus Main Thread) Installed features: [agroal, cdi, h
ibernate-orm, mutiny, narayana-jta, resteasy, smallrye-context-propagation]
```

Figure 7.3 – Quarkus starting

As you can see, some spectacular ASCII art is shown and some interesting information, including the lightning-fast startup time of fewer than 1.5 seconds.

But what are the most important benefits of Quarkus?

Better performances

One of the most famous benefits of Quarkus is its optimization. The framework was created with a *container-first* philosophy, and for this reason, it is heavily optimized both for startup time and memory usage. In order to achieve these objectives, Quarkus uses various techniques:

- **Less usage of reflection**: Reflection can be impactful in terms of performance. Quarkus reduces the use of reflection as much as possible.

- **Move as much as possible to build time**: Quarkus does as much work as possible at build time. This means that all the things that can be done in advance, such as class path scanning and configuration loading, are done at build time and persisted as bytecode. In this way, not only will the application boot faster (because it has fewer things to do), but it will also be smaller in terms of memory footprint because of all the infrastructure that is not needed at runtime; that is, the ones *precompiled* at build time are not part of the final artifact.

- **Native executables**: Optionally, Quarkus applications can be directly compiled as Linux executables thanks to support from **GraalVM** (and the `Substrate` module). This allows further optimizations, further reducing the startup time and memory footprint.

But better performance is not the only benefit of Quarkus.

Developer joy

The thing that I like the most about Quarkus is its ergonomics. As is common to hear from people working with it, Quarkus feels new and familiar at the same time. The language is extremely friendly if you come from a JEE background. It offers a ton of tools and facilities, not to mention all the *syntactic sugar* that makes even the most advanced features easy to use. In the Quarkus world, this is referred to as **developer joy**.

One of such facilities is the **developer mode**, which allows you to immediately see the changes in your application without needing a full recompile/repackage. It works like a charm when you change something (such as the source code, configuration file, and resources) and can immediately see the effect of such changes (such as simply refreshing the browser or recalling the API). I know this feature was already provided by other frameworks and libraries (with **JRebel** being one of the most famous), but the way it works out of the box is just magic for me, and, honestly, it's a great boost in terms of developer productivity.

But that's not the only *developer joy* feature. Each dependency added to Quarkus (which are more properly called **extensions**) is crafted to nicely fit the Quarkus world and use the framework's capabilities, first of all in terms of performances.

You will find a lot of facilities and conventions over configuration and *intelligent defaults*, such as the way the configuration is treated (including environment management), a simple way to use both the imperative and reactive paradigms (and make them coexist), and the way it interacts with databases (by using the `Panache` extension). *But where to start?*

Quarkus – hello world

Quarkus has a wizard for generating applications (both with **Maven** and **Gradle** support) located at `code.quarkus.io`. Follow these steps to create a new application:

1. You can create a new application with the command line by using the Maven Quarkus plugin. In the current version, this means using the following command:

    ```
    mvn io.quarkus:quarkus-maven plugin:1.12.2.Final
    :create
    ```

2. The plugin will then ask for all the required information, such as the artifact name and the dependency to start with. The following screenshot illustrates this (please note the cool emoticons too):

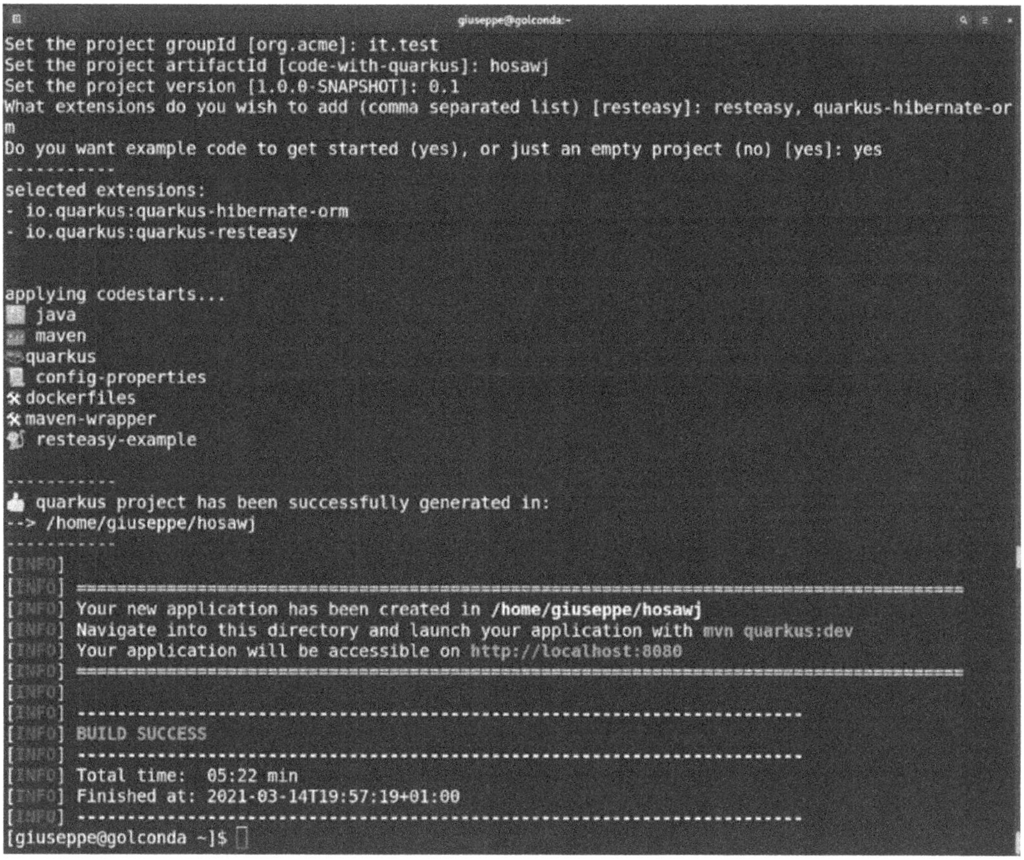

```
Set the project groupId [org.acme]: it.test
Set the project artifactId [code-with-quarkus]: hosawj
Set the project version [1.0.0-SNAPSHOT]: 0.1
What extensions do you wish to add (comma separated list) [resteasy]: resteasy, quarkus-hibernate-or
m
Do you want example code to get started (yes), or just an empty project (no) [yes]: yes
-----------
selected extensions:
- io.quarkus:quarkus-hibernate-orm
- io.quarkus:quarkus-resteasy


applying codestarts...
   java
   maven
   quarkus
   config-properties
   dockerfiles
   maven-wrapper
   resteasy-example

-----------
   quarkus project has been successfully generated in:
--> /home/giuseppe/hosawj
-----------
[INFO]
[INFO] ======================================================================
[INFO] Your new application has been created in /home/giuseppe/hosawj
[INFO] Navigate into this directory and launch your application with mvn quarkus:dev
[INFO] Your application will be accessible on http://localhost:8080
[INFO] ======================================================================
[INFO]
[INFO]
[INFO] ----------------------------------------------------------------------
[INFO] BUILD SUCCESS
[INFO] ----------------------------------------------------------------------
[INFO] Total time:  05:22 min
[INFO] Finished at: 2021-03-14T19:57:19+01:00
[INFO] ----------------------------------------------------------------------
[giuseppe@golconda ~]$
```

Figure 7.4 – The Quarkus Maven plugin

3. After the plugin execution, you will have an application scaffold that you can use as a starting point (in this example, containing `resteasy` and `hibernate` dependencies). In order to run it and experiment with the developer mode, you can use the following command:

`./mvnw compile quarkus:dev`

This command uses a Maven wrapper script (in this case, `mvnw`, because I'm running on a Linux box, but a `mvnw.cmd` file is provided for Windows environments) to run the application in development mode. Since you are using RESTEasy, by default the application will answer with a `Hello RESTEasy` string on the following endpoint: `http://localhost:8080/hello-resteasy`.

In order to try the developer mode, you can change the source code (in this case, the `GreetingResource` class) to change the response. After you do that, you can refresh your browser and see the result without needing to recompile or repackage the code.

Building techniques with Quarkus

The development mode, needless to say, supports the development phase. In order to build and distribute Quarkus applications, you have other options.

Quarkus is currently supported to run on OpenJDK (see the official website at `quarkus.io/get-started` for more information about the supported versions). In order to package your application, you can run the usual Maven command:

```
mvn clean package
```

By default, Quarkus will build a so-called fast-jar. This is basically a package optimized for boot time performance and a small memory footprint. In order to execute an application packaged in this way, you will need to copy the whole `quarkus-app` folder (in the `target` folder), which contains all the libraries and resources needed to run the application. You can then run it with a similar command to this:

```
java -jar ./quarkus-app/quarkus-run.jar
```

You can also package the application in an UberJar form (be conscious of all the limitations of this approach, as discussed in *Chapter 1, Designing Software Architectures in Java – Methods and Styles*). To do so, one easy way is to pass the `quarkus.package.uber-jar=true` property to the Maven command:

```
mvn clean package -Dquarkus.package.uber-jar=true
```

This property can also be set in the `pom.xml` file or in the configuration file of Quarkus (the `application.properties` file, by default).

Last, but not least, as mentioned at the beginning of this section, Quarkus can be compiled into a native Linux executable without the JVM needing to be executed. To do so, you can simply use the following command:

```
./mvnw package –Pnative
```

What Quarkus does under the hood is look for a GraalVM installation that is used for native compilation. The following screenshot shows what happens if we start Quarkus when the `GRAALVM_HOME` variable is not configured:

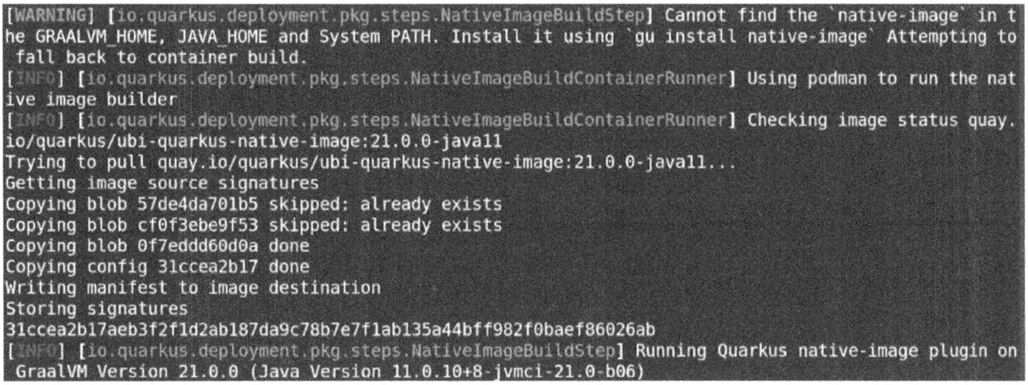

```
[WARNING] [io.quarkus.deployment.pkg.steps.NativeImageBuildStep] Cannot find the `native-image` in t
he GRAALVM_HOME, JAVA_HOME and System PATH. Install it using `gu install native-image` Attempting to
 fall back to container build.
[INFO] [io.quarkus.deployment.pkg.steps.NativeImageBuildContainerRunner] Using podman to run the nat
ive image builder
[INFO] [io.quarkus.deployment.pkg.steps.NativeImageBuildContainerRunner] Checking image status quay.
io/quarkus/ubi-quarkus-native-image:21.0.0-java11
Trying to pull quay.io/quarkus/ubi-quarkus-native-image:21.0.0-java11...
Getting image source signatures
Copying blob 57de4da701b5 skipped: already exists
Copying blob cf0f3ebe9f53 skipped: already exists
Copying blob 0f7eddd60d0a done
Copying config 31ccea2b17 done
Writing manifest to image destination
Storing signatures
31ccea2b17aeb3f2f1d2ab187da9c78b7e7f1ab135a44bff982f0baef86026ab
[INFO] [io.quarkus.deployment.pkg.steps.NativeImageBuildStep] Running Quarkus native-image plugin on
 GraalVM Version 21.0.0 (Java Version 11.0.10+8-jvmci-21.0-b06)
```

Figure 7.5 – Quarkus building a native executable through Podman

The GRAALVM_HOME variable is used to look up the install path of GraalVM. If not present, Quarkus will try a container build. This basically means that, if a container runtime (Podman or Docker) is installed on the local machine, Quarkus will download a container image to use for native building, so you can create a native executable without needing a local GraalVM installation.

Configuration management in Quarkus

One lovely characteristic of Quarkus, in line with the developer joy idea, is the way it manages configurations.

Quarkus implements the MicroProfile config specification. We will see more about the MicroProfile functionalities in Quarkus, but since config is central to all the other extensions, it's worth having a look at it now.

In order to get and use a configurable value, you can use the following annotation:

```
@ConfigProperty(name = "test.myProperty",
   defaultValue="myDefault")
String myProperty;
```

As you can see, you can provide a default value directly into the annotation.

The configuration can be loaded by a number of different sources. Quarkus looks into the following sources (listed according to decreasing priority):

- System properties (as in passing a command-line argument to the Java process, such as –DmyProperty="myValue")
- Environment variables

- `.env` files (files containing a set of environment variables) in the working directory
- An `application.properties` file (with the usual properties syntax, as in `key=value`) placed in a `config` subdirectory in the working directory
- An `application.properties` file placed in `src/main/resources`

Quarkus supports the use of profiles in the configuration properties. This allows us to have different environments (or simply different sets of configurations) in the same configuration repository (such as in the same `application.properties` file). In order to do so, you can use a prefix in the configuration key, such as this:

```
%{profile}.mykey=value
```

By default, Quarkus provides `dev`, `test`, and `prod` profiles. `dev` is activated when running in developer mode (`./mvnw compile quarkus:dev`, as seen in the previous section), `test` is activated when running tests, and `prod` is activated in all other scenarios.

You can define as many configuration profiles as you need and activate them by using the `quarkus.profile` system property or the `QUARKUS_PROFILE` environment variable.

So far, we have seen the basics of Quarkus, the most relevant benefits (including performances and language goodies), and how to build a basic `hello world` example. In the next section, we will have a look at the most common Quarkus extensions that are useful for building cloud-native applications.

Most common Quarkus extensions

Quarkus is aiming at cloud-native applications and microservices but shares some features and functionalities with the JEE world. This is thanks to its adherence to the MicroProfile specification.

Such features are implementing common use cases and are very handy, as they allow you to use existing skills and, in some cases, existing JEE code.

In this section, we will go through a quick overview of the Quarkus extensions shared with the JEE specification.

Content Dependency Injection

CDI is a structured way to wire and compose the objects of your application. CDI in Quarkus is based on the Contexts and Dependency Injection for **Java 2.0** specification, which defines CDI for both Java SE and Java EE.

The Quarkus CDI implementation leverages the ArC framework and is not fully compliant with the CDI specification, even if it provides support for the most common CDI use cases such as DI (of course), qualifiers, life cycle callbacks, and interceptors. There are some known limitations, on some specific use cases (such as the use of decorators, and the conversation scope). Following the Quarkus optimization mantra, ArC moves the discovery and injection operations at build time to achieve better performances.

REST services with JAX-RS

In order to develop REST services, Quarkus provides a **JAX-RS** extension that mimics the Jakarta EE implementation almost completely. RESTEasy is commonly added by default in new Quarkus projects. However, in order to add these features to an existing Quarkus project, you can simply use this command:

```
./mvnw quarkus:add-extension -
  Dextensions="io.quarkus:quarkus-resteasy"
```

As said, the JAX-RS implementation looks almost like the Jakarta EE implementation, so all the concepts we have seen in the previous section are still relevant (as previously mentioned, this will allow you to recycle skills and even existing code).

WebSockets

Quarkus includes the **WebSocket** support (as we have seen in the *Understanding the most common JEE APIs* section, under the *WebSockets* section). In detail, the WebSocket functionality is provided by the `undertow-websockets` extension.

Undertow is a highly performant web server technology written in Java. It can use both blocking and non-blocking APIs. Other than the WebSocket functionality (used by Quarkus), it provides other interesting web functionalities, such as full servlet API support. For this reason, Undertow is embedded into WildFly to provide web functionalities in full compliance with the JEE specification. Undertow has replaced Tomcat as the embedded web container in WildFly since **version 8**.

In order to add the WebSocket functionality to an existing Quarkus project, you can use the following command:

```
./mvnw quarkus:add-extension -Dextensions="undertow-
  websockets"
```

With this extension, you can use the WebSocket technology in the same way we saw in the *Understanding the most common JEE APIs* section, in the *WebSockets* section.

Messaging

JMS messaging is currently a preview technology in the Quarkus world. It is currently provided by two extensions, `quarkus-artemis-jms` and `quarkus-qpid-jms`.

The two dependencies are mostly equivalent from a functional point of view. Technically speaking, the `quarkus-artemis-jms` extensions use the `artemis jms` client to connect to the JMS broker, while `quarkus-qpid-jms` uses the AMQP standard as its wire protocol to connect to AMQP-compatible brokers.

Unlike the JEE version, the Quarkus framework does not provide an injectable `JMSContext` object. But it does provide a JMS `ConnectionFactory` object, so you can easily get a producer from it, such as the following:

```
@Inject
ConnectionFactory connectionFactory;
...
JMSContext context = connectionFactory.
   createContext(Session.AUTO_ACKNOWLEDGE)
context.createProducer().send(context.createQueue("test"),"
   myTestMessage");
```

Moreover, Quarkus does not provide the EJB subsystem, as it's provided in the JEE specification, so you cannot use MDBs, the classic way provided by JEE to consume messages. A quick and easy way to do so is to create a consumer (against a `JMSContext` object, as per the producer) and use the `receive()` method. Since it's a blocking call, you will have to create a new thread to encapsulate the receive logic without blocking the entire application. You'll need something like this:

```
JMSContext context = connectionFactory.
   createContext(Session.AUTO_ACKNOWLEDGE)) {
JMSConsumer consumer = context.createConsumer
   (context.createQueue("test"));
while (true) {
      Message message = consumer.receive();
      message.getBody(String.class);
}
```

The basic configurations for getting a producer and consumer to work are the server endpoint, user, and password. Those configs are stored in `quarkus.qpid-jms.url`, `quarkus.qpid-jms.username`, and `quarkus.qpid-jms.password` when using `quarkus-qpid`, and in `quarkus.artemis.url`, `quarkus.artemis.username`, and `quarkus.artemis.password` when using `quarkus-artemis`.

That's it! Now you can send and receive JMS messages with Quarkus.

Persistence

Persistence and ORM in Quarkus are provided by Hibernate, making it basically JPA compliant. This means that you can then use the same syntax that we saw in the *Understanding the most common JEE APIs* section, in the *Persistence* section. You can annotate your Java objects (commonly referred to as POJOs) with `@Entity` and the other annotations that we have seen in order to specify mappings with database tables, and you can inject the `EntityManager` object in order to retrieve and persist objects to the database.

To use Hibernate with Quarkus, you have to add the `quarkus-hibernate-orm` extension and a JDBC driver extension. The supported JDBC drivers are currently db2, Derby, H2, MariaDB, Microsoft SQL Server, MySQL, and PostgreSQL.

The basic properties to configure the database connection are `quarkus.datasource.db-kind` (configuring the type of database used), `quarkus.datasource.username`, `quarkus.datasource.password`, and `quarkus.datasource.jdbc.url`.

Although you can directly use Hibernate's `EntityManager`, Quarkus provides you with a more productive abstraction on top of it. This abstraction is Panache.

Accelerated ORM development with Panache

Panache is an amazing technology provided with Quarkus. It allows us to build ORM applications without redundant boilerplate code. It's a boost for creating **Create Read Update Delete** (**CRUD**)-like applications in no time.

In order to develop CRUD applications using Panache on top of Hibernate, you need to add the `quarkus-hibernate-orm-panache` extension.

Once you have this functionality enabled, you can use it in two main patterns, **Active Record** and the **repository**.

Panache Active Record

Active Record is an architectural pattern. It was described by Martin Fowler in his *Patterns of Enterprise Application Architecture* book.

In this pattern, one class completely represents a database table. An object created from this class represents a row (with fields mapping columns), while methods of the class map the interaction with the database, such as `persist`, `delete`, and `find`.

In Quarkus, to implement this pattern, you must make your JPA entity (annotated with `@Entity`) extend the `PanacheEntity` class. You can then use all the methods inherited from this class in order to interact with the database, including features such as `persist`, `delete`, `find` and `list`:

```
@Entity
public class MyPojo extends PanacheEntity {
@Id
private int id;
private String myField;
...
MyPojo pojo = new MyPojo();
        pojo.setMyField ("This is a test");
        pojo.persist();
MyPojo anotherPojo = MyPojo.findById(someId);
```

The obvious advantage here is that you don't have to directly interact with the `EntityManager` class anymore, and you have a number of methods ready to use for common use cases. But if you don't like the Active Record pattern, you can achieve pretty similar results with the repository approach.

Panache repository

The repository pattern is an alternative to the Active Record one. Basically, instead of having both the entities and the methods (to find, update, delete, and persist objects) implemented in the same class, you split such responsibilities and have entities with no behavior and dedicated repository classes to implement database interactions.

In the Quarkus world, this means that your entities are standard JPA entities (with no need to extend the `PanacheEntity` class), while your designated repository will need to implement the `PanacheRepository` interface. You can then use the same methods that we have seen before (`persist`, `delete`, `find`, and `list`) against the repository class:

```
@Entity
public class MyPojo{
@Id
private int id;
private String myField;
...
public class MyPojoRepository implements
  PanacheRepository<MyPojo> {
    public Person findByMyField(String myField){
        return find("myField", myField).firstResult();
    }
...
@Inject
MyPojoRepository myPojoRepository;
MyPojo pojo = new MyPojo();
        pojo.setMyField ("This is a test");
        myPojoRepository.persist(pojo);
MyPojo anotherPojo = myPojoRepository.findById(someId);
```

As you can see, the repository pattern is analogous to the Active Record one.

More complex relationships, such as one-to-many and many-to-many, can be modeled on an entity with the relevant annotations (in a similar way to what is doable with JPA), and can be retrieved and persisted with both the Active Record and Repository approaches. Moreover, Panache provides support for **Hibernate Query Language** (**HQL**) for complex queries. So far, we have learned about some of the Quarkus extensions and implemented basic APIs, similar to what we saw in the *Understanding the most common JEE APIs* section. Let's now see how Quarkus adds more features to those APIs by implementing the MicroProfile standard.

Quarkus and the MicroProfile standard

In this section, we are going to look at the MicroProfile standard and how Quarkus implements it. MicroProfile, as previously mentioned many times during this chapter (and in particular in the *Introducing MicroProfile* section), is a nice way to implement cloud-native microservices applications while adhering to a standard and hence avoiding vendor lock-in.

Quarkus, in the current version, is compatible with the **3.2 version** of the MicroProfile specification. As we have seen, MicroProfile embraces and extends the JEE specification while providing features that are useful for cloud-native and microservices development.

In the **3.2 version**, the most notable APIs in MicroProfile are as follows:

- **MicroProfile Config**, which is implemented by the Quarkus configuration, which we saw a couple of sections ago

- CDI and JAX-RS, which we saw in the *The most common Quarkus extensions* and *Understanding the most common JEE APIs* sections

- MicroProfile Fault Tolerance, OpenAPI, Health, OpenTracing, and Metrics, which we will see in *Chapter 9, Designing Cloud-Native Architectures*

- Other APIs, such as JWT authentication, Common Annotations, JSON-B, and JSON-P, which we will not cover

This has completed our overview on traditional JEE middleware, such as WildFly, and cloud-native alternatives, such as Quarkus. Let's have a look at some examples now.

Case studies and examples

In this section, we will model a very small subset of our mobile payment application. To follow up on the concepts we have seen, we will see some examples created for WildFly and Quarkus. For both technologies, since we will be interacting with a database, we will use H2, which is an easy-to-use open source database. You will find all the code in the GitHub repository located at `https://github.com/PacktPublishing/Hands-On-Software-Architecture-with-Java`.

Setting up the database

As we've said, the prerequisite for our application is to have a database up and running. To set it up, execute the following steps:

1. First, download the latest version of H2 from the www.h2database.com website.

2. You can then simply run the executable script for your platform, located under the bin directory. In my case, it was h2.sh. It will require a correctly installed JVM.

3. After the database starts, the default browser will be opened, and the embedded H2 web console will be available. If you're using H2 for the first time, it will, by default, try to connect in the embedded mode and create a test database in your home directory. You can log in with the default login credentials (which are sa as the username, with no password). The following screenshot shows the web console login screen, with all the needed configurations:

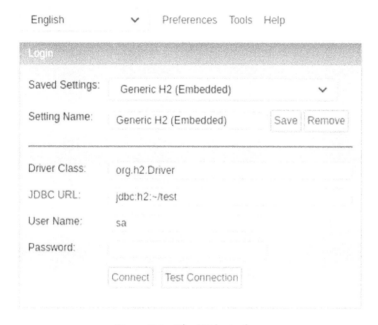

Figure 7.6 – The H2 login form

After you log in, you will be presented with a form to manipulate your newly created database. In order to create the payment table, you can copy and paste this SQL code into the SQL input form:

```sql
CREATE TABLE payment (
    id  uuid default random_uuid() primary key,
    date DATE  NOT NULL,
```

```
    currency VARCHAR(20) NOT NULL,
    sender  uuid NOT NULL,
    recipient  uuid NOT NULL,
    signature VARCHAR(50) NOT NULL,
    amount DECIMAL  NOT NULL
);
```

In this example, we are using H2 in embedded mode. This means that only one connection at a time will be allowed. So, before continuing with our examples, we will need to stop the H2 Java process to allow WildFly to connect to the database. You can then reconnect with the web console by simply relaunching H2 and using a different JDBC URL to connect in server mode. In my case, the string is as follows, and this allows more than one concurrent connection:

```
jdbc:h2:tcp://localhost/~/test
```

Also, another option is to completely skip this part and leave the table creation to Hibernate by leveraging the hbm2ddl configuration. I don't love this option, but it's still a viable alternative.

Moreover, consider that this is, of course, a simple example. In a real-world application, we would need some more tables (such as a user table). We would probably need to double-check our SQL statements with a DBA to check our data types against potential performance issues, depending on the expected volumes, or, most likely, we would have to interact with a database that's already been created for us. Now that we have a simple database, let's see how to interact with it by using WildFly.

JPA and REST with JEE and WildFly

In order to start developing our JEE application, you will need to start from an empty project (in our case, with Maven support). There are many ways to do that. The easiest one is to clone the project related to this chapter on GitHub and reuse the pom.xml and the project structure.

As an alternative, you can install and use (see the Readme file from GitHub) the WildFly Maven archetype located at https://github.com/wildfly/quickstart.

The first step of our example is accessing the table we just created via JPA. To do so, you will have to create an entity mapping to the table. As we saw in previous sections, the syntax is pretty easy:

```
@Entity
public class Payment {
```

```
    @Id
    private String id;
    private Date date;
    private String currency;
    private String sender;
    private String recipient;
    private String signature;
    private float amount;
...
```

As you see, we are using the same names defined in the database (both for identifying the table name, which corresponds to the class name, and the column names, which are linked to the class field names). Different mapping is possible with the proper annotations.

In order to manipulate our entity, we are going to use the repository pattern. Hence, we will create a `PaymentRepository` class, inject `EntityManager`, and use it for JPA operations. For the sake of simplicity, we will simply implement the `create` and `find` functionalities, but of course, these can be extended to cover all other possible requirements, such as finding by column:

```
@PersistenceContext(unitName = "hosawjPersistenceUnit")
    private EntityManager em;
     public Payment create(Payment payment)
    {
        em.persist(payment);
        return payment;
    }
    public Payment find(String id)
    {
        Payment payment=em.find(Payment.class, id);
        return payment;
    }
```

The last piece in this basic example is exposing the application using RESTful Web Services. To do so, we will need to create our `PaymentResource` class and annotate it accordingly, as we saw in the *REST services with JAX-RS* section:

```
@Path("/payments")
```

```
@Consumes(MediaType.APPLICATION_JSON)
@Produces(MediaType.APPLICATION_JSON)
public class PaymentResource {
    @Inject
     PaymentRepository repository;
    @GET
    @Path("/find/{id}")
    public Response find(@PathParam("id") String id) {
        Payment payment=repository.find(id);
        if(payment==null)
            throw new WebApplicationException
               (Response.Status.NOT_FOUND);
        else
          return Response.ok(payment).build();
    }
@POST
    @Path("/create")
    public Response create(Payment payment) {
        return Response.ok(repository.create(payment))
           .build();
    }
```

The notable thing here is that the PaymentRepository class, which we created previously, is injected using CDI and used from within the other methods. The two other methods, implementing REST capabilities (find and create), are annotated with @GET and @POST. The parameters for the find method are passed as @PathParam("id"), using the relevant annotation. The parameter for the create method is passed as a Payment object. The JSON serialization and deserialization are handled out of the box.

In order to activate the REST subsystem, as mentioned, the simplest way is to create a class that extends javax.ws.rs.core.Application and annotate it by defining the root application path, as follows:

```
@ApplicationPath("rest")
public class RestApplication extends Application
...
```

Finally, we need to configure the connection between the application server and the database.

WildFly ships with an example data source on H2, which is already configured in the default `standalone.xml` file. In order to configure the WildFly server to use the H2 database that we created in the previous section, we will have to change the `jdbc` connection string from `jdbc:h2:mem:test` to `jdbc:h2:tcp://localhost/~/test`.

Moreover, we didn't set a password in the H2 server for the database connection, so you will need to remove it.

To make our example application use such data source, you will need to change the `persistence.xml` JNDI name to the following:

`java:jboss/datasources/ExampleDS`

In the same file, you will also need to set the `hibernate` dialect to H2:

```
<property name="hibernate.dialect"
  value="org.hibernate.dialect.H2Dialect" />
```

Now everything is ready for deployment. First of all, we will start WildFly (in this case, by simply running `/bin/standalone.sh`). Then, we will package the application using a simple `mvn clean package` command. For development purposes, we can then deploy the compiled `.war` file to WildFly by copying it into the `/standalone/deployments` directory in the WildFly installation folder.

If everything worked correctly, you can then interact with REST services with this sample application. As an example, by using `curl` at the command line, you can create a payment like this:

```
curl -X POST -H 'Content-Type:application/json'
-d '{"id":"1ef43029-f1eb-4dd8-85c4-1c332b69173c",
"date":1616504158091, "currency":"EUR", "sender":"giuseppe@
test.it", "recipient":"stefano@domain.com",
"signature":"169e8dbf-90b0-4b45-b0f9-97789d66dee7",
"amount":100.0}'  http://127.0.0.1:8080/hosawj/rest/payments/
create
```

You can retrieve it like this:

```
curl -H 'Content-Type:application/json' http://127.0.0.1:8080/
hosawj/rest/payments/find/1ef43029-f1eb-4dd8-85c4-1c332b69173c
```

We have now created a simple but complete JEE example of a REST application interacting with a database using JPA. We will use the same application and see what will change when we use Quarkus.

JPA and REST (and more) with Quarkus

To create a skeleton application with Quarkus, using all the technology that we need, we can simply go to code.quarkus.io and select the technology that we need, which in our case is JAX-RS, **Jackson** (for JSON binding), and Hibernate (we will pick the version powered by Panache). Then, we can enter the group and artifact IDs, and click to download a .zip file with the right scaffold to start from.

Another alternative is to use the Maven command line, as follows:

```
mvn io.quarkus:quarkus-maven-plugin:1.12.2.Final:create
-DprojectGroupId=it.test -DprojectArtifactId=hosawj
-DclassName="it.test.rest.PaymentResource.java" -Dpath="/
payments" -Dextensions="io.quarkus:quarkus-resteasy","io.
quarkus:quarkus-resteasy-jackson","io.quarkus:quarkus-
hibernate-orm-panache","io.quarkus:quarkus-jdbc-h2"
```

This command is invoking the Quarkus Maven plugin, asking to create a new project, and defining the group ID and artifact ID to use. It specifies the name of a class exposing REST services and the path under which the services will be published. It also defines a number of extensions to be included, such as RESTEasy and Hibernate.

Once the new project is created, you can copy and paste the code developed for JEE into this project. In particular, you can override the content of /src/main/java.

For the CDI to work, we need to configure the database connection. You have to add the following properties in application.properties:

```
quarkus.datasource.db-kind=h2
quarkus.datasource.jdbc.url=jdbc:h2:tcp://localhost/~/test;DB_
CLOSE_DELAY=-1;DB_CLOSE_ON_EXIT=FALSE
quarkus.datasource.username=sa
quarkus.hibernate-orm.database.generation=drop-and-create
quarkus.hibernate-orm.packages=it.test.model
```

That's it! This is the bare-minimum change required to make the application work in Quarkus. You can launch it with this:

```
./mvnw clean compile quarkus:dev
```

These are the test methods that are exposed as a REST service (take into account that the Quarkus application is deployed as the root context, so you will have to remove the name of the application – in our case, `hosawj` – from the REST endpoints). Of course, you can also package the application in any other way that we have seen (for example, in a native executable or as a fat JAR).

But that's the simplest way to move a simple application from JEE to Quarkus. You are not using any advanced feature of Quarkus.

A simple enhancement is to expose the OpenAPI and Swagger UI. It's trivial to enable these features. You just need to add the relevant extension:

```
./mvnw quarkus:add-extension -Dextensions="quarkus-smallrye-openapi"
```

The OpenAPI for your application will now be exposed here:

```
127.0.0.1:8080/q/openapi
```

Swagger UI will now be exposed here:

```
127.0.0.1:8080/q/swagger-ui/
```

Last, but not least, it's advisable to simplify the ORM part by using Panache. To do so, you can use the existing repository and simply make it extend `PanacheRepository<Payment>`. Then, you will automatically have a lot of convenient ORM methods available, and you don't have to explicitly manage `EntityManager`. Your repository will look like this:

```
@ApplicationScoped
public class PaymentRepository implements
  PanacheRepository<Payment>{
    private Logger log =
      Logger.getLogger(this.getClass().getName());
    @Transactional
    public Payment create(Payment payment)
    {
        log.info("Persisting " + payment );
        persist(payment);
        return payment;
    }
    public Payment find(String id)
```

```
    {
        log.info("Looking for " + id );
        Payment payment=find("id", id).firstResult();
        log.info("Found " + payment );
        return payment;
    }
```

It will be very easy to simply extend using methods provided by Panache. If you prefer, it will be also very easy to get rid of the repository and implement an Active Record pattern, as discussed in the *Accelerated ORM development with Panache* section.

This will close our example section.

Summary

In this chapter, we have seen a very quick overview of the JEE specification and some very interesting alternatives, such as MicroProfile and Quarkus, which are certified MicroProfile implementations.

We have learned about the JEE standard and why it's so popular. We also learned about the basic usage of the WildFly application server, along with some widely used JEE APIs, including RESTful Web Services, JMS messaging, and JPA persistence. We also learned about the MicroProfile standard, a modern alternative to JEE, and the Quarkus framework, which implements the MicroProfile standard. We also learned about some Quarkus extensions, including RESTful Web Services, JMS messaging, and persistence with Panache.

We will see more Quarkus cloud-native features in *Chapter 9, Designing Cloud-Native Architectures*.

In the next chapter, instead, we will continue our discussion on the concept of middleware by having a look at the world of application integration.

Further reading

- Snyk, *JVM Ecosystem Report* (`res.cloudinary.com/snyk/image/upload/v1623860216/reports/jvm-ecosystem-report-2021.pdf`)

- David Delabassee, *Opening Up Java EE* (`blogs.oracle.com/theaquarium/opening-up-ee-update`)

- Dimitris Andreadis, *JBoss AS7 Reloaded* (`www.slideshare.net/dandreadis/jboss-as7-reloaded`)

- Red Hat, *The WildFly Community Official Documentation* (`docs.wildfly.org`)
- Eclipse Foundation, *The Jakarta EE Tutorial*, (`eclipse-ee4j.github.io/jakartaee-tutorial`)
- Red Hat, *Undertow* (`undertow.io`)
- Martin Fowler, *Patterns of Enterprise Application Architecture*
- The Linux Foundation, *OpenApi* (`openapis.org`)
- SmartBear, *Swagger* (`swagger.io`)

8
Designing Application Integration and Business Automation

In the previous chapter, we explored the concept of middleware in an application server. That's probably the most traditional meaning of middleware: you are using a layer providing some features to your code in order to standardize and avoid *reinventing the wheel*.

That's, of course, a concept inherent to the *middleware* term: something in between your code and the rest of the world (whether it's a database, the operating system resources, and so on). But middleware has a broader meaning in the enterprise world. One such meaning is related to the concept of **application integration**. In this sense, the middleware sits in between your application and the rest of the world, meaning other applications, legacy systems, and more.

In this chapter, we will look at some typical topics related to application integration. We will then have a look at another important related middleware aspect, which is **business automation**, more related to workflows and business rules. We will discuss the following topics in detail:

- Integration – point-to-point versus centralized

- Digging into enterprise integration patterns

- Exploring communication protocols and formats

- Introducing data integration

- Messaging

- Completing the picture with business automation

- Integration versus automation – where to draw the line

- Case studies and examples

After reading this chapter, you will be able to design and implement the most common integration, messaging, and business automation patterns, to be used in wider solution architecture design for your applications.

So, let's start with some reasoning about different integration approaches.

Integration – point-to-point versus centralized

Before digging into patterns and implementation techniques for application architecture, it's important to define that integration capabilities, as in making one application talk to another one, including different protocols and data formats, can be roughly split into two approaches:

- **Point-to-point**, where the integration capabilities are provided within each application component and components directly talk to each other

- **Centralized**, where a central integration layer plays a mediation role, hiding (partially or completely) the technological details of every component, hence facilitating the communication of components with each other

It's worth noticing that there is an important comparison to be made. We've already discussed, in *Chapter 7*, *Exploring Middleware and Frameworks*, that Java Enterprise Edition evolved into componentization with the goal of breaking monolithic approaches. This kind of architectural evolution is independent of software layers. This also means that other than the applications per se, the other architectural components (such as the integration layers) are impacted by such considerations, and so you may have a monolithic approach (as in centralized integration) and a modular approach (as in point-to-point).

The goal of this section is to give an overview of different integration approaches, starting from centralized, then modularized (point-to-point or cloud-native), touching on emerging topics (such as citizen integration), and in general providing a number of different architectural points of view on how to implement application integration.

To start, let's talk about a traditional, centralized integration approach: **Service-Oriented Architecture** (**SOA**).

Understanding service-oriented architecture

SOA is a broad term. It is more of an industry trend than a standard per se. It basically defines an architectural standard, somewhat similar to microservices (and different as well—more on this in *Chapter 9*, *Designing Cloud-Native Architectures*).

This whole concept is about creating reusable services. To do that, SOA relies on a number of different technologies, such as SOAP web services, an **Enterprise Service Bus** (**ESB**), and sometimes other components, such as a service registry (**Universal Description Discovery and Integration** (**UDDI**), which used to be a standard for this area), security, governance, and repositories.

The ESB is the relevant component for this chapter. Very often, SOA has been loosely adopted and ultimately abandoned in enterprise contexts (for reasons such as scalability and complexity), while the ESB has survived such architectures.

Enterprise service bus – what and why?

The ESB technology is commonly considered to have been born together with SOA, even though some of its concepts predate SOA technology.

Some commonly used ESB products include the following:

- **Red Hat Fuse** (`https://www.redhat.com/it/technologies/jboss-middleware/fuse`), distributed by Red Hat, and made using Apache Camel, which is the framework that we are going to see in this chapter.

- **Tibco BusinessWorks** (`https://www.tibco.com/`), distributed by Tibco. This is a widespread solution among many enterprise customers.

- **MuleSoft** (`https://www.mulesoft.com`), distributed by Salesforce, particularly suited to integrating SaaS applications.

While SOA focuses on supporting the construction of composable services and modular architecture (by stressing standard protocol usage, common security and governance policies, and a machine-readable registry of exposed services), ESB does some heavy lifting behind the scenes. An ESB provides all the glue needed for making the communication between different technologies transparent. The idea is we want to standardize services (such as SOAP) to make ESB interoperable and ultimately reusable to create new applications. We can integrate existing applications and services by using an ESB. An ESB revolves around the concept of a message, being the basic unit of information managed in each integration.

There are a number of ways to represent a message, but they normally include the following:

- A **header**, including a variable amount of metadata (usually in the form of key-value pairs), which may include information such as a unique ID, the message creation timestamp, and the original sender identifier (which is the system that generated the message).

- A **body**, which includes the message data (or payload). The data may be structured, meaning that it can be validated against a schema (such as `.xsd` for `.xml` files).

Given that the message represents the information flowing into our integration system, an ESB is then further composed of the following kinds of logical building blocks, dealing with such information:

- **Connectors**, providing interoperability (sending and receiving messages) with different technologies, such as databases, filesystems, and SaaS components.

- **Formats**, providing compliance with different message types, such as `.json`, `.xml`, and `.csv`. These are used to validate messages (to ensure the format is correct) or to convert a message between formats (to make the integration between different systems possible). We will see some widespread message formats in detail in the upcoming sections.

- **Patterns**, providing well-known integration behaviors, solving common integration problems such as content-based routing, splitting, and aggregating.

In this book, we will refer to integrations defined as routes. A **route**, in the context of integration, is composed of the following:

- One or more **sources** (or endpoints), which are basically systems generating messages. This is usually a connector implementing a specific technology (such as receiving REST calls, reading files, or getting data from a database).

- One or more **destinations** (or endpoints), which are the systems that receive the messages. Also, in this case, this is commonly a connector for a specific technology (such as inserting data into a SaaS system, writing files, or calling a web service).

- One or more **integration steps**, which are the business logic of the integration itself. Integration steps can imply changing the data format by calling a third-party system (using a connector) in order to retrieve (or send) data or even to implement a specific pattern (as per the previous section, so content-based routing, splitting, and so on).

This is what an integration route schematically looks like: a source, a destination, and a number of steps in between. The messages flow in such a way, following the required steps:

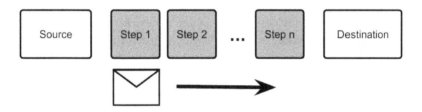

Figure 8.1 – Integration route

Please note that, usually, the steps are executed sequentially (straight through integration routes). However, according to specific patterns, it may be possible to have optional steps (skipped in some cases) or steps executed in parallel (for performance purposes). Now, when hearing about messages, you may get fooled into thinking that the concept of integration is inherently asynchronous. But in this context, this is not necessarily true. Conversely, integration may be (and usually is) a synchronous interaction, meaning that the initiator of such an integration process waits for the execution to complete.

Asynchronous integrations are behaviorally different. The initiator of such a process sends the message to the integration route and doesn't wait for the completion. It's usually enough to get an *acknowledgment* from the integration infrastructure, meaning that the system has taken charge of the message.

To implement such logic, usually, it's enough to use a message broker. In this way, you can publish the messages into a dedicated *parking space* (which is the broker) and have one or more consumers take it and execute the integration logic against it. Then, the integration logic may or may not signal the result of integration in some way (by using another message or synchronously calling an endpoint, such as a REST service). With this approach, you will have producers and consumers decoupled. We will see more about message brokers in the upcoming sections.

However, while most (if not all) of the principles of integration still hold valid today, ESBs have evolved and play a different role (and with different names) in the modern, cloud-native world.

Integration in the cloud-native world

With microservices and cloud-native architectures becoming popular, many started to question the role of ESBs and **integration**. The most common reason behind this is the lack of scalability. The microservices architectural approach heavily relies on the concept of product teams, each developing and having responsibility for a well-defined piece of software (implementing a subset of use cases).

A central ESB is simply against such an idea: in order to have service A talk to service B, you will need an integration route in the ESB, which means that both service A and service B are coupled to the system, both from a technical and an organizational point of view. You will have to pay attention to changes in your service that may break the compatibility with the central ESB (and the services dependent on it). Also, as a further side effect, you will introduce a single point of failure in the platform. Moreover, in the worst case, you'll have to raise a ticket to a specific team, which you'll need to implement yourself. This kind of complex synchronization and tight coupling between different projects is not the best in a fast-moving, self-service-oriented, cloud-native world.

But what happens if you remove the concept of the ESB from your architecture altogether?

Well, the problems that an ESB tries to solve will still exist, so you will need to solve them anyway. In order to integrate service A with service B (especially if service A and B use different technologies and protocols to communicate with each other), you will need to implement some glue. So, commonly, integration ends up buried in your services. While this is a somewhat widespread practice (more on this in *Chapter 9, Designing Cloud-Native Architectures*), I still think this has some downsides to be considered:

- You end up polluting your business logic with technological glue that needs to be encapsulated and isolated from your domain model (as per the patterns seen in *Chapter 6, Exploring Essential Java Architectural Patterns*).

- You will likely have many different implementations for the same use case (think about SOAP to REST or XML to JSON). This is inherently inefficient and may increase the occurrence of bugs.

- You will hardly reach the complete decentralization of integration capabilities. Supporting infrastructures for things such as service discovery, observability, and security will likely be needed, and are more difficult to distribute (and decentralizing such capabilities may be just wrong).

As usual, when we look at these kinds of considerations, there is not a complete answer that's good for everybody. Of course, relying on a complex and extensive centralized ESB may be a bottleneck (both technical and organizational), while trying to decentralize such capabilities may lead to repetition and a lack of governance. A common approach to resolving this kind of dilemma is basically to still rely on centralization but make it lighter and smarter. Some approaches to reduce coupling and implement more flexible integration include the following:

- It may be that your ESB becomes a set of reusable integration components (organized around capabilities) that you basically re-instantiate (and maybe modify) in your project context (hence, depending on the team providing such components, in a way).

- Such components may also not even technically be artifacts. It may be that you simply share the best practices and code samples (or even the complete code) with the project teams working with related projects. In this way, you still have some (light) control over what's going on, but each team has more freedom in understanding the component, building it, evolving it (if needed), and maybe reverting changes into the main collection via a pull request. Hence, this creates an open community behind integration capabilities across different projects.

- Another approach is to still use an ESB but limit it to one small boundary. So, instead of having a single, huge integration bus for the whole company, we can have smaller ones by department or project. They could be logical tenants of the same ESB (hence, reusing skills and best practices) or even completely different ones, based on different technologies. Once again, this is kind of a trade-off: you may still end up having repetition and/or bottlenecks, so the downsides may outweigh the benefits if you don't manage it properly.

So, even though ESBs are often viewed badly in modern architectures, the need for integration is still there, and it's important to properly study your environment in order to make good choices and evolve it correctly.

Citizen integration

One last trend that is worth highlighting is **citizen integration**. This is a trend highly studied by consulting firms and considered to be a game-changer in some scenarios. Basically, citizen integration is about having non-technical users (such as business analysts, managers, and other similar roles) being able to create integrations on their own, without having to rely on developers and other technical teams. To do so, our citizen integrators rely on highly expressive and user-friendly interfaces, usually simply accessible from the browser, and provide integration capabilities with wizards and drag and drop. Such interfaces are part of what's commonly called an **Integration Platform as a Service** (**IPaaS**).

As you can imagine, this is too good to be true: IPaaS and citizen integration is, of course, not a silver bullet. It's hard to solve every possible use case with such tools that commonly work very well on a specified subset of the infinite integration problems. There are technical implications too. IPaaS is a platform that needs to be configured and connected to backend systems, which can be a challenge (also from the security point of view), especially if you consider that such platforms are commonly hosted on the cloud.

So, I think that the whole concept of citizen integration is still relevant and deserves to be thoroughly considered in your integration strategy but usually does not solve all the integration needs a complex enterprise may have and should be targeted at a well-defined subset of them.

In this section, we explored the basic components and characteristics of integration, including the concept of an integration route, steps, and messages. We also discussed what an ESB is and how such a concept is evolving, starting from centralized SOA and going toward more modern, decentralized, self-service approaches.

Beyond the semantic difference and historical evolution of the integration technologies, there is a common sharing of knowledge about the integration patterns used. We will look at them in the next section.

Digging into enterprise integration patterns

The most complete and widely used collection of integration patterns is enterprise integration patterns. **Enterprise integration patterns** are a list of recipes for implementing well-known solutions to well-known problems in integration. Indeed, very often, the issues that occur when implementing an integration solution fall into some recognizable categories. According to common groupings, such categories include the following:

- **Message routing**, which includes all the issues and solutions about message dispatching, with topics such as filtering, routing, and aggregating messages

- **Message transformation**, which is more focused on the message content, including all kinds of message manipulation techniques, such as enriching, filtering, and *uniforming* the message content

- **System management**, which is a category including known techniques for managing and operating the integration system as a whole, including wiretaps, message archiving, and tracing

In this section, we will see a curated list of these patterns.

Message routing

The **message routing** family of integration patterns is a set of integration techniques aimed at programmatically defining the destination of an integration message. In this way, you can sort messages or define complex integration logic by chaining different integration steps designed for different types of messages. The most commonly used routing patterns are the following:

- **Message filter**: This is probably the easiest routing pattern. Here, a message filter simply discards the messages that don't comply with a specified policy. Such a policy can be a rule as complex as needed, which takes the message as input and outputs a Boolean value. The message is discarded according to that value. Common implementations of such a pattern include the comparison of some message attributes against a defined set of values. An example of a message filter is shown here:

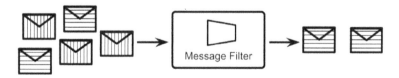

Figure 8.2 – Message filter

As you can see in the diagram, the message filter applies a policy to input messages and discards the messages that are not compliant with such a policy.

- **Content-based router**: This is slightly more complex than the filter pattern. Content-based router dispatch uses logic similar to the message filter. As a result, the message can be delivered to two or more different destinations (including other integration steps, queues, or other kinds of endpoints). Of course, unlike the message filter use case, the criteria here don't output a Boolean value, but two or more different results mapping to the destination endpoint:

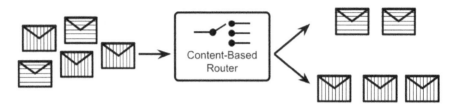

Figure 8.3 – Content-based router

We will further discuss the content-based router approach in *Chapter 9, Designing Cloud-Native Architectures*, as it will conceptually support some interesting cloud-native behaviors in the area of release management.

- **Aggregator**: The aggregator is an interesting pattern because, unlike the others described in this list, it is a stateful one. In the aggregator pattern, the incoming messages are collected (according to some defined policy) and composed as a more complex message. Being stateful is relevant here because you may want to understand what happens if such components crash when some messages are currently in flight, and how to react to such situations:

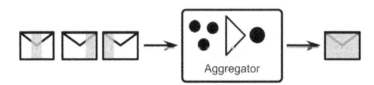

Figure 8.4 – Aggregator

- **Splitter**: This complements the aggregator pattern. A complex message is taken as an input and is divided into two or more different messages. Then, it may be followed by a content-based router to help dispatch each message to a different path to implement different business logic:

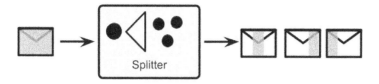

Figure 8.5 – Splitter

- **Routing slip**: This is a slightly different pattern, useful to model complex integration logic, unpredictable beforehand. With this pattern, you basically attach metadata to each of your messages and identify the next integration step (if any) that needs to be applied against such a message. This metadata can be calculated using any policy relevant to your use case. You will then need to have a component (similar to a registry) that associates the key present in this metadata with a defined destination (being another component or other endpoints):

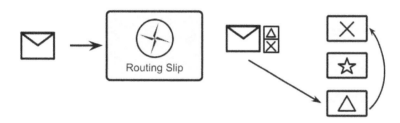

Figure 8.6 – Routing slip

In the previous diagram, the objects with a shape (a cross, star, or triangle) represent the available integration steps. By implementing the **routing slip** integration pattern, each message obtains a list of integration steps, which is attached as metadata to the message itself and calculated starting from the message content. In this particular case, our message will then go through the steps represented by the *triangle* and the *cross mark*, while skipping the step represented by the *star*.

Now let's move on to another family of patterns, focused on message transformation.

Message transformation

As it's easy to imagine, message transformation patterns focus on changing the data format of the message body. This is useful when connecting systems based on different data models or formats (think about connecting a database to a REST service or a legacy application to a SaaS solution). The pattern used for message transformation is generically referred to as message translator and simply operates on the message body, manipulating it to change the format. Apart from this generic description, there are some specific, recognizable types of message translators. Some examples are the following:

- **Content filter**: A content filter is somewhat analogous to the message filter. But instead of dropping the message as a whole when the content doesn't comply with a set of rules, it operates within the message data, discards part of the content, and only keeps the part of the message that is relevant (by checking it against a set of conditions):

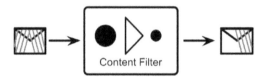

Figure 8.7 – Content filter

- **Content enricher**: This complements the content filter. A content enricher adds some new data to the message content. To do that, it relies on an external repository (such as a database). The enrichment algorithm may use a replacement (each value corresponds to another one, like when changing a ZIP code for a city name), a fixed value (adds the same value to each message), or more complex logic. Here is a diagram of it:

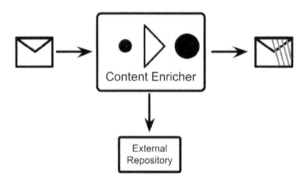

Figure 8.8 – Content enricher

- **Canonical data model**: This is a common approach in ESBs. Basically, in order to decouple the message format of all the participants of the system, a *neutral* format is defined to be used in the ESB. This is usually a superset of all the messages, or simply a different format. In order to implement this approach, each system is plugged into the ESB with a special **Message Translator** component, which translates the native format of each system to the canonical data model, and vice versa:

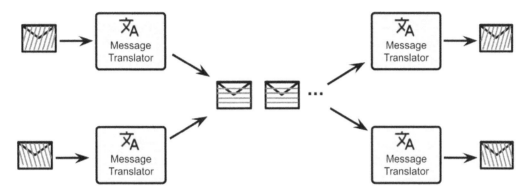

Figure 8.9 – Canonical data model

- **Normalizer**: This is a special case of the canonical data model approach. In order to maintain the common data format inside the ESB, but use a single endpoint for each external system, you can use the router component (as per the *Message routing* section). The only purpose of such a component will be to look into the messages, recognize the message format (by looking into the body or header), and route it to a specific message translator, which must be able to translate the message format to the common data format:

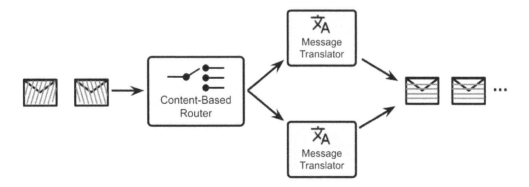

Figure 8.10 – Normalizer

These are just some well-known examples, but the message translators are usually something very specific to the business logic, including custom approaches, such as the merging of different fields, string formatting, and calculations. In the next section, we will talk about system management patterns.

System management

System management patterns are essentially positioned as a way to monitor and manage integration routes in production. So, in this sense, they are useful for the operation of the platform and ensuring the service level for the customer. However, there are several patterns that could also be useful for implementing logic that solves specific use cases (besides being useful for monitoring and management). Such patterns include the following:

- **Detour**: A detour is a practical technique for ensuring a particular treatment for some messages. In practice, you will have a content-based router triggering a specific path when some condition happens. The content-based router may be triggered by certain content in the incoming messages (as usual) or may be based on specific external conditions (such as special messages coming in, and maybe even on specific channels different from the one on which the rest of the traffic comes). When activated, the detour will route the messages to a different path that may be used for debugging, testing, or validating such messages.

  The detour opens a lot of interesting (and modern) use cases, such as the concept of the circuit breaker and other cloud-native patterns (we'll see more about this in *Chapter 9, Designing Cloud-Native Architectures*). In the following diagram, there's an example of a detour: each message is inspected and, depending on the content (using the content-based routing pattern), it is routed to the normal path or a special path (if some conditions are met). In this way, you can activate special handling for some specific messages:

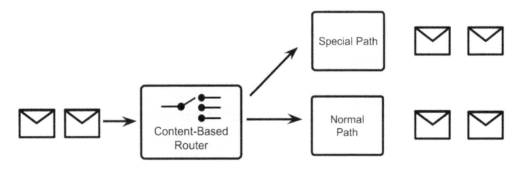

Figure 8.11 – Detour

- **Wiretap**: Wiretap is a pretty simple pattern. You basically add a step to the integration route that duplicates all the incoming messages and sends a copy to an alternative channel (while a copy continues to travel on the usual integration route). In this way, you can monitor the incoming messages (such as counting them or inspecting them) and analyze the system behavior with real data:

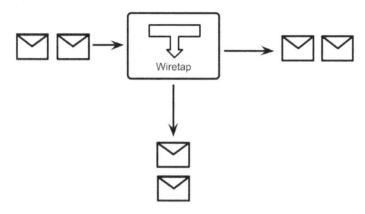

Figure 8.12 – Wiretap

- **Message history**: Message history is a simple and structured way to understand the path that each message flows through. Think about an integration route with multiple paths (such as conditional ones, which are diverted by content-based routers and similar patterns). It may be useful for debugging purposes or even mandated for regulation purposes (such as audit logging) to have a registry of every step spanned by the message. Message history suggests doing so by attaching some data at each step. This is commonly done by adding a unique key for each system in a specific message header. At the end of the integration route, you will have a list of keys identifying each integration step. Even in this case, this is not so different from tracking a cloud-native pattern needed for heavily distributed architectures (such as microservices). Here is a diagram for visualizing message history:

Figure 8.13 – Message history

In this diagram, we see the integration steps are represented by a symbol (a *cross mark* and a *triangle*). Each time a message passes into an integration step, the message is marked with an identifier corresponding to it. So, at the end of the integration route, you know exactly the path that each message has followed (if it skipped any step, went through optional paths, and so on).

- **Message store**: There are use cases in which you want to know exactly the content of each message, including intermediate transformation. This can be required for a subset of messages (such as for troubleshooting purposes) or all messages (as we saw in the message history pattern, that may be for audit logging requirements). The message store pattern suggests implementing this case by attaching a wiretap to each integration and diverting every message (or some messages, conditionally) to a shared message store (such as a database).

  It may be necessary to add some complementary metadata, such as a timestamp, an identifier for each step, and maybe a signature (for checking the data integrity). In some cases, the message store may need to implement specific technologies for non-repudiation, such as **Write Once, Read Many** (**WORM**), in terms of special anti-tampering hardware. The following diagram visualizes the workings of the message store:

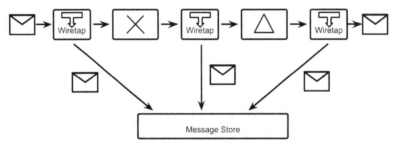

Figure 8.14 – Message store

- **Test message**: This is a simple health check for integration routes. Basically, in order to understand the message flow (such as whether there is any intermediate component losing messages or taking too long to process them), you inject some special test messages into the integration route. You will then need a content-based router at the end of the integration route in order to identify such special messages (such as looking for a particular pattern in the data or a special key in a header). Then, you'll need to route it to a monitoring system, which can then check whether every message is returned (so that there is no message dropping) or calculate the elapsed time, and so on.

  Bear in mind that every intermediate step may need to be aware of or at least resistant to this kind of test message. This means that if you are calling external systems or writing data to a database, you may want to instruct a specific step to skip in the case of test messages. In the next diagram, we can see a graphical representation of this pattern, that is, a content-based router that identifies a special test message and routes it to a monitoring system, instead of the standard integration flow:

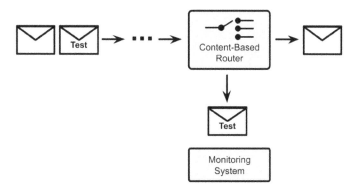

Figure 8.15 – Test message

The group of system management patterns is different from what we have seen so far. They are less focused on application logic and data and more on the monitoring, maintenance, and operation of the integration infrastructure. This does not mean that you cannot use them to implement some use cases (think about the Wiretap pattern, which can be a way to implement multiple different behaviors on the same message), but that's for sure not the main usage.

As we said, all the patterns that we have seen so far are useful both for synchronous and asynchronous integration. However, when it comes to async use cases, a whole new set of considerations arises in terms of messaging brokers and integration with them. This is partially related to enterprise integration patterns and partially implicit in the technology itself (which may be referred to as **message-oriented middleware**, or more commonly, **queue managers**). In the next section, we will have a look at those cases.

The Camel integration framework

Apache Camel is likely the most famous open source integration framework. It was created in the years after 2000 and it has been evolving constantly since then, mostly because of the very active community behind it. At the time of writing, Camel has hundreds of contributors and thousands of stars on GitHub.

Camel isn't exactly an ESB but can be used as one. It is more like a core engine containing integration capabilities. Indeed, Camel implements the enterprise integration patterns by design (and other patterns, including some techniques for cloud-native applications). Moreover, Camel includes hundreds of connectors for specific technologies (such as queues, databases, and applications) and data formats (such as JSON and XML). Camel can be run standalone or on top of a selection of runtimes (including Quarkus, which we saw in the previous chapter). It can be deployed as an ESB (centralizing all the integration capabilities at one point) or embedded in your applications (distributing such capabilities where it's needed).

The Camel DSL

Camel exactly implements the concept of routes as we have seen it so far, as a sequence of specific steps to run against each message (intended as a piece of data). In order to specify each route with Camel, you can use a .xml file or the Java **Domain-Specific Language** (**DSL**), which is basically a dialect of Java made for the purpose of expressing concepts specific to the Camel world. For the purpose of this section, we will use the Java DSL, which allows the definition of routes using a Java-fluent API.

This is what a simple integration route that converts JSON to XML looks like:

```
from(platformHttp("/camel/hello"))
.unmarshal()
.json(JsonLibrary.Jackson, MyClass.class)
.marshal()
.jacksonxml()
.to(file("/myfilePath?fileName=camelTest.xml"));
```

As you will see, there is `from`, which is the endpoint starting the integration route (in our case, by exposing an HTTP REST endpoint, by using a component called `platformHttp`), and `to`, which writes the final result to a file (by using the `file` component). In between, you can see an example of data transformation, including the mapping (`unmarshal`) of a JSON object to a Java object, and then mapping back (`marshal`) of such a **Plain Old Java Object** (**POJO**) to XML.

We will see a more complete example in the *Case studies and examples* section. Now, let's have an overview of the messaging concepts.

Messaging

Messaging is a core concept in the integration world. In the previous section, we discussed messages as the basic unit of data flowing inside each integration step. Let's now focus a bit more on the concepts specific to messaging, such as message brokers, asynchronous interactions, producers, and consumers. First, we will start with the broker concept.

Defining the broker concept

A **broker** is a common, elementary concept in IT. It can be intended as an architectural solution as well as a technology.

From an architectural standpoint, a broker allows producers to push messages into an intermediate system (a broker itself), which dispatches it to one or more consumers. The message broker concept is described in the homonymous enterprise integration pattern.

Beyond this simple description, a huge number of variants and other concepts can be elaborated on, influenced by the underlying technology and the use case we are trying to model. Examples of broker technology include Apache ActiveMQ, Kafka, and RabbitMQ.

Now, let's dig into some basic messaging concepts.

Queues versus topics

The first categorization that is common in a Java programmer's mind is queues versus topics. This differentiation has been made famous by the **Java Message Service (JMS)**, which is the API defining messaging practices under the Java Enterprise standard.

In the JMS world, a **queue** is defined in the message broker, which takes care of messages sent by producers and dispatches them to a consumer. If there are no consumers available, the queue stores them until one connects, trying to avoid the loss of messages. This is referred to as the **store and forward** approach. The queue can also be used for point-to-point connections (one producer and one consumer) as the **point-to-point channel enterprise integration pattern**.

A common usage of queues is to have one or more producers and a number of consumers that may also vary with time, depending on the number of messages to work effectively (an example of horizontal scaling). In this case, each consumer takes a message in an exclusive way, usually with some sort of transactional semantic. This pattern is named **Competing Consumer** in the enterprise integration patterns world.

A **topic** has a slightly different semantic. In a topic, the messages sent by producers are propagated to all the consumers connected in that particular moment. This is similar to the concept of a broadcast, commonly used in networking. Consumers usually lose all the messages sent before they were connected with that particular topic.

Queues and topics are two high-level concepts that encompass, in recognizable names, a number of different characteristics of the messages, producers, and consumers involved (and may include different variants). In the enterprise integration pattern world, a queue is defined as a point-to-point channel including the Competing Consumer pattern. The topic is instead defined by the concept of the **publish-subscribe channel**, in which you have one or more producers, and not every consumer is competing, but instead receives a copy of each message, in a broadcast fashion, where everybody receives every message.

Message quality of service

An important concept, often related to the underlying messaging technology, is the **quality of service** (also known as **QoS**). QoS, in the context of messages, refers to the *commitment* that the broker takes on when it comes to delivering our message to consumers. This refers to what happens after the producer puts a message into the system and gets an acknowledgment from the broker. Then, based on the configuration of the system, three delivery scenarios are possible:

- **At most once**, which means that the message may not be delivered at all, but if it's indeed delivered, it will not be delivered more than once. Here, the use case is about *best-effort* messages (so, we can lose some), where duplication is to be avoided (because it *pollutes* our downstream systems). A real-world example of this is currency exchange rates. These are values that change very often, and in some scenarios (such as high-frequency trading), you would rather lose one value (which is valid for a very short period of time and overridden by a new one) than just having a *ghost* value caused by a message duplicate. Here is a diagram to illustrate this:

Figure 8.16 – At most once message delivery

- **At least once**, which implies that messages will never get lost, but may be sent more than once to consumers. Here, the use case is, of course, the opposite to the previous one. In particular, it's more important to not lose any messages. In the real world, this could be an **Internet of Things** (**IoT**) scenario: imagine collecting field data from an industrial machine. You may prefer to have all messages (which, for example, may highlight an imminent failure), even if this means that you may have duplicates (which could be discarded in downstream systems or simply be considered as harmless). The following diagram exemplifies this:

Figure 8.17 – At least once message delivery

- **Exactly once**, which is the ideal scenario that you can imagine when approaching a messaging system. Needless to say, here the broker guarantees that your message will be delivered and no duplicates will exist. This, of course, may be a mandatory requirement in some kinds of use cases. Typically, in the real world, this is related to financial services: once you have entered a payment transaction, you cannot afford to lose it, nor execute it twice. The following diagram demonstrates this:

Figure 8.18 – Exactly once message delivery

Now, you might be wondering, why don't we simply stick with the exactly once delivery scenario every time, and simplify our lives? The answer is simple and expected: exactly once is the most expensive of the three. Since the system will need to lock at some point (to check for duplicates), providing there are the same number of messages and the same hardware, exactly once would probably be the worst choice in terms of performance. This may not be noticeable with low traffic, but it may be crucial if you are designing with high-traffic peaks in mind.

Zero message loss

In messaging, it's a common requirement to guarantee the zero loss of messages (and as we have seen, this is a combination of at least once and exactly once QoS). In order to provide such requirements, messaging systems usually use two kinds of solutions:

- **Message persistence**, which is usually on a filesystem or database. This means that a producer will get an acknowledgment for putting a message in a queue only after the message is serialized on the persistent storage. In this way, in the event of a system crash, it is guaranteed that the situation can be recovered by reading from the journal. Here is a diagram for demonstration:

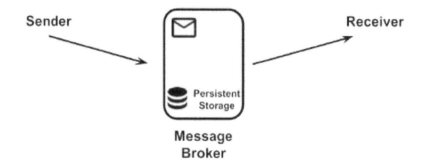

Figure 8.19 – Message persistence

- **Message copies**, which are sent to different instances of the message broker. The producer gets the acknowledgment for putting a message in the queue after a copy of the message is propagated (over the network) to one or more (usually configurable) backup instances of the messaging system. This guarantees that, in the case of our messaging system crashing, the backup instances can take over and deliver the message. Of course, in this scenario, you are reducing but not eliminating risks. You may still end up with all the instances down in the case of catastrophic failures, and you should plan accordingly (such as using different physical locations, where possible), as shown in the following diagram:

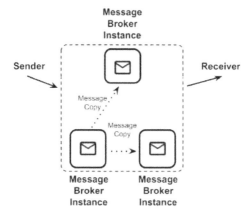

Figure 8.20 – Message copies

Zero message loss scenarios almost always have performance impacts.

Other messaging concepts

As has been said, depending on the underlying implementation technology, there are a number of use cases that can be implemented in the messaging world. Here is a list of the most useful ones:

- **Dead Letter Queue** (**DLQ**): This is pretty common in any messaging system. A DLQ is basically a special location, as shown in the following diagram, to redirect messages when certain conditions happen (such as no consumers are available after a certain amount of time, as we will see in the next point about time to live) or simply when the broker doesn't know what to do with a message (for example, for runtimes or configuration errors). It's a common behavior to persist and monitor the DLQ as an indicator if something goes wrong and if the messages contain any recoverable data.

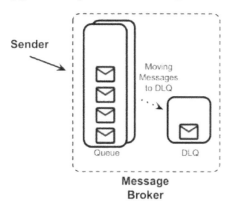

Figure 8.21 – Dead letter queue

- **Time to live**: When used, time to live is an attribute associated with each message when it is inserted into the queue. It will define the expiry of the message: if a message is still in the queue after the expiration has occurred (because there are no consumers or they aren't fast enough), it could be discarded or moved to a special queue (such as the DLQ).

  It's an elegant way to model some use cases: there are some kinds of data that are just useless after a certain amount of time has passed (maybe because some more recent information has become available by that time). In this way, you avoid putting overhead on the consumers. However, if you have too many messages expiring, there is probably a need for something else to be tuned (such as the availability and performance of the consumers).

- **Duplicate checks**: Some broker implementations can check messages against duplicates. This is usually a delicate matter to handle. There are different possible implementations, but the most common one involves the presence of a unique identifier for the message (which can be provided externally, such as a database key, or calculated by the broker, such as a hash) and storing such messages in a proper data structure (such as a database or a key-value store). Each message is then checked against such a structure, and if a duplicate is found, the message is discarded. The message store commonly has a fixed maximum size or an expiration to avoid indefinite growth.

- **Priority**: This is a common requirement for some use cases. Basically, it is the possibility to identify some messages as having a higher priority than others (usually setting a specific header), to inform the broker to have it delivered before the other messages in the queue (if present).

- **Bridge**: This is an infrastructure for multiple queue management that basically passes messages from one broker to another, as shown in the following diagram. It can copy the messages or just move them to another queue and broker. It's useful to interface with different technologies and existing systems, or even to provide reliability (such as a multi-site messaging system):

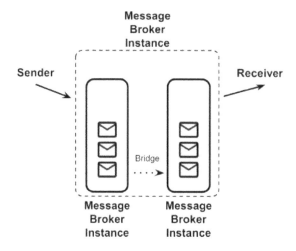

Figure 8.22 – Bridge infrastructure

- **Filters**: This is a common functionality of brokers, which mimics the content-based router pattern that we have already seen. It's basically a configuration instructing the broker to move messages between different queues when some conditions happen (such as if a special header is present or a condition is met in the message payload).

- **Chunking**: It may happen that a queue is used to transfer data of a consistent size. In order to avoid hogging the broker and handle very big messages, a broker can implement chunking. As it's easy to imagine, a big message is then chunked into smaller parts before being delivered, as shown in the following diagram. However, some mechanism is needed to reconstruct the message on the consumer's side. A common one is to tag each chunk with an identifier and a sequence number:

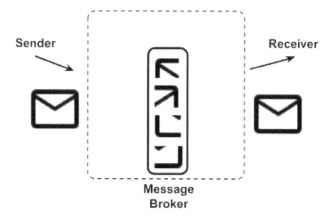

Figure 8.23 – Message chunking

- **Schema**: It's sometimes useful to perform some validation on the messages inserted into the broker. A smart way to do that is to define a data schema (such as an XSD). The messages that are not compliant with such a schema are then discarded or moved to special queues (such as the DLQ), as follows:

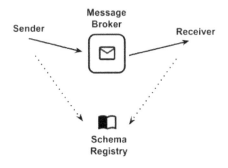

Figure 8.24 – Data schema in messaging

This list completes our considerations about messaging. In this section, we have seen many constructs (such as brokers, queues, and topics) and configurations (such as QoS and zero message loss) that can be used to model a lot of different use cases. In the next section, we will focus on the protocols and formats of data.

Exploring formats

As we have seen in the previous sections, integration works with flows (synchronous or asynchronous) of small information bites (in the form of messages) to be acted upon. Such messages are usually formatted into well-known shapes. Let's have a quick overview of the most common ones.

XML

Ostracized for being verbose and cumbersome, and often considered old and past it, **eXtensible Markup Language** (**XML**) is simply here to stay. And for good reason, since, as we will see, it has a number of powerful and useful features. To start, XML files are expressive and structured, and there is a lot of tooling supporting them.

This is what a simple XML file looks like:

```xml
<?xml version="1.0" encoding="UTF-8"?>
<myTag>
   <mySubTag myAttribute="myValue" >my content</mySubTag>
</myTag>
```

I'm sure everybody is familiar with XML; however, just to set common ground, the characteristics of a proper `.xml` file are as follows:

- It is text-based.

- There is a special tag at the beginning, called a **prolog**, specifying the version and encoding of the document (such as `<?xml version="1.0" encoding="UTF-8"?>`).

- There is a root tag including all the other tags in the document (excluding the prolog, which is considered to be a special element of the document).

- Each tag of the document can include text content (`<myTag>my content</myTag>`) or other tags (`<myTag> <mySubTag>...</mySubTag> </myTag>`). This is called an **element**.

- Each tag may include one or more key-value pairs, called **attributes** (such as `<myTag myKey="myValue" ...>...</myTag>`).

- Each tag must be opened and closed properly (such as `<myTag>...</myTag>`). The shorthand form is allowed if the tag is empty (`<myTag/>`). Tags must be properly nested: if you open a tag, you can open other tags inside it, but you need to close the parent tag before closing the child tags (`<myTag><myOtherTag></myOtherTag></myTag>` is allowed, while `<myTag><myOtherTag></myTag></myOtherTag>` is not).

- Special characters, such as <, >, and ", must be replaced with special entity references, such as `&lt;`, `&gt;`, and `&quot;`, which are commonly called **escape sequences** and are basically one-to-one mappings between each special character and the related entity reference.

Most likely, such rules are just taken for granted: after all, you have probably already edited a `.html` file (which is a sibling of the `.xml` file) or a configuration file in the XML format.

As mentioned, detractors of XML say that it is long and hardly human-readable, not to mention the frustration of parsing errors when you try to manually edit it: one single character off will often corrupt the whole file.

However, due to this simple but powerful syntax, XML provides some interesting features:

- **It allows for automatic validation by using an XML schema (XSD)**: An XSD is considered a class when the `.xml` file is considered to be the instance. An XSD can be applied to a given `.xml` file to ensure it is compliant with such a specification. That's crucial in machine-to-machine interactions and may reduce the number of runtime errors. By defining an XSD, you are basically creating an XML dialect suitable for your own problem.

- **It can be searched (using queries)**: By using technologies such as XPath and XQuery, you can define patterns that will allow you to find a specific portion of a .xml document. That's particularly interesting in the context of integration (think about the content filter or content-based router patterns that we have seen), so most of the available ESB technology provides support for this kind of feature.

- **It can be automatically transformed**: By using the XSLT dialect, you can define transformations for .xml files. In this way, you can set rules allowing a processor to change a .xml file from one definition to another, mapping and transforming tags in the source files to something different in the target files. Also, in this case, it's an interesting feature in the integration world that can basically cover most of the message transformation patterns.

Talking about XML is like talking about Java: there is plenty of criticism around calling it an old and outmoded standard. However, while more modern approaches have, of course, come along and deserve attention, XML, like Java, provides proper support for a wide range of use cases to date, due to a structured set of rules and the extensive availability of supporting tools and technology.

Working with XML in Java

The translation of .xml files from and to Java objects is a pretty common task. There are basically two ways to do so:

- The first (and now less common) way to parse XML is to use **streaming**. This is useful if you don't know the structure of the .xml document upfront that you are going to parse. So, you rely on a streaming approach, in which XML is traversed from the beginning to the end, and each element triggers events, such as the start element and the end element.

  Each event contains the data for the particular elements (contents and attributes). While it is not particularly widespread today and has some practical disadvantages (the creation of Java objects is cumbersome and random access to elements is not allowed), this kind of parsing has the advantage of usually being very efficient, especially in terms of memory usage. The most famous implementation of XML streaming in Java is **SAX** (www.saxproject.org).

- The most common way to implement XML serialization and deserialization is to use **direct mapping**. With this approach, there is a direct link between elements (and attributes) of the XML content and fields of the POJO. Such linking is defined by a proper mapping, which could be defined in configuration files or, more conveniently, by using annotations. Part of the mapping can also be implicit (such as fields mapped to homonymous XML elements and vice versa).

Nested elements are commonly mapped using collections or other complex subobjects. This approach is heavily used in integration (but not only that), as XML content is mapped to Java objects that are then used for business logic, checks, and other interactions. The most common implementation of XML mapping in Java is provided by **Jakarta XML Binding** (**JAXB**), which is part of the JEE specification. It is also worth knowing that `Jackson`, a JSON library that we saw in *Chapter 7, Exploring Middleware and Frameworks*, in the *JPA and REST (and more) with Quarkus* section, can also be used as a framework for REST serialization for XML mapping (and supporting other data formats too).

Whatever the approach is for parsing, mapping XML to Java is a pretty common use case in the enterprise world, as XML is a widely used format for data interchange (used in many different industries, including banking and healthcare).

In the next section, we are going to see a challenger of XML in the field of web services: JSON notation.

JSON

We have already seen and used JSON, in *Chapter 7, Exploring Middleware and Frameworks*, in the *Jakarta RESTful web services* section. Now, it's time for a bit of theory about it.

JSON is the acronym for **JavaScript Object Notation**. It is a text representation for representing data. The technology was born in the context of web development when the AJAX application became widespread. We will see more about AJAX and web development in *Chapter 10, Implementing User Interaction*, but for now, it's enough to know that it's now a common technology that started to be used around 1999 and is about web pages dynamically requesting data from the backend after the page is downloaded by the browser. To do so, the JavaScript language is used on the client side for both requesting and parsing such data.

While it is possible to use XML to serialize such data, JSON emerged as an effective and simpler alternative. JSON is indeed native to JavaScript, and the serialization/deserialization of JavaScript objects to JSON is done without the need for external libraries. This is what a simple JSON file looks like:

```
{
    "myKey":"myValue",
    "myOtherKey": 42,
    "mySubObject":
```

```json
{
    "mySubKey": "mySubValue",
    "myArray":[ "value1", "value2", "value3" ]
}
}
```

JSON is basically made of primitive types (such as strings, Booleans, and numbers), objects, which have one or more key-value pairs enclosed in curly brackets, and arrays, which are collections of other objects, arrays, or primitive types, enclosed in square brackets. The thing that made JSON popular, other than being native to JavaScript, is that it is less verbose and more human-readable than XML.

The major criticism of JSON is that it's less *structured* than XML, which has produced a number of other concepts and technologies in terms of validation (XSD, as we saw in the previous section), web services (SOAP), querying (the aforementioned XPath and XQuery), and more (such as security and other features associated with the SOAP standard).

However, JSON nowadays covers some (if not all) of those features, both natively and via third-party implementation. It's worth mentioning that JSON Schema is a technology available for syntactic validation, and other implementations, such as JSONPath, are used for querying JSON documents. Moreover, JSON is commonly used as a base technology in NoSQL document databases (we'll see more on this in *Chapter 11, Dealing with Data*). In the next couple of sections, we are going to see the interactions between JSON and YAML (which is a widely used data format nowadays), and, of course, JSON and Java.

JSON and YAML

YAML Ain't Markup Language (YAML) is an alternative data serialization language created in 2001 that became widespread with the popularity of Kubernetes because it's used as a format to encode resources and configurations (we'll see more on Kubernetes in *Chapter 9, Designing Cloud-Native Architectures*). YAML is also widely used in frameworks such as Quarkus and Spring Boot for managing configurations of microservices. YAML is designed to be easily human-readable and is heavily based on key-value-like structures (and more complex objects), which are organized using a syntax similar to the Python language, which relies on spaces to define hierarchies.

This is what a simple YAML file looks like:

```yaml
---
myKey: myValue
myOtherKey: 42
```

```
mySubObject:
  mySubKey: mySubValue
  myArray:
  - value1
  - value2
  - value3
```

It's interesting to note that, since YAML can (but does not enforce doing so) use a syntax based on curly brackets, it is indeed a proper superset of JSON. This means that YAML provides some additional features that are not present in JSON (such as comments and richer data type management).

A YAML parser, in other words, can parse JSON documents. Moreover, if the additional features are not used, a YAML document can be directly translated to JSON (and vice versa) without losing any data. Indeed, the example for YAML that we have seen is the exact representation of the example for JSON that we saw in the section before.

Working with JSON in Java

As we already know, the parsing of JSON files is native in JavaScript, while in Java the already mentioned `Jackson` library is a common way to work with JSON. The mapping, as we saw in *Chapter 7*, *Exploring Middleware and Frameworks*, is made by associating (explicitly by using an annotation, or implicitly by relying on the name) each field of the POJO to each key of the `.json` file, similar to the approach of JAXB for XML mapping. This kind of mapping is particularly useful when dealing with REST web services.

Protobuf

Protocol Buffers (**Protobuf**) is a slightly different way to store data. It was created by Google as an internal tool (widely used within their infrastructure) and then was open sourced. The most notable difference from the other technologies seen so far is that Protobuf is a binary protocol. As per the other technologies seen so far, it is language-independent, so you can use it as a way to communicate from Java to other technologies.

Google (and the other organizations and contributors involved in the open source community) provides tools for serializing, deserializing, and in general working with Protobuf, including an SDK for Java. The SDK contains a compiler (protoc) that acts as a source code generator. Basically, when given a specific configuration (in a `.proto` file), it creates all the needed scaffolding for serializing and deserializing POJOs to and from byte arrays (and they then can be sent over the network, persisted to a file, or used as a message). Since the output is in a binary format, it is very efficient and optimized.

The configuration is basically a declaration of all the fields contained in the POJO you want to serialize, plus some metadata:

```
syntax = "proto3";

option java_outer_classname = "MyPojoProto";
option java_package = " it.test";

message MyPojo {
    string myField = 1;
    repeated string myList = 2;
    int32 myNumber = 3;
}
```

Here are some details about the preceding block of code:

- `syntax` refers to the version of Protobuf used. **Proto3** is the current version at the time of writing.

- The two `option` keywords are specific to Java. They will configure the name of the class and the package containing all the autogenerated facilities.

- `message` is the description of each field. Other than the name of the object (`MyPojo`), it defines the name of each field and the primitive type (`string`, `int32`, and so on). The field can be prefixed by the `repeated` keyword, meaning that a specific field can be present multiple times in a valid message. If that keyword is not present, it can be present zero or one times (not more than once). Last but not least, each field is attached to a numerical index (`1`, `2`, `3`, and so on), which Protobuf uses as a unique identifier for the fields in a message.

Running the protoc compiler against the `.proto` file will generate a class (in our case, named `MyPojoProto`). This file will contain an inner class that will be used to represent our POJO (a message, in Protobuf jargon, which in our case is named `MyPojo`). In the class, there will also be a number of utility methods, including a builder to create such messages, and methods to serialize and deserialize to and from byte arrays.

In this section, we have seen a number of widely used data formats, such as XML, which is a traditional, old, and widely used technology; JSON, which has become more and more popular also, thanks to JavaScript and web technologies; and Protobuf, a less-used alternative with a different approach and aiming to reach cases where a binary format is needed.

Exploring communication protocols

In the previous sections, we focused on the data formats used for storing and exchanging information in a standard way. The next step is identifying the ways to exchange such information, in other words, the most commonly used communication protocols.

SOAP and REST

SOAP and **REST** are two widely used communication protocols. Even if they have been mentioned many times in previous chapters (and in this chapter too), I think it's still relevant to provide a quick summary of them, as this can be the key to understanding the role of communication protocols in integration systems:

- **SOAP**: As mentioned before, this used to be a key component of the so-called SOA. Being based on the XML data format, it's usually used over the HTTP protocol. The documents are exchanged via formatted XML files included in a root tag called `envelope`, containing a header and a body. Being regulated by a lot of substandards, SOAP is used to define the methods, the exchanged data, and optionally other specifications, such as the security, to be used. Last but not least, SOAP provides a well-structured way for defining method signatures and performing validations, called **WSDL**. SOAP is less popular currently for the same reasons as the XML technology: it is verbose and less flexible than most modern alternatives.

- **REST**: This is considered to be a less formal, more flexible alternative to SOAP. In this sense, it's improperly defined as a protocol; it's more of an architectural style. REST is basically a definition of a set of operations (based on the HTTP verbs, such as `GET`, `PUT`, `POST`, and `DELETE`). Such operations are performed against resources, which are identified by the URIs. The threatened resources can be formatted in many different ways, but JSON is a widely used way to do so. REST is way more lightweight than SOAP. For this reason, some of the features embedded in SOAP (such as security, session handling, and validation) are not natively part of REST and are usually implemented by using external tools, libraries, and extensions.

Of course, that's just a very high-level introduction to SOAP and REST, but since they are widely used, well-defined protocols, there is a lot of relevant material available that can be used for getting more information. Having said that, it should be clear by now that SOAP and REST are ways to allow different systems (across different languages and technologies) to communicate with each other, and basically implement APIs for both querying data and invoking remote operations. Now, let's see a couple of more modern, alternative approaches commonly used today for achieving similar goals.

gRPC

gRPC Remote Procedure Call (gRPC) is a modern, open source framework developed originally by Google, and then released in open source as part of the CNCF projects umbrella. It defines a complete way for implementing interoperability between different systems. In order to do so, it provides a number of client libraries for all major languages, including Java, PHP, and Python.

gRPC natively implements a lot of useful mechanisms that are often missing or implemented externally in SOAP and REST. Such mechanisms include bidirectional streaming and notifications (full-duplex communication), security, synchronous and asynchronous patterns, and flow control. Another key characteristic is that gRPC natively uses Protobuf as a serialization technique, hence providing more stability and fewer issues with cross-language communication. For all of those reasons, gRPC is now considered to be a good alternative to REST and SOAP for the communication between microservices and has proven to be most useful, in production and in many well-known contexts (such as Netflix, Spotify, and Dropbox), in providing low-footprint, high-performance communications.

From a practical standpoint, in order to use gRPC communication, it is of course necessary to retrieve the relevant library for the language that we are going to use. As said, Java is a great choice. Once the dependency is provided, you have a component acting as a server and another component acting as a client. Once the server has been started, the client can connect to it and from that point, fully bidirectional communication is established.

Let's see a practical example of a server and a client implementation, using the official Java gRPC library. Here is a basic server implementation:

```
    . . .
    int port = 9783;
    server = ServerBuilder.forPort(port)
        .addService(new PingImpl())
        .build()
        .start();
    logger.info("Server started, listening on " + port+"
        ...");
    server.awaitTermination();
    . . .
  static class PingImpl extends PingGrpc.PingImplBase {
    @Override
    public void send(PingRequest req,
      StreamObserver<PingReply> responseObserver) {
```

```
        logger.info("Received request " + req.getMsg() + "
            ...");
        PingReply reply = PingReply.newBuilder().setMsg("pong
            " + req.getMsg()).build();
        responseObserver.onNext(reply);
        responseObserver.onCompleted();
    }
  }
...
```

In this simple example, you can see a Java class launching and an embedded gRPC server. The `main` method creates the server using the `ServerBuilder` class provided by the library. In order to build the server, a port is passed (`9783`, in this case), then a `static` class is passed, which defines the implementation of the server method defined by the RPC (in this case, a `send` method, answering to a simple request by passing a string). The server is then built and started in the same chain of method calls in the `ServerBuilder` utility. Lastly, the `awaitTermination` method is called, and basically blocks the execution while waiting for connections and handling them.

Let's now see how a simple gRPC client can be implemented to contact this server:

```
...
String message = "Ciao!";
String target = "localhost:9783";
ManagedChannel channel =
  ManagedChannelBuilder.forTarget(target)
.usePlaintext()
.build();
blockingStub = PingGrpc.newBlockingStub(channel);
logger.info("Trying to ping with message " + message + "
   ...");
PingRequest request =
  PingRequest.newBuilder().setMsg(message).build();
PingReply response;
response = blockingStub.send(request);
logger.info("Received response: " + response.getMsg());
...
```

As you can see, in the previous simple example, `ManagedChannel` is built, passing some parameters (the host and port to contact the server, in this case, locally). Then, a stub is instantiated. A `request` object is built, and a message is set inside (in this case, the `Ciao` string). The `send` method is then invoked against this stub, passing the `request` object. The response is then collected and logged.

As mentioned before, gRPC relies on Protobuf by default for defining serialization. That's where the request and reply objects are defined, and the signature for the `send` method is declared. Here is a sample `.proto` definition for our example:

```
syntax = "proto3";

option java_multiple_files = true;
option java_package = "it.test";
option java_outer_classname = "GrpcTestProto";
option objc_class_prefix = "HLW";


package grpctest;

service Ping {
  // Sends a greeting
  rpc Send (PingRequest) returns (PingReply) {}
}

message PingRequest {
  string msg = 1;
}

message PingReply {
  string msg = 1;
}
```

That's all for our primer about gRPC. Of course, in the real world, more things need to be taken into account, such as correctly shutting down the server, handling exceptions, and any other features (such as retries, flow control, or load balancing) that you may want to use. In the next section, we are going to see another protocol that is commonly compared and used alongside REST: GraphQL.

GraphQL

GraphQL is a technology for defining complete API systems in order to query and manipulate data. It has some similarities with the REST and SQL technologies, but it's really a unique idea, as it defines APIs that are structured while providing freedom to the clients, who can specify what kind of data they are requesting. GraphQL was originally implemented by Facebook, which then released the governance of the project to an open source community under the Linux Foundation.

As mentioned previously, an aspect that is really interesting (and unique) of GraphQL is that the client is controlling the kind of data that is sending requests to the server, thus making this technology well suited for mobile applications and, in general, optimizing the communication, because only the data needed is transferred. In order to do so, GraphQL defines a special way to make queries that explicitly define the kind of data we are requesting to the server. As an example, take a look at the following query:

```
query {
  payments{
    date
    amount
    recipient
  }
}
```

This is a simple definition asking for payments and three specific fields of each payment. Of course, some conditions for querying can be passed, such as the following:

```
query {
  getPayments(recipient: "giuseppe") {
    amount
    data
  }
}
```

Of course, there are a lot of other options that can be explored. GraphQL supports complex, nested types. You can specify queries with multiple conditions. It is possible to use other interesting features, such as pagination, sorting, and caching.

In order to implement and expose GraphQL APIs in your projects, there are at least two different options:

- You can implement a server, embedded in your backend code. In this case, it can be useful to use a framework such as the Domain Graph Service framework built by Netflix (`github.com/netflix/dgs-framework`). Other options include GraphQL Spring Boot (`github.com/graphql-java-kickstart/graphql-spring-boot`) and graphql-java (`github.com/graphql-java/graphql-java`).

- Another option is to use a standalone server. In this case, instead of embedding the GraphQL functionalities in your code, you will configure an external application that provides data through GraphQL APIs and retrieves it from a data store (such as a SQL database). A couple of popular implementations of such an approach are Apollo (`apollographql.com`) and Hasura (`hasura.io`).

In order to consume and query GraphQL APIs, your best bet is to use a client for your language. There are a number of semi-official implementations for a lot of languages. Due to the protocol being heavily used for web and mobile applications, JavaScript, Android, and iPhone clients are very common. Of course, there are also a couple of libraries for Java, such as graphql-java (seen before for its server capabilities), which can be used as a client too.

In this section, we have seen a number of different technologies in the scope of APIs. We glanced at API technologies, briefly looking at SOAP and REST, and then some modern alternatives, such as gRPC and GraphQL. In the next section, we are going to dig a bit more into the world of data and integration in such a layer.

Introducing data integration

Data integration is a very widespread technique, or rather, consists of a range of techniques.

Under this umbrella terminology, there are a lot of different approaches aiming to consolidate, enrich, filter, and in general work on data, potentially in a range of different formats, to generate different results. Basically, while the integration techniques seen in the *Digging into enterprise integration patterns* section are about transient data (being part of a method call, as a web service, or an asynchronous interaction, such as a message), data integration focuses on data at rest, so when it's persisted on a data store, such as a database or a file. Better again, data integration starts and ends with data persisted (at rest), usually with a big amount of data (such as databases and `.csv` files).

I have to admit that this is not my favorite approach, and I advise against indiscriminate use, especially in greenfield applications. Indeed, data integration can generate a lot of side effects, including stale data (if something goes wrong in the process), an unpredictable amount of time taken to complete the processes, and scalability issues. Moreover, you may end up having less *trust* in the data you deal with, as you may not know who the master is and which data is the most recent or reliable.

Given this warning, I also have to say that, in a more or less structured way, data integration is very widespread in enterprise contexts, especially in the context of data warehouses and batch processing. The more common data integration techniques include the following:

- **Extract, Transform, and Load** (**ETL**): This is a generic term indicating the process of reading data from one or more sources, transforming it (enriching, joining, filtering, and other techniques, more or less similar to what we saw in the *Message transformation* section), and loading it to one or more target storage systems (as a database). This can be done by using specialized software (proprietary or open source) or custom developments (such as SQL queries or custom-written software).

- **Data virtualization**: This is an approach that tries to minimize the downsides of ETL. It basically involves the same steps as ETL but without replicating the data. To do so, the last step (load) is replaced by the virtualization of a target system (usually a database). This means that, instead of loading the data in a target database, there is a *fake* database simulated by the data virtualization technology of choice (which can be an open source or proprietary product). This translates the requests into queries or other ways to collect data from the source systems.

    If it sounds complicated and cumbersome, it's because it is complicated and cumbersome. There can be caching in between (to enhance performance), as the generated queries (or whatever will be needed for collecting data from source systems, which can also be files or other data stores) are usually not so optimized. In general, an approach that can work very well in some scenarios could go awfully in other cases (depending on the source data and the transformations needed).

- **Change data capture**: This is an alternative technique for aligning different data sources. There is a process of listening for changes in a data source and propagating such changes to the systems that are interested in them. The listening for changes is usually specific to each source technology but is commonly done by polling the system (such as with a scheduled query running repeatedly) or by parsing the system metadata (usually the so-called **transaction log**). It is indeed a log maintained by some databases keeping a track of changes. The events detected in this way are then usually propagated in queues (Kafka is particularly widespread as a technology for such use cases). Last but not least, one or more consumers will then listen for some or all the events generated and use them to create a target data store with the desired format.

In this section, we had an overview of data virtualization techniques. In the next section, we will talk about another important piece of enterprise middleware systems, business automation, which includes rules and workflow engines.

Completing the picture with business automation

This section is focused on another big use case of enterprise middleware. While the previous section was about integrating applications with each other by translating data formats and protocols, in this section, we are going to see how to decouple the business logic from the application code.

What do we mean by that? Basically, in each application, there is a part of the behavior that may be subjected to a change periodically. We are not talking about the business logic *as a whole*, but about the specific subsections that are likely known in advance as being required to change due to some specific conditions, such as new business requirements. This kind of logic is usually grouped into two categories:

- **Rules**, which include all kinds of calculations and algorithms that are specific to the business domain and can be isolated, changed, and fine-tuned in the application life cycle. We already introduced the concept of business rules, in *Chapter 3, Common Architecture Design Techniques*, in the *Decision model and notation* section, which is standard notation for business rules.

- **Workflows**, which are modeled around the concept of business processes by mapping a use case as a set of sequential steps. We already introduced the concept of business workflows, in *Chapter 3, Common Architecture Design Techniques*, in the *Business process model and notation* section, which is standard notation for business processes.

Why should you use such separation of logic and implementation in your applications? Well, there are at least two important reasons, which are as follows:

- Having the business logic encapsulated in a rule or workflow will make it quicker, cheaper, and safer to change the business logic itself, in order to fix bugs or adhere to changing external conditions. Depending on the technology you are going to use, it may be supported for a hot reload of the logic, meaning that you can change the behavior of the application with minimal or no downtime. Even if hot reload is not supported, changes in the business logic will still have a very limited impact (such as changing a text file or a database), with minimal consequences on the rest of the application. This means that you can run a smaller set of tests, and the risk of introducing bugs and regressions elsewhere is limited.

- Depending on the language used for the business logic, it can be validated or even directly edited by the business owners (or by a non-technical person anyway). Some technologies for business rules and workflows, such as the aforementioned **Decision Model and Notation** (**DMN**) and **Business Process Model and Notation** (**BPMN**), indeed are basically human-readable as there are tools available to provide a graphical representation of the logic included. Also, the concepts used (such as the task, the item, or the decision table) require no technical knowledge and are intended to have a direct mapping to business concepts.

Business rules, as well as workflows, can be basically deployed in a centralized or embedded way. The considerations about it are similar to the ones that we saw in the integration area:

- When deployed in a centralized way, all your rules or workflows sit on a server (or a cluster of servers), and you interact with them remotely (such as via a REST service call or a message). In this way, everything is organized, and you have a central view and management of all the business artifacts. The downside is, as usual, that this may become a bottleneck and a single point of failure. As the performance slows down, a crash or a maintenance window will impact all the applications dependent on this central component.

- If deployed in an embedded mode, you include the business engine powering workflows and rules in each component that needs such functionalities. Of course, the rules and workflows per se will still be deployed separately (usually being loaded from external text files or artifacts). The embedded mode will allow better scalability as each component will have full control over the decision capabilities. On the other hand, you will lack central administration and governance capabilities.

Let's now have a look at those two technologies in detail and learn when you should use them in your application.

Business rules

Business rules are a way to express and isolate logic from the implementation, or better, the algorithm that ultimately leads to a decision from the technical details behind it. The examples here are very different scenarios. A common one is the concept of promotions in an e-commerce environment: the products, the price, and the rest of the behavior stay the same, but you may change the amount of discount calculated based on the time of year, the number of items in stock, or simply new requirements coming from the business.

Another widespread use of business rules is regarding anti-fraud. Basically, before accepting a payment request, you make several checks to ensure that said payment is not a fraudulent one. The number and type of checks you perform may vary with time, as you may discover more fraudulent cases and add other controls in order to detect them. In the next section, we will extend the concept of business rules by introducing the temporal dimension. This concept is called **Complex Event Processing** (**CEP**).

Complex event processing

CEP is a concept related to business rules. The most widely accepted distinction between business rule processing and CEP is that in CEP, the concept of time and event flow is the core of the computation.

With CEP, each decision can be influenced by previous events that occurred, both in a time window (such as in the last hour) or an event window (such as in the last 10 events).

Let's get back to our anti-fraud example. A business rule checking for fraud will make a decision based on the data related to the specific incoming payment, such as the amount or the profile of the sender. CEP-based checking for fraud will add the temporal dimension to it, so the evaluated information will include past payment transactions. You may want to check whether any of the last 10 transactions have been suspected of fraud, or you may want to check whether, in the last hour, other payment transactions have occurred in a very distant location (such as a different country).

Now that we have cleared the basics of business rules and CEP, let's have a look at the Drools project, which is a very widespread implementation of such concepts.

The Drools project

Drools is a widespread open source business rules engine, created in the early 2000s. It is part of a broader community called **Knowledge Is Everything** (**KIE**), which included related features, such as a workflow manager implementation (**jBPM**—more about that in a few sections) and the related tooling (such as rules modeling and graphical interfaces). Drools ships a lot of interesting capabilities, such as great performance, a small footprint, and compatibility with different rule languages, including DRL and the already mentioned DMN. Moreover, Drools can be deployed in various configurations, including embedded and server mode (supporting a number of different runtimes, including Quarkus).

The decision model and notation

As already mentioned in *Chapter 3*, *Common Architecture Design Techniques*, DMN is a language for modeling business decisions in a way that's understandable to both technical and non-technical people. The language is based on XML, so while it is text-based (hence easily versionable in a source code repository), it's hardly human-readable because it contains all the properties and coordinates for visualizing the components onscreen. However, there are plenty of free and commercial tools to edit and visualize such files. You can download some from the Kogito tooling page (`github.com/kiegroup/kogito-tooling`), or you can have the same experience online for free (at the `dmn.new` page).

This is what a simple `hello world` rule looks like in the editor:

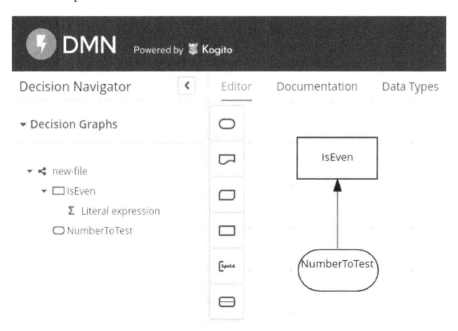

Figure 8.25 – A simple DMN rule

This rule will check whether a number in input is even, returning `Yes` or `No`.

The rounded component is a `DMN Input Data` element, which contains the `NumberToTest` input variable, while the `Is Even ?` rectangle component is a `DMN Decision` containing the algorithm. In this case, if we look in the panel on the left, the component contains a so-called **literal expression**. By editing the component, we can see this expression:

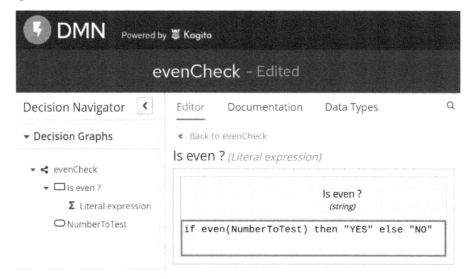

Figure 8.26 – A simple DMN rule

Of course, this is a very simple example expression. In a real-world application, you could have much more complex logic, such as a combination of different expressions, a function, and a decision table. The decision would likely have other components, such as different input, complex types, and reusable decisions.

Let's now extend our reasoning with business workflows.

Business workflows

Business workflows can be seen, conceptually, as an extension of the concept of business rules. By using workflows, indeed, you aim to logically separate the business logic from the implementation and application logic.

While with rules you isolate a decision (which may be simple or complex at will), with workflows you model an entire process. You will still start from a set of information and arrive at a final outcome, but the process will involve a number of different steps and it will usually be *passivated* every now and then (usually on a database) while waiting for each step to complete.

The steps can be fully automated (such as when calling a REST service or sending a message to a queue or similar things), or simply represent a human task happening outside of our application (such as the signature on a paper document or other manual tasks) and need an explicit confirmation (which may be via email or completing a web form) in order to signal the completion (and let the process continue). So, to keep it simple, while a business rules model represents a single calculation, a business process involves a set of different steps and each decision may go through a set of different paths.

Indeed, one core function of workflows, other than modeling and isolating business processes, is to give insights into the process performance and statistics from both a business and a technical point of view.

Let's suppose that we implement a business process to represent a loan request. You will have a first step for getting the data for the request, including the amount of money requested, the name of the requestor, and their age. You will then most likely have a set of validation, such as a background check of the requestor, verification of their salary, and the history of the payments made by the requestor. Each of these steps can be modeled as an item in a workflow, which can be completed automatically (such as calling an external system asking for information) or by asking an operator (such as sending an email and waiting for the reply). It's a common practice to model some of those steps as business rules, according to what we have seen so far in this chapter.

Let's suppose you have such a system in production. You can then extract the historical data from the past process instances and understand how well your workflow is performing. *How many loan requests get approved at the end? How much time do you spend on each task? What's the average age of each requestor?*

From this valuable data, the business can get insights to change the process (such as simplifying, changing, or removing some steps), to create different promos (such as a special loan with different interest rates for a specific audience), and so on. This is basically the reason why you want to isolate decisions (whether rules or processes). You now can easily know what's happening in your application and fine-tune such behavior while having a limited impact on the other functionalities.

The most widespread approach to technically model such business processes is to rely on the BPMN notation, which is a standard. Also, as DMN is based on XML, it is human-readable and editable by using graphical tools. For more information on BPMN, please refer to *Chapter 3, Common Architecture Design Techniques*.

The jBPM project

jBPM is another project under the KIE umbrella, providing a lightweight and extensible engine for running BPMN workflows. Similar to Drools, jBPM can be deployed in many different ways, including embedded in your applications and standalone, and it can rely on many different runtimes (such as **JBoss WildFly** and Quarkus), implementing both traditional and cloud-native scenarios.

As per the Drools project, jBPM provides some free tools to visualize and edit BPMN files at `github.com/kiegroup/kogito-tooling` and you can use the online editor on the `bpmn.new` page.

This is what a simple workflow looks like:

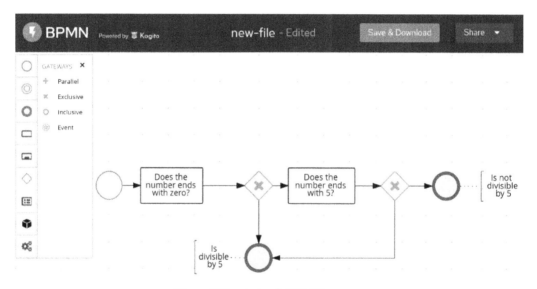

Figure 8.27 – A simple BPMN process

In this simple process, we check the divisibility of a number by 5. There is a start, then a couple of checks (whether the number ends with zero or five), and a logic gateway that leads to an end in both cases: is or is not divisible.

This section completes our overview of business automation. In the next section, we will compare the architectural roles of integration and automation.

Integration versus automation – where to draw the line

A common discussion when designing complex software architectures is where to define the boundary between integration and automation.

After all, there is a bit of overlap: both a business workflow and an integration route can call a number of external systems sequentially or while going through conditions (which may be represented, in both cases, as business rules).

Of course, there is not a fixed answer for every behavior. I personally prefer to avoid polluting the business automation with too many technical integrations (such as connectors for specific uncommon technologies, everything that is not a call to a web service or a message in a queue) and the integration routes with conditions that are dependent on specific business requirements (such as modeling a business process as an integration route). But other than this high-level, common-sense advice, there are a few considerations that can help in understanding whether a particular feature should stay in the automation or integration layer:

- If the flow that represents our use case involves a significant number of human tasks (such as when human interaction is needed), it will likely be a business workflow.

- If you have to deal with technologically driven behavior, such as retries in the case of errors in service calls or other details about the protocols used by the external system, it is most likely something to encapsulate in an integration route.

- If the process has business relevance, so every step is something significant to a business person, who could be interested in the performance of the process (meaning how many processes are stuck in a step, or how much time is needed for a particular path), it is likely to be a business workflow.

- If no passivation is needed, meaning that the majority of instances are straight-through processes, or in other words, a set of steps performed one after the other without needing to be persisted in a data store (waiting for a signal or other events to restart), it is likely an integration route.

That is my personal advice on how to consider whether a particular feature should be in an integration or business automation layer. Indeed, in many cases, you will need a combination of both layers:

- An integration layer is used to encapsulate technical details into interactions with third-party and other external systems (such as databases) and expose such functionalities as a higher-level, composite API (such as a REST service).

- A workflow layer orchestrates the calls to those high-level APIs, adds human tasks (if needed) and processes instance persistence, and in general models business processes and gets metrics and insights on such process execution.

This completes our overview of business automation. In the next section, we will have a look at examples of integration and automation using the aforementioned technologies.

Case studies and examples

In this section, we are going to go on with our payment use case, to see some examples of integration and business automation.

For example purposes, we will use Camel, jBPM, and Drools. Our target runtime will be Quarkus, which we already saw in the previous chapter.

But many of the concepts and implementations are applicable to other runtimes, such as embedded ones (as in using the runtime as a dependency of your Java application), deployed on JBoss WildFly, and deployed on Spring Boot.

Integrating payment capabilities

Our first use case to implement integration is the connection of payment capabilities with a legacy backend. Let's suppose that we have developed our microservices payment application and it is working correctly. A new business requirement is to integrate a legacy platform for settlement purposes (which is a kind of accounting operation done after payments).

It's fairly easy for our core application to call a REST service for this purpose, but the legacy system used for settlement works with .xml files placed in a shared folder. That's a perfect fit for integration: there is no business logic apart from some plumbing to make the two systems talk to each other, and it will be fairly easy to implement with a rich and expressive framework such as Camel.

There are a number of ways to create a Camel project on top of Quarkus. As we saw in the previous chapter, we can create it with the mvn command, or go to the Quarkus website and use a web wizard to download an empty project scaffold.

The dependencies that we are going to use are in the camel-quarkus family. Here are the details:

- camel-quarkus-platform-http is basically a bridge to make the existing HTTP server of the runtime (Quarkus, in our case) usable from Camel.

- `camel-quarkus-jackson` is the component to marshal POJO to JSON and vice versa.

- `camel-quarkus-jacksonxml` works on the same concept, but for XML serialization. As we have seen, `Jackson` is a library that can be used for JSON, XML, and other formats.

- `camel-quarkus-file` is the default component for reading and writing files.

We have already created a `Payment` Java class for holding our payment data, in the previous chapter. You can see the fields used in the following code (*the rest of the class is getters and setters plus some more boilerplate code*):

```java
public class Payment {
    private String id;
    private Date date;
    private String currency;
    private String sender;
    private String recipient;
    private String signature;
    private float amount;
...
```

In order to define a Camel integration route using Java DSL, it's enough to create a class extending `EndpointRouteBuilder`, as follows (*imports are omitted here*):

```java
@ApplicationScoped
public class PaymentSettlement extends EndpointRouteBuilder
{
    @Override
    public void configure() throws Exception {
        from(platformHttp("/camel/settlement"))
        .unmarshal()
        .json(JsonLibrary.Jackson, Payment.class)
        .setHeader("PaymentID", simple("${body.id}"))
        .marshal()
        .jacksonxml()
        .to(file("{{settlement.path}}?fileName=
          ${header.PaymentID}.xml"));
```

```
        }
    }
```

This simple Java code models an integration route that starts by exposing an HTTP endpoint, unmarshals the requests coming as JSON objects (using the `Jackson` framework), mapping it to a Java object of the `Payment` class, sets a header, then marshals the Java object to XML (using `Jackson` again), and finally writes the XML to a file.

To call the Camel route, we have to post a request to `/camel/settlement` (`http://127.0.0.1:8080/camel/settlement`, if we are running locally), which is a JSON representation of the `Payment` object (as seen in the previous chapter). Here's an example:

```
{
    "id":"1ef43029-f1eb-4dd8-85c4-1c332b69173c",
    "date":1616504158091,
    "currency":"EUR",
    "sender":"giuseppe@test.it",
    "recipient":"stefano@domain.com",
    "signature":"169e8dbf-90b0-4b45-b0f9-97789d66dee7",
    "amount":10.0
}
```

Regarding the Camel route, we already saw a similar flow a couple of sections ago, in the *The Camel DSL* section, following our first look at the Camel framework, in the *The Camel integration framework* section. However, there are a couple of things worth noticing:

- There is a `setHeader` method, which locates the `id` field from the body of the current message flowing through the Camel route (which is the ID of the payment transaction) and sets it into the `PaymentId` header so it can be reused later as a name for the `.xml` file that we are generating. Note that the `simple` expression language is used, which can be used for navigating the payload (using the dot notation) and express conditions.

- The Quarkus properties (defined according to what we saw in the previous chapter, using the system properties defined in `application.properties` or in many other ways) are directly accessed and used with double curly brackets. In this case, the file component accesses the `{{settlement.path}}` variable to set the destination path of the settlement file.

But what if we want to avoid the generation of settlement for an amount of less than 10 €? That's easy. It is enough to implement the filter EIP and basically drop the messages that do not respect the relevant condition:

```
from(platformHttp("/camel/settlement"))
        .unmarshal()
        .json(JsonLibrary.Jackson, Payment.class)
        .setHeader("PaymentID", simple("${body.id}"))
        .setHeader("Amount", simple("${body.amount}"))
        .marshal()
        .jacksonxml()
            .filter(simple("${header.amount} > 10"))
            .to(file("{{settlement.path}}?fileName=
                ${header.PaymentID}.xml"));
```

As you can see, the Camel component is indeed called `filter`.

And what if we want to add a different behavior for the two conditions (less than 10 € or more than 10 €)? The EIP here is a content-based router, which can be implemented in Camel using the `choice` component, like this:

```
from(platformHttp("/camel/settlement"))
        .unmarshal()
        .json(JsonLibrary.Jackson, Payment.class)
        .setHeader("PaymentID", simple("${body.id}"))
        .setHeader("Amount", simple("${body.amount}"))
        .marshal()
        .jacksonxml()
            .choice()
                .when(simple("${header.amount} > 10"))
                    .to(file("{{settlement.path}}?fileName=
                        ${header.PaymentID}.xml"))
                .otherwise()
                    .log("No settlement needed")
        .end();
```

In this case, we are simply logging in the `otherwise` case, but in the real world, you may consider doing something more (such as writing a different file format or in different storage) and you can also add a number of different `when` conditions.

More generally, this is just a taste of what Camel can do. In the real world, there are a lot of technology connectors, formats, and patterns, and the routes can be modularized to call each other. On the official Camel website, you can even find a page about mapping between EIP and Camel components. This completes our integration example. Let's now have a look at automation.

Automating customer onboarding

Our payment platform will for sure need a process to onboard customers. It's pretty common to have some actions supporting the creation of profiles for new customers, including validations and the provisioning of customers on many different systems. This is a task that is a perfect fit for business automation.

Indeed, customer onboarding is commonly driven by (changing) business requirements (such as the number of steps for registering a user and facilitating the onboarding of some categories for promotional purposes). Moreover, these kinds of processes may be regulated by laws, and so it may happen that you have different workflows in different geolocations (such as some countries requiring wet signatures on paper needing to be modeled as human tasks) and changing steps with changing regulations over time.

Last but not least, the process of provisioning a user is typically stateful: you will want to have it persisted on a data store for auditing, reporting, and customer experience purposes. It may happen that a customer starts the registration process on a mobile app, then the user continues doing other steps on a computer, then lastly, finalizing it by going to a bank branch.

A workflow will allow this kind of experience (also known as *omnichannel*) by persisting each step (where needed) and providing a stateful process. Let's start with modeling one single step of the workflow: a business rule modeling the age check of a customer. That's quite easy with DMN and the visual editor provided by the **Kogito** project (maybe the online one, or a standalone version, as a plugin for an IDE, such as **VSCode**):

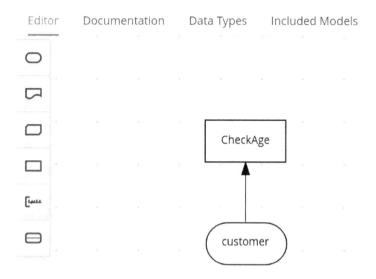

Figure 8.28 – A simple DMN validation rule

This is a very simple rule: using the **customer** data structure and providing a **CheckAge** DMN decision. Here is what's inside such a decision:

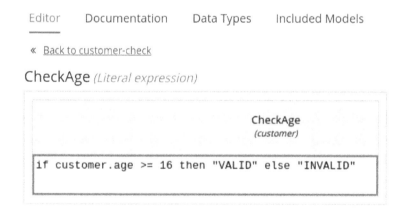

Figure 8.29 – The rule expression

In this example, we are using a very simple **Friendly Enough Expression Language (FEEL)** expression, checking on the age field of the customer structure. Here is what the structure looks like in the **Data Types** editor:

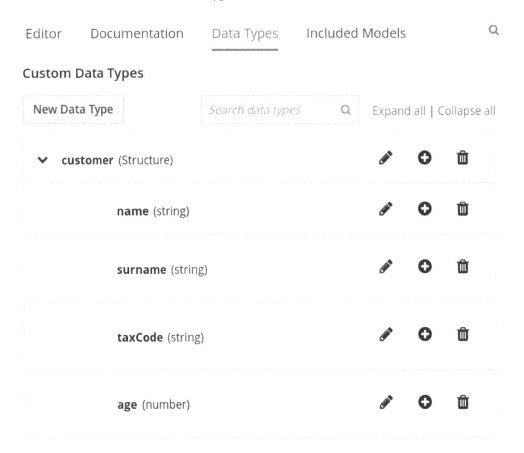

Figure 8.30 – The customer data type

In the sources of this example, you can also find the POJO representing the same structure (and it is interoperable with it). In order to invoke this rule, we need to post this REST request in JSON:

```
POST http://localhost:8080/customer-check HTTP/1.1
content-type: application/json

{
    "customer":{
        "name":"Giuseppe",
```

```
        "surname":"Bonocore",
        "age":37,
        "taxCode":"dads213213fasfasf"
    }
}
```

Here, the Kogito engine will reply with VALID or INVALID. This is, of course, pretty useful: you can easily create decision services providing business rules (usually more complex than the one seen in this example) and use them in your project. But there is more: this rule can become one step in a more complex workflow.

Let's imagine a very simple prototype of a customer onboarding process: you have the start of the process, a preparation step (which may include sending a request to a CRM or other systems), and the evaluation of the age of the customer (by using the DMN rule that we have just seen). If the age is INVALID, you may want to have some special handling (such as asking for the permission of a parent). We did exactly that in this simple BPMN workflow:

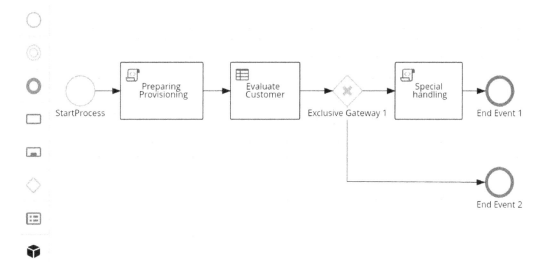

Figure 8.31 – The customer data type

In this example, the **Evaluate Customer** step is using our DMN rule, while the other steps, for example, purposes, are simple *script tasks* printing the information on the console (using a `System.out` call). Of course, you may want to have different kinds of tasks here, such as REST calls to external services, sending messages, or human tasks. Whatever your implementation is, you can trigger the workflow with a REST request, such as the following:

```
POST http://localhost:8080/customer_onboarding HTTP/1.1
content-type: application/json

{
    "customer":{
        "name":"Giuseppe",
        "surname":"Bonocore",
        "age":37,
        "taxCode":"dads213213fasfasf"
    }
}
```

Take into account that in this simple example, we have not configured a persistence layer, nor defined tasks requiring passivation of the process instance. If this is needed, you can easily do it by adding some configurations (such as in the `application.properties` file). Once you have a process requiring and using persistence, you can then query the Kogito engine, asking for the status of encapsulated (or even completed) process instances, the list of pending tasks, and so on (like you could do in a typical BPMN workflow engine). This completes our examples for this chapter.

Summary

In this chapter, we have looked at a lot of technologies, completing the middleware overview that we started in the last chapter. You have also learned what an ESB is (including connectors, patterns, and data formats).

We have looked at the enterprise integration patterns and the Camel library, which is an implementation of enterprise integration patterns. We have also looked at messaging systems to support the concept of integration in asynchronous scenarios. We then shifted our view of process automation by digging into business rules and business workflows and having a glimpse at Kogito, which is a complete business automation engine running on Quarkus.

After this chapter, you should be able to understand the basics of enterprise integration, including messaging capabilities. We have also seen what business automation is, including workflows and rules, and how to differentiate what should stay in an integration layer from what should stay in a business automation layer. By using some open source libraries, we have gone through a couple of examples of implementing these concepts in Java.

In the next chapter, we will see how to design and implement a modern distributed application by applying cloud-native architecture recommended practices.

Further reading

- The arc42 official website: `https://arc42.org/`
- Gregor Hohpe, Bobby Woolf, Enterprise Integration Patterns (`www.enterpriseintegrationpatterns.com`)
- *Enterprise Integration Patterns*, by Gregor Hohpe and Bobby Woolf, published by Pearson Education (2012)
- Apache Software Foundation: The Apache Camel project (`camel.apache.org`)
- Apache Software Foundation: The Apache Camel project – mapping to EIP (`camel.apache.org/components/latest/eips/enterprise-integration-patterns.html`)
- The official XML website (`XML.org`)
- The official JSON website (`JSON.org`)
- The Google Protobuf Java tutorial (`developers.google.com/protocol-buffers/docs/javatutorial`)
- The official gRPC website (`grpc.io`)
- The KIE project, including Drools, jBPM, and more (`www.kiegroup.org`)
- The Kogito project, providing business automation on Quarkus (`kogito.kie.org`)

9
Designing Cloud-Native Architectures

Nowadays, the microservices architectural model is mainstream. At the time of writing, we are likely in the *Trough of Disillusionment*. This widespread terminology comes from the Gartner Hype Cycle model and is a way of identifying phases in the adoption of technology, starting from bleeding edge and immaturity and basically going through to commodity.

This means, in my opinion, that even if we are starting to recognize some disadvantages, microservices are here to stay. However, in this chapter, I would like to broaden the point of view and look at the so-called cloud-native architectures. Don't get confused by the term *cloud*, as these kinds of architectures don't necessarily require a public cloud to run (even if one cloud, or better, many clouds, is the natural environment for this kind of application).

A cloud-native architecture is a way to build resistant, scalable infrastructure able to manage traffic peaks with little to no impact and to quickly evolve and add new features by following an Agile model such as **DevOps**. However, a cloud-native architecture is inherently complex, as it requires heavy decentralization, which is not a matter that can be treated lightly. In this chapter, we will see some concepts regarding the design and implementation of cloud-native architectures.

In this chapter, you will learn about the following topics:

- Why create cloud-native applications?
- Learning about types of cloud service models
- Defining twelve-factor applications
- Well-known issues in the cloud-native world
- Adopting microservices and evolving existing applications
- Going beyond microservices
- Refactoring apps as microservices and serverless

That's a lot of interesting stuff, and we will see that many of these concepts can help improve the quality of the applications and services you build, even if you are in a more traditional, less cloud-oriented setup. I am sure that everybody reading this chapter already has an idea, maybe a detailed idea, of what a cloud-native application is. However, after reading this chapter, this idea will become more and more structured and complete.

So, to start the chapter, we are going to better elaborate on what a cloud-native application is, what the benefits are, and some tools and principles that will help us in building one (and achieving such benefits).

Why create cloud-native applications?

Undoubtedly, when defining cloud-native, there are a number of different nuances and perspectives that tackle the issue from different perspectives and with different levels of detail, ranging from technical implications to organizational and business impacts.

However, in my opinion, a cloud-native application (or architecture, if you want to think in broader terms) must be designed to essentially achieve three main goals (somewhat interrelated):

- **Scalability** is, of course, immediately related to being able to absorb a higher load (usually because of more users coming to our service) with no disruption and that's a fundamental aspect. However, scalability also means, in broader terms, that the application must be able to downscale (hence, reducing costs) when less traffic is expected, and this may imply the ability to run with minimal or no changes in code on top of different environments (such as on-premises and the public cloud, which may be provided by different vendors).

- **Modularity** is about the application being organized in self-contained, modular components, which must be able to interoperate with each other and be replaced by other components wherever needed. This has huge impacts on the other two points (scalability and resiliency). A modular application is likely to be scalable (as you can increase the number of instances of a certain module that is suffering from a load) and can be easily set up on different infrastructures. Also, it may be using the different backing services provided by each infrastructure (such as a specific database or filesystem), thereby increasing both scalability and resiliency.

- **Resiliency**, as we just mentioned, can be defined as being able to cope with unpredicted events. Such events include application crashes, bugs in the code, external (or backing) services misbehaving, or infrastructure/hardware failures. A cloud-native application must be designed in a way that avoids or minimizes the impact of such issues on the user experience. There are a number of ways to address those scenarios and we will see some in this section. Resiliency includes being able to cope with unforeseen traffic spikes, hence it is related to scalability. Also, as has been said, resiliency can also be improved by structuring the overall application in modular subsystems, optionally running in multiple infrastructures (minimizing single-point-of-failure problems).

As a cloud-native architect, it is very important to see the business benefits provided by such (preceding) characteristics. Here are the most obvious ones, for each point:

- **Scalability** implies that a system can behave better under stress (possibly with no impacts), but also has predictable costs, meaning a higher cost when more traffic comes and a lower one when it's not needed. In other words, the cost model scales together with the requests coming into the system. Eventually, a scalable application will cost less than a non-scalable one.

- **Modularity** will positively impact system maintainability, meaning reduced costs as regards changes and the evolution of the system. Moreover, a properly modularized system will facilitate the development process, reducing the time needed for releasing fixes and new features in production, and likely reducing the time to market. Last but not least, a modular system is easier to test.

- **Resiliency** means a higher level of availability and performance of the service, which in turn means happier customers, a better product, and overall, a better quality of the user experience.

While cloud-native is a broad term, implying a big number of technical characteristics, benefits, and technological impacts, I think that the points that we have just seen nicely summarize the core principles behind the cloud-native concept.

Now, it's hard to give a perfect recipe to achieve each of those goals (and benefits). However, in this chapter, we are going to link each point with some suggestions on how to achieve it:

- **PaaS** is an infrastructural paradigm providing support services, among other benefits, for building scalable infrastructures.

- The **twelve-factor apps** are a set of principles that assist in building modular applications.

- **Cloud-native patterns** are a well-known methodology (also implemented in the MicroProfile specification) for building resilient applications.

In the next section, we are going to define what PaaS is and how our application is going to benefit from it.

Learning about types of cloud service models

Nowadays, it is common to refer to modern, cloud-native architectures by means of a number of different terms and acronyms. The *as a service* phrase is commonly used, meaning that every resource should be created and disposed of on-demand, automatically. *Everything as a service* is a wider term for this kind of approach. Indeed, with cloud computing and microservices, applications can use the resources of a swarm (or a cloud, if you want) of smaller components cooperating in a network.

However, such architectures are hard to design and maintain because, in the real world, the network is basically considered unreliable or at least has non-predictable performances. Even if the network behaves correctly, you will still end up with a lot of *moving parts* to develop and manage in order to provide core features, such as deploying and scaling. A common tool for addressing those issues is PaaS.

PaaS is an inflated term, or, better yet, every *as a service* term is overused, and sometimes there is no exact agreement and definition of the meaning and the boundaries between each *as a service* set of tools. This is my personal view regarding a small set of *as a service* layers (that can be regarded as common sense, and indeed is widely adopted):

- **Infrastructure as a Service** (**IaaS**) refers to a layer providing *on-demand* computational resources needed to run workloads. This implies **Virtual Machines** (**VMs**) (or physical servers) can network to make them communicate and store persistent data. It does not include anything above the OS; once you get your machines, you will have to install everything needed by your applications (such as a **Java Virtual Machine** (**JVM**), application servers, and dependencies).

- **Platform as a Service** (**PaaS**) is used for a layer that abstracts most of the infrastructure details and provides services that are useful for the application to be built and run. So, in PaaS, you can specify the runtimes (VM, dependencies, and servers) needed by your application and the platform will provide it for you to use.

  PaaS could also abstract other concepts for you, such as storage (by providing object storage, or other storage services for you to use), security, serverless, and build facilities (such as CI/CD). Last but not least, PaaS provides tools for supporting the upscale and downscale of the hosted applications. Most PaaS platforms provide their own CLIs, web user interfaces, and REST web services to provision, configure, and access each subsystem. PaaS, in other words, is a platform aiming to simplify the usage of an infrastructural layer to devs and ops. One common way of implementing PaaS is based on containers as a way to provision and present a runtime service to developers.

- **Software as a Service** (**SaaS**) is one layer up from PaaS. It is mostly targeted at final users more than developers and implies that the platform provides, on-demand, applications ready to use, completely abstracting the underlying infrastructure and implementation (usually behind an API). However, while the applications can be complex software suites, ready for users to access (such as office suites or webmail), they can also be specific services (such as security, image recognition, or reporting services) that can be used and embedded by developers into more complex applications (usually via API calls).

The following diagram shows you a comparison of **IaaS**, **PaaS**, and **SaaS**:

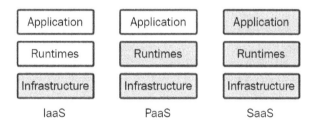

Figure 9.1 – IaaS versus PaaS versus SaaS

Now, we have a definition of boundaries between some *as a service* layers. We should get back to our initial thoughts, *how is PaaS a good way to support a heavily distributed, cloud-native architecture such as "the network is the computer"?*

PaaS simplifies the access to the underlying computing resources by providing a uniform packaging and delivering model (usually, by using containers). It orchestrates such components by deploying them, scaling them (both up and down), and trying to maintain a service level wherever possible (such as restarting a faulty component). It gives a set of administration tools, regardless of the technology used inside each component. Those tools address features such as log collection, metrics and observabilities, and configuration management. Nowadays, the most widely used tool for orchestration is **Kubernetes**.

Introducing containers and Kubernetes

Container technology has a longstanding history. It became popular around 2013 with the Docker implementation, but initial concepts have their roots in the Linux distributions well before then (such as **Linux Containers** (**LXC**), launched around 2008). Even there, concepts were already looking very similar to the modern containers that can be found in older systems and implementations (**Solaris** zones are often mentioned in this regard).

We could fill a whole book just talking about containers and Kubernetes, but for the sake of simplicity and space, we will just touch on the most important concepts useful for our overview on defining and implementing cloud-native architectures, which is the main goal of this book. First things first, let's start with what a container is, simplified and explained to people with a development background.

Defining containers and why they are important

In a single sentence, a container is a way to use a set of technologies in order to fool an application into thinking it has a full machine at its disposal.

To explain this a bit better, containers wrap a set of concepts and features, usually based on Linux technology (such as `runc` and `cgroups`), which are used to isolate and limit a process to make it play nicely with other processes sharing the same computational power (physical hardware or VMs).

To achieve those goals, the container technology has to deal with the assignment and management of computing resources, such as networking (ports, IP addresses, and more), CPU, filesystems, and storage. The supporting technology can create *fake* resources, mapping them to the real ones offered by the underlying resources. This means that a container may think to expose a service on port `80`, but in reality, such a service is bound to a different port on the host system, or it can think to access the root filesystem, but in reality, is confined to a well-defined folder.

In this way, it's the container technology that administers and partitions the resources and avoids conflicts between different applications running and competing for the same objects (such as network ports). But that's just one part of the story: to achieve this goal, our application must be packaged in a standard way, which is commonly a file specifying all the components and resources needed by our application to run.

Once we create our container (starting from such a descriptor), the result is an immutable container image (which is a binary runtime that can also be signed for integrity and security purposes). A container runtime can then take our container image and execute it. This allows container applications to do the following:

- **Maximize portability**: Our app will run where a compatible runtime is executed, regardless of the underlying OS version, or irrespective of whether the resources are provided by a physical server, a VM, or a cloud instance.

- **Reduce moving parts**: Anything tested in a development environment will look very similar to what will be in production.

- **Isolate configurations from executable code**: The configurations will need to be external and injected into our immutable runtime image.

- **Describe all the components**: You should nicely describe all the components of our application, for both documentation purposes and inspection (so you can easily understand, for example, the patch level of all of your Java machines).

- **Unify packaging, deployment, and management**: Once you have defined your container technology, you can package your applications and they will be managed (started, stopped, scaled, and more) all in the same way regardless of the internal language and technologies used.

- **Reduce the footprint**: While you could achieve most of the advantages with a VM, a container is typically way lighter (because it will carry only what's needed by the specific application, and not a full-fledged OS). For such a reason, you can run more applications using the same number of resources.

Those are more or less the reasons why container technology became so popular. While some of those are specific to infrastructural aspects, the advantages for developers are evident: think how this will simplify, as an example, the creation of a complete test or dev environment in which every component is containerized and running on the correct version (maybe a production one because you are troubleshooting or testing a fix).

So far so good: containers work well and are a nice tool for building modern applications. *What's the warning here?* The point is, if you are in a local environment (or a small testing infrastructure), you can think of managing all the containers manually (or using some scripts), such as provisioning it on a few servers and assigning the configurations required. *But what will happen when you start working with containers at scale?* You will need to worry about running, scaling, securing, moving, connecting, and much more, for hundreds or thousands of containers. This is something that for sure is not possible to do manually. You will need an orchestrator, which does exactly that. The standard orchestrator for containers today is Kubernetes.

Orchestrating containers with Kubernetes

Kubernetes (occasionally shortened to **K8s**) is, at the time of writing, the core of many PaaS implementations. As will become clear at the end of this section, Kubernetes offers critical supporting services to container-based applications. It originated from work by Google (originally known as *Borg*) aimed at orchestrating containers providing most of the production services of the company. The Kubernetes operating model is sometimes referred to as declarative. This means that Kubernetes administrators define the target status of the system (such as *I want two instances of this specific application running*) and Kubernetes will take care of it (as an example, creating a new instance if one has failed).

Following its initial inception at Google, Kubernetes was then released as an open source project and is currently being actively developed by a heterogeneous community of developers, both enterprise sponsored and independent, under the Cloud Native Computing Foundation umbrella.

Kubernetes basic objects

Kubernetes provides a set of objects, used to define and administer how the applications run on top of it. Here is a list of these objects:

- A Pod is the most basic unit in a Kubernetes cluster, including at least one container (more than one container is allowed for some specific use cases). Each **Pod** has an assigned set of resources (such as CPU and memory) and can be imagined as a **Container** plus some metadata (including network resources, application configurations, and storage definitions). Here is a diagram for illustration:

Pod

Figure 9.2 – A Pod

- Namespaces are how Kubernetes organizes all the other resources and avoids overlaps. In that sense, they can be intended as *projects*. Of course, it's possible to restrict access for users to specific namespaces. Commonly, namespaces are used to group containers belonging to the same application and to define different environments (such as dev, test, and prod) in the same Kubernetes cluster.

- Services are network load balancers and DNS servers provided by the Kubernetes cluster. Behind a **Service**, there are a number of Pods answering incoming requests. In this way, each **Pod** can access functions exposed from other Pods, thereby circumventing accessing such Pods directly via their internal IP (which is considered a bad and unreliable way). Services are, by default, internal, but they could be exposed and accessed from outside the Kubernetes cluster (by using other Kubernetes objects and configurations that aren't covered here). The following diagram illustrates the structure of a Service:

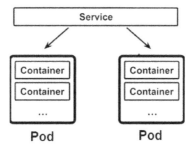

Figure 9.3 – A Service

- Volumes are a means for Kubernetes to define access to persistent storage to be provided to the Pods. Containers do indeed use, by default, ephemeral storage. If you want a container to have a different kind of storage assigned, you have to deal with volumes. The storage (like many other aspects) is managed by Kubernetes in a pluggable way, meaning that behind a volume definition, many implementations can be attached (such as different storage resources provided by cloud services providers or hardware vendors).

- ConfigMaps and Secrets are the standard way, in Kubernetes, of providing configurations to Pods. They are basically used to inject application properties (such as database URLs and user credentials) without needing an application rebuild. Secrets are essentially the same idea but are supposed to be used for confidential information (such as passwords).

  By default, Secrets are strings encoded in Base64 (and so are not really secure), but they can be encrypted in various ways. ConfigMaps and Secrets can be consumed by the application as environment variables or property files.

- ReplicaSet, StatefulSet, and DaemonSet are objects that define the way each Pod should be run. ReplicaSet defines the number of instances (replicas) of the Pods to be running at any given time. StatefulSet is a way to define the ordering in which a given set of Pods should be started or the fact that a Pod should have only one instance running at a time. For this reason, they are useful for running stateful applications (such as databases) that often have these kinds of requirements.

  DaemonSet, instead, is used to ensure that a given Pod has an instance running in each server of the Kubernetes cluster (more on this in the next section). DaemonSet is useful for some particular use cases, such as monitoring agents or other infrastructural support services.

- The Deployment is a concept related to ReplicaSet and Pods. A Deployment is a way to deploy ReplicaSets and Pods by defining the intermediate steps and strategy to perform Deployments, such as rolling releases and rollbacks. Deployments are useful for automating the release process and reducing the risk of human error during such processes.

- Labels are the means that Kubernetes uses to identify and select basically every object. In Kubernetes, indeed, it is possible to tag everything with a label and use that as a way to query the cluster for objects identified by it. This is used by both administrators (such as to group and organize applications) and the system itself (as a way to link objects to other objects). As an example, Pods that are load-balanced to respond to a specific service are identified using labels.

Now that we have had a glimpse into Kubernetes' basic concepts (and a glossary), let's have a look at the Kubernetes architecture.

The Kubernetes architecture

Kubernetes is practically a set of Linux machines, with different services installed, that play different roles in the cluster. It may include some Windows servers, specifically for the purpose of running Windows workloads (such as .NET applications).

There are two basic server roles for Kubernetes:

- **Masters** are the nodes that coordinate with the entire cluster and provide the administration features, such as managing applications and configurations.

- **Workers** are the nodes running the entire applications.

Let's see a bit about what's in each server role. We'll start with the **master node components**.

Master nodes, as said, carry out the role of coordinating workloads across the whole cluster. To do so, these are the services that a master node commonly runs:

- `etcd` **server**: One of the most important components running on a master is the `etcd` server. `etcd` is a stateful component, more specifically, a key-value database. Kubernetes uses it to store the configuration of the cluster (such as the definition of Pods, Services, and Deployments), basically representing the desired state of the entire cluster (remember that Kubernetes works in a declarative way). `etcd` is particularly suited to such needs because it has quite good performance and works well in a distributed setup (optionally switching to a read-only state if something bad happens to it, thereby allowing limited operativity even under extreme conditions, such as a server crash or a network split).

- **API server**: To interact with `etcd`, Kubernetes provides an API server. Most of the actions that happen in Kubernetes (such as administration, configuration checks, and the reporting of state) are done through calls to the API server. Such calls are basically JSON via HTTP.

- **Scheduler**: This is the component that handles selecting the right worker to execute a Pod. To do so, it can handle the basic requirements (such as the first worker with enough resources or lower loads) or advanced, custom-configured policies (such as anti-affinity and data locality).

- **Controller manager**: This is a set of processes implementing the Kubernetes logic that we have described so far, meaning that it continuously checks the status of the cluster against the desired status (defined in `etcd`) and operates the requisite changes if needed. So, for example, if a **Pod** is running fewer instances than what is configured, the **Controller** manager creates the missing instances, as shown here:

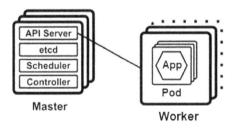

Figure 9.4 – Master and worker nodes

This set of components, running into master nodes, is commonly referred to as the **control plane**. In production environments, it's suggested to run the master nodes in a high-availability setup, usually in three copies (on the basis of the `etcd` requirements for high availability).

As is common in a master/slave setup, the master nodes are considered to be precious resources configured in a high-availability setup. Everything that is reasonably possible should be done to keep the master nodes running and healthy, as there can be unforeseen effects on the Kubernetes cluster in case of a failure (especially if all the master instances fail at the same time).

The other component in a Kubernetes cluster is the worker nodes: **worker node components**.

The worker nodes are the servers on a Kubernetes cluster that actually run the applications (in the form of Pods). Unlike masters, workers are a disposable resource. With some exceptions, it is considered safe to change the number of worker nodes (by adding or removing them) in a running Kubernetes cluster. Indeed, that's a very common use case: it is one of the duties of the master nodes to ensure that all the proper steps (such as recreating Pods and rebalancing the workload) are implemented following such changes.

Of course, if the changes to the cluster are planned, it is likely to have less impact on the application (because, as an example, Kubernetes can evacuate Pods from the impacted nodes before removing it from the cluster), while if something unplanned, such as a crash, happens, this may imply some service disruptions. Nevertheless, Kubernetes is more or less designed to handle this kind of situation. Master nodes run the following components:

- **Container runtime**: This is a core component of a worker node. It's the software layer responsible for running the containers included in each Pod. Kubernetes supports, as container runtimes, any implementation of the **Container Runtime Interface (CRI)** standard. Widespread implementations, at the time of writing, include containerd, Docker, and CRI-O.

- **kubelet**: This is an agent running on each worker. Kubelet registers itself with the Kubernetes API server and communicates with it in order to check that the desired state of Pods scheduled to run in the node is up and running. Moreover, kubelet reports the health status of the node to the master (hence, it is used to identify a faulty node).

- **kube-proxy**: This is a network component running on the worker node. Its duty is to connect applications running on the worker node to the outside world, and vice versa.

Now that we have a clear understanding of the Kubernetes objects, server roles, and related components, it's time to understand why Kubernetes is an excellent engine for PaaS, and what is lacking to define it as PaaS *per se*.

Kubernetes as PaaS

If the majority of your experience is in the dev area, you may feel a bit lost after going through all those Kubernetes concepts. Even if everything is clear for you (or you already have a background in the infrastructural area), you'll probably agree that Kubernetes, while being an amazing tool, is not the easiest approach for a developer.

Indeed, most of the interactions between a developer and a Kubernetes cluster may involve working with `.yaml` files (this format is used to describe the API objects that we have seen) and the command line (usually using kubectl, the official CLI tool for Kubernetes) and understanding advanced container-based mechanisms (as persistent volumes, networking, security policies, and more).

Those aren't necessarily the most natural skills for a developer. For such reasons (and similar reasons on the infrastructure side), Kubernetes is commonly not regarded as a PaaS *per se*; it is more like being a core part of one (an engine). Kubernetes is sometimes referred to as **Container as a Service** (**CaaS**), being essentially an infrastructure layer that orchestrates containers as the core feature.

One common metaphor used in this regard is with the Linux OS. Linux is made by a low-level, very complex, and very powerful layer, which is the kernel. The kernel is vital for everything in the Linux OS, including managing processes, resources, and peripherals. But no Linux users exclusively use the kernel; they will use the Linux distributions (such as Fedora, Ubuntu, or RHEL), which top up the kernel with all the high-level features (such as tools, utilities, and interfaces) that make it usable to final users.

To use it productively, Kubernetes (commonly referred to in this context as vanilla Kubernetes) is usually complemented with other tools, plugins, and software, covering and extending some areas. The most common are as follows:

- **Runtime**: This is related to the execution of containers (and probably the closer, more instrumental extensions that Kubernetes needs to rely on to work properly and implement a PaaS model). Indeed, strictly speaking, Kubernetes doesn't even offer a *container runtime*; but, as seen in the previous section, it provides a standard (the CRI) that can be implemented by different runtimes (we mentioned containerd, Docker, and CRI-O). In the area of runtimes, *network* and *storage* are also worth mentioning as stacks are used to provide connectivity and persistence to the containers. As in the container runtime, in both network and storage runtimes, there is a set of standards and a glossary (the aforementioned services, or volumes) that is then implemented by the technology of choice.

- **Provisioning**: This includes aspects such as automation, infrastructure as code (commonly used tools here include Ansible and Terraform), and container registries in order to store and manage the container images (notable implementations include Harbor and Quay).

- **Security**: This spans many different aspects of security, from *policy* definition and enforcement (one common tool in this area is Open Policy Agent), runtime security and threat detection (a technology used here is Falco), and image scanning (Clair is one of the implementations available) to vault and secret encryption and management (one product covering this aspect is HashiCorp Vault).

- **Observability**: This is another important area to make Kubernetes a PaaS solution that can be operated easily in production. One de facto standard here is **Prometheus**, which is a time series database widely used to collect metrics from the different components running on Kubernetes (including core components of the platform itself). Another key aspect is log collection, to centralize the logs produced.

  Fluentd is a common choice in this area. Another key point (that we already introduced in *Chapter 7, Exploring Middleware and Frameworks*, in the sections on micro profiling) is tracing, as in the capability of correlating calls to different systems and identifying the execution of a request when such a request is handled by many different subsystems. Common tools used for this include Jaeger and OpenTracing. Last but not least, most of the telemetry collected in each of those aspects is commonly represented as dashboards and graphics. A common choice for doing that is with **Grafana**.

- **Endpoint management**: This is a topic related to networking, but at a higher level. It involves the definition, inventory, discovery, and management of application endpoints (that is, an API or similar network endpoints). This area is commonly addressed with a service mesh. It offloads communication between the containers by using a network proxy (using the so-called **sidecar pattern**) so that such a proxy can be used for tracing, securing, and managing all the calls entering and exiting the container. Common implementations of a service mesh are Istio and Linkerd. Another core area of endpoint management is so-called API management, which is similar conceptually (and technically) to a service mesh but is more targeted at calls coming from outside the Kubernetes cluster (while the service mesh mostly addresses Pod-to-Pod communication). Commonly used API managers include 3scale and Kong.

- **Application management**: Last but not least, this is an area related to how applications are packaged and installed in Kubernetes (in the form of container images). Two core topics are application definition (where two common implementations are Helm and the Operator Framework) and CI/CD (which can be implemented, among others, using Tekton and/or Jenkins).

All the aforementioned technologies (and many more) are mentioned and cataloged (using a similar glossary) in the CNCF landscape (visit `https://landscape.cncf.io`). **CNCF** is the **Cloud Native Computing Foundation**, which is an organization related to the Linux Foundation, aiming to define a set of vendor-neutral standards for cloud-native development. The landscape is their assessment of technologies that can be used for such goals including and revolving around Kubernetes (which is one of the core software parts of it).

So, I think it is now clear that Kubernetes and containers are core components of PaaS, which is key for cloud-native development. Nevertheless, such components mostly address runtime and orchestration needs, but many more things are needed to implement a fully functional PaaS model to support our applications.

Looking at things the other way around, you can wonder what the best practices are that each application (or component or microservice) should implement in order to fit nicely in PaaS and behave in the best possible way in a cloud-native, distributed setup. While it's impossible to create a magical checklist that makes every application a cloud-native application, there is a well-known set of criteria that can be considered a good starting point. Applications that adhere to this list are called twelve-factor applications.

Defining twelve-factor applications

The **twelve-factor applications** are a collection of good practices suggested for cloud-native applications. Applications that adhere to such a list of practices will most likely benefit from being deployed on cloud infrastructures, face *web-scale* traffic peaks, and resiliently recover from failures. Basically, twelve-factor applications are the closest thing to a proper definition of microservices. PaaS is very well suited for hosting twelve-factor apps. In this section, we are going to have a look at this list of factors:

- **Codebase**: This principle simply states that all the source code related to an app (including scripts, configurations, and every asset needed) should be versioned in a single repo (such as a Git source code repository). This implies that different apps should not share the same repo (which is a nice way to reduce coupling between different apps, at the cost of duplication).

Such a repo is then the source for creating Deployments, which run instances of the application, compiled (where relevant) by a CI/CD toolchain and launched in a number of different environments (such as production, test, and dev). A Deployment can be based on different versions of the same repo (as an example, a dev environment could run experimental versions, containing changes not yet tested and deployed in production, but still part of the same repo).

- **Dependencies**: A twelve-factor app must explicitly declare all the dependencies that are needed at runtime and must isolate them. This means avoiding depending on implicit and system-wide dependencies. This used to be a problem with traditional applications, as with Java applications running on an application server, or in general with applications *expecting* some dependencies provided by the system.

  Conversely, twelve-factor apps specifically declare and isolate the applications needed. In this way, the application behavior is more repeatable, and a dev (or test) environment is easier to set up. Of course, this comes at the cost of consuming more resources (disk space and memory, mostly). This requirement is one of the reasons for containers being so popular for creating twelve-factor apps, as containers, by default, declare and carry all the necessary dependencies for each application.

- **Config**: Simply put, this factor is about strictly separating the configurations from the application code. Configurations are intended to be the values that naturally change in each environment (such as credentials for accessing the database or endpoints pointing to external services). In twelve-factor apps, the configurations are supposed to be stored in environment variables. It is common to relax this requirement and store a little configuration in other places (separated from code), such as config files.

  Another point is that the twelve-factor apps approach suggests avoiding grouping configurations (such as grouping a set of config values for prod, or one for test) because this approach does not scale well. The advice is to individually manage each configuration property, associating it with the related Deployment. While there are some good rationalizations beyond this concept, it's also not uncommon to relax at this point and have a grouping of configurations following the naming of the environment.

- **Backing services**: A twelve-factor app must consider each backing service that is needed (such as databases, APIs, queues, and other services that our app depends on) as attached resources. This means that each backing service should be identified by a set of configurations (something such as the URL, username, and password) and it should be possible to replace it without any change to the application code (maybe requiring a restart or refresh).

By adhering to such factors, our app will be loosely coupled to an external service, hence scaling better and being more portable (such as from on-premises to the cloud, and vice versa). Moreover, the testing phase will benefit from this approach because we can easily swap each service with a mock, where needed.

Last but not least, the resiliency of the app will increase because we could, as an example, swap a faulty service with an alternative one in production with little to no outage. In this context, it's also worth noticing that in the purest microservices theory, each microservice should have its own database, and no other microservice should access that database directly (but to obtain the data, it should be mediated by the microservice itself).

- **Build, release, and run**: The twelve factor approach enforces strict separation for the build, release, and run phases. Build includes the conversion of source code into something executable (usually as the result of a compile process) and the release phase associates the executable item with the configuration needed (considering the target environment).

  Finally, the run phase is about executing such a release in the chosen environment. An important point here is that the whole process is supposed to be stateless and repeatable (such as using a CI/CD pipeline), starting from the code repo and configurations. Another crucial point is that each release must be associated with a unique identifier, to map and track exactly where the code and config ended up in each runtime. The advantages of this approach are a reduction in moving parts, support for troubleshooting, and the simplification of rollbacks in case of unexpected behaviors.

- **Processes**: An app compliant with the twelve factors is executed as one or more processes. Each process is considered to be stateless and shares nothing. The state must be stored in ad hoc backing services (such as databases). Each storage that is local to the process (being disk or memory) must be considered an unreliable cache.

  It can be used, but the app must not depend on it (and must be capable of recovering in case something goes wrong). An important consequence of the stateless process model is that sticky sessions must be avoided. A sticky session is when consequent requests must be handled by the same instance in order to function properly. Sticky sessions violate the idea of being stateless and limit the horizontal scalability of applications, and hence should be avoided. Once again, the state must be offloaded to relevant backing services.

- **Port binding**: Each twelve-factor app must directly bind to a network port and listen to requests on such a port. In this way, each app can become a backing service for other twelve-factor apps. A common consideration around this factor is that usually, applications rely on external servers (such as PHP or a Java application server) to expose their services, whereas twelve-factor apps embed the dependencies needed to directly expose such services.

  That's basically what we saw with JEE to cloud-native in the previous chapter; Quarkus, as an example, has a dependency to undertow to directly bind on a port and listen for HTTP requests. It is common to then have infrastructural components routing requests from the external world to the chosen port, wherever is needed.

- **Concurrency**: The twelve-factor app model suggests implementing a horizontal scalability model. In such a model, concurrency is handled by spinning new instances of the affected components. The smallest scalability unit, suggested to be scaled following the traffic profiles, is the process. Twelve-factor apps should rely on underlying infrastructure to manage the process's life cycle.

  This infrastructure can be the OS process manager (such as `systemd` in a modern Linux system) or other similar systems in a distributed environment (such as PaaS). Processes are suggested to span different servers (such as VMs or physical machines) if those are available, in order to use resources correctly. Take into account the fact that such a concurrency model does not replace other internal concurrency models provided by the specific technology used (such as threads managed by each JVM application) but is considered a kind of extension of it.

- **Disposability**: Applications adhering to the twelve-factor app should be disposable. This implies that apps should be fast to start up (ideally, a few seconds between the process being launched and the requests being correctly handled) and to shut down. Also, the shutdown should be handled gracefully, meaning that all the external resources (such as database connections or open files) must be safely deallocated, and every inflight request should be managed before the application is stopped.

  This *fast to start up/safe to shut down* mantra will allow for horizontal scalability with more instances being created to face traffic peaks and the ones being destroyed to save resources when no more are needed. Another suggestion is to create applications to be tolerant to hard shutdown (as in the case of a hardware failure or a forced shutdown, such as a process kill). To achieve this, the application should have special procedures in place to handle incoherent states (think about an external resource improperly closed or requests partially handled and potentially corrupted).

- **Dev/prod parity**: A twelve-factor app must reduce the differences between production and non-production environments as much as possible. This includes differences in software versions (meaning that a development version must be released in production as soon as possible by following the *release early, release often* mantra).

  But the mantra also referred to when different teams are working on it (devs and ops should cooperate in both production and non-production environments, avoiding the handover following the production release and implementing a DevOps model). Finally, there is the technology included in each environment (the backing services should be as close as possible, trying to avoid, as an example, different types of databases in dev versus production environments).

  This approach will provide multiple benefits. First of all, in the case of a production issue, it will be easier to troubleshoot and test fixes in all the environments, due to those environments being as close as possible to the production one. Another positive effect is that it will be harder for a bug to find its way into production because the test environments will look like the production ones. Last but not least, having similar environments will reduce the hassle of having to manage multiple variants of stacks and versions.

- **Logs**: This factor points to the separation of log generation and log storage. A twelve-factor app simply considers logs as an event stream, continuously generated (usually in a textual format) and sent to a stream (commonly the standard output). The app shouldn't care about persisting logs with all the associated considerations (such as log rotation or forwarding to different locations).

  Instead, the hosting platform should provide services capable of retrieving such events and handling them, usually on a multitier persistence (such as writing recent logs to files and sending the older entries to indexed systems to support aggregation and searches). In this way, the logs can be used for various purposes, including monitoring and business intelligence on platform performance.

- **Admin processes**: Many applications provide supporting tools to implement administration processes, such as performing backups, fixing malformed entries, or other maintenance activities. The twelve-factor apps are no exception to this. However, it is recommended to implement such admin processes in an environment as close as possible to the rest of the application.

Wherever possible, the code (or scripts) providing those features should be executed by the same runtime (such as a JVM) and using the same dependencies (the database driver). Moreover, such code must be checked out in the same code repo and must follow the same versioning schema as the application's main code. One approach to achieving this outcome is to provide an interactive shell (properly secured) as part of the application itself and run the administrative code against such a shell, which is then approved to use the same facilities (connection to the database and access to sessions) as the rest of the application.

Probably, while reading this list, many of you were thinking about how those twelve factors can be implemented, especially with the tools that we have seen so far (such as Kubernetes). Let's try to explore those relationships.

Twelve-factor apps and the supporting technology

Let's review the list of the twelve factors and which technologies covered so far can help us to implement applications adhering to such factors:

- **Codebase**: This is less related to the runtime technology and more to the tooling. As mentioned before, nowadays, the versioning tool widely used is Git, basically being the standard (and for good reason). The containers and Kubernetes, however, are well suited for supporting this approach, providing constructs such as containers, Deployments, and namespaces, which are very useful for implementing multiple deploys from the same codebase.

- **Dependencies**: This factor is, of course, dependent on the programming language and framework of choice. However, modern container architectures solve the dependency issue in different stages. There is usually one dependency management solution and declaration at a language level (such as Maven for Java projects and npm for JavaScript), useful for building and running the application prior to containerization (as in the dev environment, on a local developer machine).

  Then, when containers come into play, their filesystem layering technology can further separate and declare the dependencies from the application (which constitutes the very top layer of a container). Moreover, the application technology is basically able to formalize every dependency of the application, including the runtime (such as JVM) and the OS version and utilities (which is an inherent capability of the container technology).

- **Config**: This factor has plenty of ways of being implemented easily. I personally very much like the way Kubernetes manages it, defining `ConfigMap` objects and making them available to the application as environment variables or configuration files. That makes it pretty easy to integrate into almost every programming language and framework, and makes the configuration easy to be versioned and organized per environment. This is also a nice way to standardize configuration management regardless of the technology used.

- **Backing services**: This factor can be mapped one to one to the Kubernetes Services object. A cloud-native application can easily query the Kubernetes API to retrieve the service it needs, by name or by using a selector. However, it's worth noticing that Kubernetes does not allow a Pod to explicitly declare the dependencies to other Services, likely because it delegates the handling of corner cases (such as a service missing or crashing, or the need for reconnection) to each application. There are, however, multiple ways (such as using Helm charts or operators) in which to set up multiple Pods together, including the Services to talk to each other.

- **Build, release, and run**: This is pretty straightforward in containers and the Kubernetes world. A build can be intended as the container image build (and the application is, from there, regarded as immutable). The release can be defined with the creation of the build and other objects (including `config`) needed to import the containerized application into Kubernetes. Last but not least, Kubernetes handles (using the container runtime of choice) the running of the application.

- **Processes**: This factor is also quite well represented in Kubernetes. Indeed, each container is, by default, confined in its own runtime, while sharing nothing between each other. We know that containers are stateless too. A common strategy for handling the state is by using external resources, such as connections to databases, services, or persistent volumes. It's worth observing that Kubernetes, by using DaemonSets and similar constructs, allows exceptions to this behavior.

- **Port binding**: Even in this case, Kubernetes and containers allow all the requisite infrastructure to implement apps adhering to the port binding factor. Indeed, with Kubernetes, you can declare the port that your node will listen to (and Kubernetes will manage conflicts that could potentially arise between Pods asking for the same port). With Services, you can add additional capabilities to it, such as port forwarding and load balancing.

- **Concurrency**: This is inherent to the Kubernetes containers model. You can easily define the number of instances each Pod should run at any point in time. The infrastructure guarantees that all the required resources are allocated for each Pod.

- **Disposability**: In Kubernetes, the Pod life cycle is managed to allow each application to shut down gracefully. Indeed, Kubernetes can shut down each Pod and prevent new network traffic from being routed to that specific Pod, hence providing the basics for zero downtime and zero data loss. Then, Kubernetes can be configured to run a pre-stop hook, to allow a custom action to be done before the shutdown.

  Following that, Kubernetes sends a **SIGTERM** signal (which is a standard Linux termination signal) to communicate with the application with the intention of stopping it. The application is considered to be responsible for trapping and managing such a signal (disposing of resources and shutting down) if it's still up and running. Finally, after a timeout, if the application has not yet stopped, Kubernetes forcefully terminates it by using a **SIGKILL** signal (which is a more drastic signal than SIGTERM, meaning that it cannot be ignored by the application that will be terminated). Something similar can be said for the startup: Kubernetes can be configured to do some actions in case the start of a Pod goes wrong (as an example, taking too long).

  To do this, each application can be instrumented with probes, to detect exactly when an application is running and is ready to take new traffic. So, even in this case, the infrastructure provides all the necessary pieces to create an application compliant with this specific factor.

- **Dev/prod parity**: Similar to the other factors in this list, this is more about the processes and disciplines practiced in your particular development life cycle, meaning that no tool can ensure adherence to this factor if there is no willingness to do so. However, with Kubernetes natively being a declarative environment, it's pretty easy to define different environments (normally mapping to namespaces) that implement each development stage needed (such as dev and prod) and make such environments are as similar as possible (with frequent deploys, implementing checks if the versions and configuration differ too much, and using the same kind of backing services).

- **Logs**: These play a big part in Kubernetes architecture as there are many alternatives to manage them. The most important lesson, however, is that a big and complex infrastructure based on Kubernetes is mandated to use some log collection strategy (usually based on dealing with logs as an event stream). Common implementations of such an approach include using Fluentd as a log collector or streaming log lines into a compatible event broker (such as Kafka).

- **Admin processes**: This is perhaps a bit less directly mapped to Kubernetes and container concepts and more related to the specific language, framework, and development approaches that are used. However, Kubernetes allows containers to be accessed using a shell, so this way it can be used if the Pod provides the necessary administrative shell tools. Another approach can be to run specific Pods that can use the same technologies as our applications, just for the time needed to perform administrative processes.

As I've said many other times, there is no magic recipe for an application to be cloud-native (or microservices-compliant, or simply performant and well written). However, the twelve factors provide an interesting point of view and give some food for thought. Some of the factors are achievable by using features provided by the hosting platform or other dependencies (think about config or logs), while others are more related to the application architecture (backing services and disposability) or development model (codebase and dev/prod parity).

Following (and extending, where needed) this set of practices will surely be beneficial for the application's performance, resiliency, and cloud readiness. To go further in our analysis, let's look at some of the reasoning behind well-known patterns for cloud-native development and what supporting technologies we can use to implement them.

Well-known issues in the cloud-native world

Monolithic applications, while having many downsides (especially in the area of Deployment frequency and scalability), often simplify and avoid certain issues. Conversely, developing an application as cloud-native (hence, a distributed set of smaller applications) implies some intrinsic questions to face. In this section, we are going to see some of those issues. Let's start with fault tolerance.

Fault tolerance

Fault tolerance is an umbrella term for a number of aspects related to resiliency. The concept basically boils down to protecting the service from the unavailability (or minor failures) of its components. In other words, if you have chained services (which is very common, maybe between microservices composing your application or when calling external services), you may want to protect the overall application (and user experience), making it resilient to the malfunction of some such services.

By architecting your application in this way, you can avoid overstressing downstream components that are already misbehaving (such as giving exceptions or taking too long) and/or implementing a graceful degradation of the application's behavior. Fault tolerance can be obtained in various ways.

To keep this section practical and interesting, I am going to provide an introduction to each pattern and discuss how this can be implemented in Java. As a reference architecture, we are keeping the MicroProfile (as per what we saw in *Chapter 7, Exploring Middleware and Frameworks*), so we can have a vendor-independent implementation.

Circuit breaker

The most famous fault-tolerance technique is the **circuit breaker**. It became famous thanks to a very widespread implementation, which is **Netflix Hystrix** (now no longer actively developed). Resilience4j is widely accepted and commonly maintained as an alternative.

The circuit breaker pattern implies that you have a configurable threshold when calling another service. If such a call fails, according to the threshold, the circuit breaker will open, blocking further calls for a configurable amount of time. This is similar to a circuit breaker in an electrical plant, which will open in case of issues, preventing further damages.

This will allow the next calls to fail fast and avoid further calling to the downstream service (which may likely be already overloaded). The downstream system then has some time to recover (perhaps by manual intervention, with an automatic restart, or autoscaling). Here is an example where a circuit breaker is not implemented:

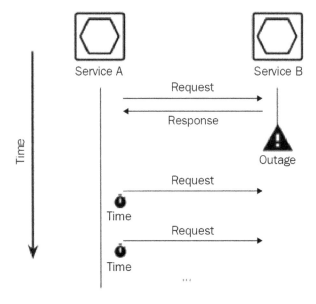

Figure 9.5 – Without a circuit breaker

As we can see in the preceding diagram, without a circuit breaker, calls to a failed service (due to a crash or similar outages) keep going in timeout. This, in a chain of calls, can cause the whole application to fail. In the following example, we'll implement a circuit breaker:

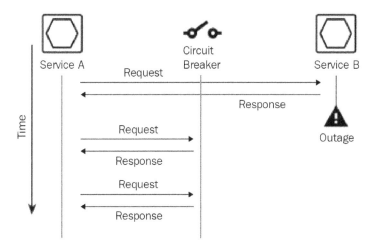

Figure 9.6 – With a circuit breaker

Conversely, in an implementation using a circuit breaker, in the case of a service failing, the circuit breaker will immediately identify the outage and provide the responses to the service calling (**Service A**, in our case). The response sent can simply be an error code (such as HTTP 500) or something more complex, such as a default static response or a redirection to an alternative service.

In a MicroProfile, you can configure a circuit breaker as follows:

```
@CircuitBreaker(requestVolumeThreshold = 4, failureRatio =
   0.5, delay = 10000)
public MyPojo getMyPojo(String id){...
```

The annotation is configured in a way that, if you have a failure ratio of 50% over four requests (so two failures in four calls), the circuit breaker will stay open for 10 seconds, failing immediately on calls in such a time window (without directly calling the downstream instance). However, after 10 seconds, the next call will be attempted to the target system. If the call succeeds, `CircuitBreaker` will be back to closed, hence working as before. It's worth noticing that the circuit breaker pattern (as well as other patterns in this section) can be implemented at a service mesh level (especially, in a Kubernetes context). As we saw a couple of sections ago, the service mesh works at a network level in Pod-to-Pod communication and can then be configured to behave as a circuit breaker (and more).

Fallback

The **fallback** technique is a good way to implement a *plan B* against external services not working. This will allow for a graceful fallback if the service fails, such as a default value or calling an alternative service.

To implement this in a MicroProfile, you can simply use the following annotation:

```
@Fallback(fallbackMethod = "myfallbackmethod")
public MyPojo getMyPojo(String id){...
```

In this way, if you get an exception in your getMyPojo method, myfallbackmethod will be called. Such methods must, of course, have a compatible return value. The fallback method, as said, may be an alternative implementation for such default values or different external services.

Retries

Another powerful way to deal with non-working services is to **retry**. This may work well if the downstream service has intermittent failures, but it will answer correctly or fail in a reasonable amount of time.

In this scenario, you can decide that it's good enough to retry the call in the event of a failure. In a MicroProfile, you can do that using the following annotation:

```
@Retry(maxRetries = 3, delay = 2000)
public MyPojo getMyPojo(String id){...
```

As you can see, the maximum number of retries and the delay between each retry are configurable with the annotation. Of course, this kind of approach may lead to a high response time if the downstream system does not fail fast.

Timeout

Last but not least, the **timeout** technique will precisely address the problem that we have just seen. Needless to say, a timeout is about timeboxing a call, imposing a maximum amount of time for it to be completed before an exception is raised.

In a MicroProfile, you can simply annotate a method and be sure that the service call will succeed or fail within a configured amount of time:

```
@Timeout(300)
public MyPojo getMyPojo(String id){...
```

In such a configuration, the desired service will have to complete the execution within 300 ms or will fail with a timeout exception. In this way, you can have a predictable amount of time in your chain of services, even if your external service takes too much time to answer.

All the techniques discussed in this section aim to enhance the resiliency of cloud-native applications and address one very well-known problem of microservices (and, more generally, distributed applications), which is failure cascading. Another common issue in the cloud-native world concerns transactionality.

Transactionality

When working on a classical monolithic Java application, **transactions** are kind of a resolved issue. You can appropriately mark your code and the container you are running in (be it an application server or other frameworks) to take care of it. This means that all the things that you can expect from a transactional **Atomicity**, **Consistency**, **Isolation**, **Durability** (**ACID**)-compliant system are provided, such as rollbacks in the case of failures and recovery.

In a distributed system, this works differently. Since the components participating in a transaction are living in different processes (that may be different containers, optionally running on different servers), traditional transactionality is not a viable option. One intuitive explanation for this is related to the network connection between each participant in the transaction.

If one system asks another to complete an action (such as persisting a record), no answer might be returned. *What should the client do then?* It should assume that the action has been successful and the answer is not coming for external reasons (such as a network split), or it should assume that the action has failed and optionally retry. Of course, there is no easy way to face these kinds of events. The following are a couple of ideas for dealing with data integrity that can be adopted.

Idempotent actions

One way to solve some issues due to distributed transactionality is **idempotency**. A service is considered idempotent if it can be safely called more than once with the same data as input, and the result will still be a correct execution. This is something naturally obtained in certain kinds of operations (such as read operations or changes to specific information, such as the address or other data of a user profile), while it must be implemented in some other situations (such as money balance transactions, where a double charge for the same payment transaction is, of course, not allowed).

The most common way to correctly handle idempotency relies on a specific key associated with each call, which is usually obtained from the payload itself (as an example, calculating a hash over all the data passed). Such a key is then stored in a repository (this can be a database, filesystem, or in-memory cache). A following call to the same service with the same payload will create a clash on such a key (meaning that the key already exists in the repository) and will then be handled as a no-operation.

So, in a system implemented in this way, it's safe to call a service more than once (as an example, in case we got no response from the first attempt). In the event that the first call was successful and we received no answer for external reasons (such as a network drop), the second call will be harmless. It's common practice to define an expiration for such entries, both for performance reasons (avoiding growing the repository too much, since it will be accessed at basically every call) and for correctly supporting the use case of your specific domain (for instance, it may be that a second identical call is allowed and legitimate after a specific timeout is reached).

The Saga pattern

The **Saga pattern** is a way to deal with the problem of transactions in a distributed environment more fully. To implement the Saga pattern, each of the systems involved should expose the opposite of each business operation that includes updating the data. In the payments example, a charge operation (implying a write on a data source, such as a database) should have a twin *undo* operation that implements a top-up of the same amount of money (and likely provides some more descriptions, such as the reason for such a cancellation).

Such a complementary operation is called compensation, and the goal is to undo the operation in the event of a failure somewhere else. Once you have a list of services that must be called to implement a complex operation and the compensation for each of them, the idea is to call each service sequentially. If one of the services fails, all the previous services are notified, and the undo operation is called on them to put the whole system in a state of consistency. An alternative way to implement this is to call the first service, which will be in charge of sending a message to the second service, and so on. If one of the services fails, the messages are sent back to trigger the requisite compensation operations. There are two warnings regarding this approach:

* The signaling of the operation outcome after each write operation (which will trigger the compensations in case of a failure) must be reliable. So, the case in which a failure happens somewhere and the compensations are not called must be avoided as far as possible.

- The whole system must be considered as eventually consistent. This means that there are some specific timeframes (likely to be very short) in which your system is not in a consistent state (because the downstream systems are yet to be called, or a failure just happened and the compensations are not yet executed).

An elegant way to implement this pattern is based on the concept of change data capture. **Change data capture** is a pattern used for listening to changes on a data source (such as a database). There are many different technologies to do that, including the polling of the data source or listening for some specific events in the database transaction logs. By using change data capture, you can be notified when a write happens in the data source, which data is involved, and whether the write has been successful. From such events, you can trigger a message or a call for the other systems involved, continuing your distributed transaction or *rolling back* by executing the compensation methods.

The Saga pattern, in a way, makes us think about the importance of the flow of calls in a microservices application. As seen in this section (and also in the *Why create cloud-native applications?* section regarding resiliency), the order (and the way) in which we call the services needed to compose our functionality can change the transactionality, resiliency, and efficiency (think about parallel versus serial, as discussed in *Chapter 6, Exploring Essential Java Architectural Patterns*, under the *Implementing for large-scale adoption* section). In the next section, we are going to elaborate a bit more on this point.

Orchestration

The Saga pattern highlights the sequence (and operations) in which every component must be called in order to implement eventual consistency. This is a topic that we have somewhat taken for granted.

We have talked about the microservice characteristics and ways of modeling our architectures in order to be flexible and define small and meaningful sets of operations. *But what's the best way to compose and order the calls to those operations, so as to create the higher-level operations implementing our use case?* As usual, there is no easy answer to this question. The first point to make concerns the distinction between **orchestration** and another technique often considered an alternative to it, which is **choreography**. There is a lot of debate ongoing about the differences between orchestration and choreography. I don't have the confidence to speak definitively on this subject, but here is my take on it:

- **Orchestration**, as in an orchestra, implies the presence of a conductor. Each microservice, like a musician, can use many services (many sounds, if we stay within the metaphor), but it looks for hints from the conductor to make something that, in cooperation with the other microservices, simply works.

- **Choreography**, as in a dance, is studied beforehand and requires each service (dancer) to reply to an external event (other dancers doing something, music playing, and so on). In this case, we see some similarities with what we saw in the *Event-driven and reactive* section of *Chapter 6, Exploring Essential Java Architectural Patterns*).

In this section, we are going to focus on orchestration.

Orchestrating in the backend or the frontend

A first, simple approach to orchestration implies a **frontend aggregation** level (this may be on the client side or server side). This essentially means having user experience (as in the flow of different pages, views, or whatever the client technology provides) dictate how the microservice functions are called.

The benefit of this approach is that it's easy and doesn't need extra layers, or other technology in your architecture, to be implemented.

The downsides, in my opinion, are more than one. First of all, you are tightly coupling the behavior of the application with the technical implementation of the frontend. If you need to change the flow (or any specific implementation to a service), you are most likely required to make changes in the frontend.

Moreover, if you need to have more than one frontend implementation (which is very common, as we could have a web frontend and a couple of mobile applications), the logic will become sprawled in all of those frontends and a change must be propagated everywhere, thereby increasing the possibility of making mistakes. Last but not least, directly exposing your services to the frontend may imply having a mismatch of granularity between the amount of data microservices offer with the amount of data the frontend will need. So, choices you may need to make in the frontend (as pagination) will need to slip into the backend microservices implementation. This is not the best solution, as every component will have unclear responsibilities.

The obvious alternative is moving the orchestration to the backend. This means having a component between the microservices implementing the backend and the technologies implementing the frontend, which has the role of aggregating the services and providing the right granularity and sequence of calls required by the frontend.

Now the fun begins: *How should this component be implemented?*

One common alternative to aggregating at the frontend level is to pick an API gateway to do the trick. **API gateways** are pieces of software, loosely belonging to the category of integration middlewares, that sit as a man in the middle between the backend and frontend, and proxy the API calls between the two. An API gateway is an infrastructural component that is commonly equipped with additional features, such as authentication, authorization, and monetization.

The downside is that API gateways are usually not really tools designed to handle aggregation logic and sequences of calls. So sometimes, they are not capable of handling complex orchestration capabilities, but simply aggregate more calls into one and perform basic format translation (such as SOAP to REST).

A third option is to use a **custom aggregator**. This means delegating one (or more than one) microservices to act as an orchestrator. This solution provides the maximum level of flexibility with the downside of centralizing a lot of functionalities into a single architectural component. So, you have to be careful to avoid scalability issues (so it must be appropriately scalable) or the solution becoming a single point of failure (so it must be appropriately highly available and resilient). A custom aggregator implies a certain amount of custom code in order to define the sequence of calls and the integration logic (such as formal translation). There are a couple of components and techniques that we have discussed so far that can be embedded and used in this kind of component:

- A **business workflow** (as seen in *Chapter 8*, *Designing Application Integration and Business Automation*) can be an idea for describing the sequence of steps orchestrating the calls. The immediate advantage is having a graphical, standard, and business-understandable representation of what a higher-level service is made of. However, this is not a very common practice, because the current technology of business workflow engines is designed for a different goal (being a stateful point to persist process instances).

  So, it may have a performance impact and be cumbersome to implement (as BPMN is a business notation, while this component is inherently technological). So, if this is your choice, it is worthwhile considering a lightweight, non-persistent workflow engine.

- **Integration patterns** are to be considered with a view of implementing complex aggregation (such as protocol translation) and composition logic (such as sequencing or parallelizing calls). Even in this case, to keep the component scalable and less impactful from a performance standpoint, it is worthwhile considering lightweight integration platforms and runtimes.

- The **fault-tolerance** techniques that we saw a couple of sections ago are a good fit in this component. They will allow our composite call to be resilient in case of the unavailability of one of the composing services and to fail fast if one of them is misbehaving or simply answering slowly. Whatever your choice for the aggregation component, you should consider implementing fault tolerance using the patterns seen.

Last but not least, a consideration to be made about orchestration is whether and how to implement the **backend for frontend** pattern (as briefly seen in *Chapter 4, Best Practices for Design and Development*). To put it simply, different frontends (or better, different clients) may need different formats and granularity for the higher-level API. A web UI requires a different amount of data (and of different formats) than a mobile application, and so on. One way to implement this is to create a different instance of the aggregation component for each of the frontends. In this way, you can slightly change the frontend calls (and the user experience) without impacting the microservices implementation in the backend.

However, as with the frontend aggregation strategy, a downside is that the business logic becomes sprawled across all the implementations (even if, in this case, you at least have a weaker coupling between the frontend and the orchestration component). In some use cases, this may lead to inconsistency in the user experience, especially if you want to implement omnichannel behavior, as in, you can start an operation in one of the frontends (or channels) and continue with it in another one. So, if you plan to have multiple aggregation components (by means of the backend for frontend pattern), you will likely need to have a consistency point somewhere else (such as a database persisting the state or a workflow engine keeping track of the current and previous instances of each call).

This section concludes our overview of microservices patterns. In the next section, we are going to consider when and how it may make sense to adopt microservices and cloud-native patterns or evolve existing applications toward such paradigms.

Adopting microservices and evolving existing applications

So, we had an overview of the benefits of microservices applications and some of their particular characteristics. I think it is now relevant to better consider why you should (or should not) adopt this architectural style. This kind of consideration can be useful both for the creation of new applications from scratch (in what is called *green-field development*) and modernization (termed *brown-field applications*). Regarding the latter aspect, we will discuss some of the suggested approaches for modernizing existing applications in the upcoming sections.

But back to our main topic for this section: *why should you adopt the microservices-based approach?*

The first and most important reason for creating microservices is the **release frequency**. Indeed, the most famous and successful production experiences of microservices applications are related to services heavily benefitting from being released often.

This is because a lot of features are constantly released and experimented with in production. Remembering what we discussed in relation to the Agile methodologies, doing so allows us to test what works (what the customers like), provide new functionalities often (to stay relevant to the market), and quickly fix issues (which will inevitably slip into production because of the more frequent releases).

This means that the first question to ask is: *Will your application benefit from frequent releases?* We are talking about once a week or more. Some well-known internet applications (such as e-commerce and streaming services) even push many releases in production every day.

So, if the service you are building will not benefit from releasing this often – or worse, if it's mandated to release according to specific timeframes – you may not need to fully adhere to the microservices philosophy. Instead, it could turn out to be just a waste of time and money, as of course, the application will be much more complicated than a simple monolithic or *n*-tier application.

Another consideration is **scalability**. As stated before, many successful production implementations of microservices architectures are related to streaming services or e-commerce applications. Well, that's not incidental. Other than requiring constant experimentation and the release of new features (hence, release frequency), such services need to scale very well. This means being able to handle many concurrent users and absorbing peaks in demand (think about Black Friday in an e-commerce context, or the streaming of live sporting events). That's supposed to be done in a cost-effective way, meaning that the resource usage must be minimized and allocated only when it is really needed.

So, I think you get the idea: microservices architectures are supposed to be applied to projects that need to handle thousands of concurrent requests and that need to absorb peaks of 10 times the average load. If you only need to manage much less traffic, once again microservices could be overkill.

A less obvious point to consider is **data integrity**. As we mentioned a few sections ago, when talking about the Saga pattern, a microservices application is a heavily distributed system. This implies that transactions are hard or maybe impossible to implement. As we have seen, there are workarounds to mitigate the problem, but in general, everybody (especially business and product managers) should be aware of this difficulty.

It should be thoroughly explained that there will be features that may not be updated in real time, providing stale or inaccurate data (and maybe some missing data too). The system as a whole may have some (supposedly short) timeframes in which it's not consistent. Note that it is a good idea to contextualize and describe which features and scenarios may present these kinds of behaviors to avoid bad surprises when testing.

At the same time, on the technical design side, we should ensure we integrate all possible mechanisms to keep these kinds of misalignments to a minimum, including putting in place all the safety nets required and implementing any reconciliation procedure that may be needed, in order to provide a satisfactory experience to our users.

Once again, if this is not a compromise that everybody in the project team is willing to make, maybe microservices should be avoided (or used for just a subset of the use cases).

As we have already seen in *Chapter 5*, *Exploring the Most Common Development Models*, a prerequisite for microservices and cloud-native architectures is to be able to operate as a DevOps team. That's not a minor change, especially in big organizations. But the implications are obvious: since each microservice has to be treated as a product with its own release schedule, and should be as independent as possible, then each team working on each microservice should be self-sufficient, breaking down silos and maximizing collaboration between different roles. Hence, DevOps is basically the only organizational model known to work well in supporting a microservices-oriented project. Once again, this is a factor to consider: if it's hard, expensive, or impossible to adopt this kind of model, then microservices may not be worth it.

An almost direct consequence of this model is that each team should have a supporting technology infrastructure that is able to provide the right features for microservices. This implies having an automated release process, following the CI/CD best practices (we will see more about this in *Chapter 13*, *Exploring the Software Life Cycle*). And that's not all: the environments for each project should also be easy to provision on-demand, and possibly in a self-service fashion. Kubernetes, which we looked at a couple of sections ago, is a perfect fit for this.

It is not the only option, however, and in general, cloud providers offer great support to accelerate the delivery of environments (both containers and VMs) by freeing the operations teams from many responsibilities, including hardware provisioning and maintaining the uptime of some underlying systems.

In other words, it will be very hard (or even impossible) to implement microservices if we rely on complex manual release processes, or if the infrastructure we are working on is slow and painful to extend and doesn't provide features for the self-service of new environments.

One big advantage of the microservices architecture is the **extensibility** and **replaceability** of each component. This means that each microservice is related to the rest of the architecture via a well-defined API and can be implemented with the technology best suited for it (in terms of the language, frameworks, and other technical choices). Better yet, each component may be evolved, enhanced, or replaced by something else (a different component, an external service, or a SaaS application, among others). So, of course, as you can imagine, this has an impact in terms of integration testing (more on that in *Chapter 13*, *Exploring the Software Life Cycle*), so you should really consider the balance between the advantages provided and the impact created and resources needed.

So, as a summary for this section, microservices provide a lot of interesting benefits and are a really cool architectural model, worth exploring for sure.

On the other hand, before you decide to implement a new application following this architectural model, or restructuring an existing one to adhere to it, you should consider whether the advantages will really outweigh the costs and disadvantages by looking at your specific use case and whether you will actually use these advantages.

If the answer is no, or partially no, you can still take some of the techniques and best practices for microservices and adopt them in your architecture.

I think that's definitely a very good practice: maybe part of your application requires strong consistency and transactionality, while other parts have less strict requirements and can benefit from a more flexible model.

Or maybe your project has well-defined release windows (for external constraints), but will still benefit from fully automated releases, decreasing the risk and effort involved, even if they are not scheduled to happen many times a day.

So, your best bet is to not be too dogmatic and use a mix-and-match approach: in this way, the architecture you are designing will be better suited to your needs. Just don't adopt microservices out of FOMO. It will just be hard and painful, and the possibility of success will be very low.

With that said, the discussion around new developmental and architectural methodologies never stops, and there are, of course, some ideas on what's coming next after microservices.

Going beyond microservices

Like everything in the technology world, microservices got to an inflection point (the *Trough of Disillusionment*, as we called it at the beginning of this chapter). The reasoning behind this point is whether the effort needed to implement a microservices architecture is worth it. The benefit of well-designed microservices architectures, beyond being highly scalable and resilient, is to be very quick in deploying new releases in production (and so experiment with a lot of new features in the real world, as suggested by the adoption of Agile methodology). But this comes at the cost of having to develop (and maintain) infrastructures that are way more complex (and expensive) than monolithic ones. So, if releasing often is not a primary need of your particular business, you may think that a full microservices architecture constitutes overkill.

Miniservices

For this reason, many organizations started adopting a compromise approach, sometimes referred to as **miniservices**. A miniservice is something in the middle between a microservice and a monolith (in this semantic space, it is regarded as a macroservice). There is not a lot of literature relating to miniservices, mostly because, it being a compromise solution, each development team may decide to make trade-offs based on what it needs. However, there are a number of common features:

- Miniservices may break the dogma of one microservice and one database and so two miniservices can share the same database if needed. However, bear in mind that this will mean tighter coupling between the miniservices, so it needs to be evaluated carefully on a case-by-case basis.

- Miniservices may offer APIs of a higher level, thereby requiring less aggregation and orchestration. Microservices are supposed to provide specific APIs related to the particular domain model (and database) that a particular microservice belongs to. Conversely, a miniservice can directly provide higher-level APIs operating on different domain models (as if a miniservice is basically a composition of more than one microservice).

- Miniservices may share the deployment infrastructure, meaning that the deployment of a miniservice may imply the deployment of other miniservices, or at least have an impact on it, while with microservices, each one is supposed to be independent of the others and resilient to the lack of them.

So, at the end of the day, miniservices are a *customized* architectural solution, relaxing on some microservices requirements in order to focus on business value, thereby minimizing the technological impact of a full microservices implementation.

Serverless and Function as a Service

As the last point, we cannot conclude this chapter without talking about serverless. At some point in time, many architects started seeing serverless as the natural evolution of the microservices pattern. **Serverless** is a term implying a focus on the application code with very little to no concern regarding the underlying infrastructure. That's what the *less* part in serverless is about: not that there are no servers (of course), but that you don't have to worry about them.

Looking from this point of view, serverless is truly an evolution of the microservices pattern (and PaaS too). While serverless is a pattern, a common implementation of it takes the container as the smallest computing unit, meaning that if you create a containerized application and deploy it to a serverless platform, the platform itself will take care of scaling, routing, security, and so on, thereby absolving the developer of responsibility for it.

A further evolution of the serverless platform is referred to as **Function as a Service** (**FaaS**). In serverless, in theory, the platform can manage (almost) every technology stack, provided that it can be packaged as a container, while with FaaS, the developer must comply with a well-defined set of languages and technologies (usually Java, Python, JavaScript, and a few others). The advantage that balances such a lack of freedom is that the dev does not need to care about the layers underlying the application code, which is really just writing the code, and the platform does everything else.

One last core characteristic, common to both serverless and FaaS, is the scale to zero. To fully optimize platform usage, the technology implementing serverless and FaaS can shut the application down completely if there are no incoming requests and quickly spin up an instance when a request comes. For this reason, those two approaches are particularly suitable for being deployed on a cloud provider, where you will end up paying just for what you need. Conversely, for implementing the scale to zero, the kind of applications (both the framework and the use case) must be appropriate. Hence, applications requiring a warmup or requiring too long to start are not a good choice.

Also, the management of state in a serverless application is not really an easy problem to solve (usually, as in microservices, the state is simply offloaded to external services). Moreover, while the platforms providing serverless and FaaS capabilities are evolving day after day, it is usually harder to troubleshoot problems and debug faulty behaviors.

Last but not least, there are no real standards (yet) in this particular niche, hence the risk of lock-in is high (meaning that implementations made to run on a specific cloud stack will be lost if we want to change the underlying technology). Considering all the pros and cons of serverless and FaaS, these approaches are rarely used for implementing a full and complex application. They are, instead, a good fit for some specific use cases, including batch computations (such as file format conversions and more) or for providing *glue* code connecting different, more complex functions (such as the ones implemented in other microservices).

In the next section, we are going to discuss a very hot topic on a strategy for evolving existing applications toward cloud-native microservices and other newer approaches such as serverless.

Refactoring apps as microservices and serverless

As we discussed a couple of sections earlier, software projects are commonly categorized as either green- or brown-field.

Green-field projects are those that start from scratch and have very few constraints on the architectural model that could be implemented.

This scenario is common in start-up environments, for example, where a brand-new product is built and there is no legacy to deal with.

The situation is, of course, ideal for an architect, but is not so common in all honesty (or at least, it hasn't been so common in my experience so far).

The alternative scenario, brown-field projects, is where the project we are implementing involves dealing with a lot of legacy and constraints. Here, the target architecture cannot be designed from scratch, and a lot of choices need to be made, such as what we want to keep, what we want to ditch, and what we want to adapt. That's what we are going to discuss in this section.

The five Rs of application modernization

Brown-field projects are basically application modernization projects. The existing landscape is analyzed, and then some decisions are made to either develop a new application, implement a few new features, or simply enhance what's currently implemented, making it more performant, cheaper, and easier to operate.

The analysis of what's existing is often an almost manual process. There are some tools available for scanning the existing source code or artifacts, or even to dynamically understand how applications behave in production. But often, most of the analysis is done by architects, starting from the data collected with the aforementioned tools, using existing architectural diagrams, interviewing teams, and so on.

Then, once we have a clear idea about what is running and how it is implemented, choices have to be made component by component.

There is a commonly used methodology for this that defines five possible outcomes (the *five Rs*). It was originally defined by Gartner, but most consultancy practices and cloud providers (such as Microsoft and AWS) provide similar techniques, with very minor differences.

The five Rs define what to do with each architectural component. Once you have a clear idea about how a brown-field component is implemented and what it does, you can apply one of the following strategies:

- **Rehost**: This means simply redeploying the application on more modern infrastructure, which could be different hardware, a virtualized environment, or using a cloud provider (in an IaaS configuration). In this scenario, no changes are made to the application architecture or code. Minor changes to packaging and configuration may be necessary but are kept to a minimum. This scenario is also described as lift-and-shift migration and is a way to get advantages quickly (such as cheaper infrastructure) while reducing risks and transformation costs. However, of course, the advantages provided are minimal, as the code will still be old and not very adherent to modern architectural practices.

- **Refactor**: This is very similar to the previous approach. There are no changes to architecture or software code. The target infrastructure, however, is supposed to be a PaaS environment, possibly provided by a cloud provider. In this way, advantages such as autoscaling or self-healing can be provided by the platform itself while requiring only limited effort for adoption. CI/CD and release automation are commonly adopted here. However, once again, the code will still be unchanged from the original, so it may be hard to maintain and evolve.

- **Revise**: This is a slightly different approach. The application will be ported to a more modern infrastructure (PaaS or cloud), as with the Rehost and Refactor strategies. However, small changes to the code will be implemented. While the majority of the code will stay the same, crucial features, such as session handling, persistence, and interfaces, will be changed or extended to derive some benefits from the more modern underlying infrastructure available. The final product will not benefit from everything the new infrastructure has to offer but will have some benefits. Plus, the development and testing efforts will be limited. The target architecture, however, will not be microservices or cloud-native, rather just a slightly enhanced monolith (or *n* tier).

- **Rebuild**: Here, the development and testing effort is way higher. Basically, the application is not ported but instead is rewritten from scratch in order to use new frameworks and a new architecture (likely microservices or microservices plus something additional). The rebuilt architecture is targeted for hosting on a cloud or PaaS. Very limited parts of the application may be reused, such as pieces of code (business logic) or data (existing databases), but more generally, it can be seen as a complete green-field refactoring, in which the same requirements are rebuilt from scratch. Of course, the effort, cost, and risk tend to be high, but the benefits (if the project succeeds) are considered worthwhile.

- **Replace**: In this, the existing application is completely discarded. It may be simply retired because it's not needed anymore (note that in some methodologies, *Retire* is a *sixth R*, with its own dedicated strategy). Or it may be replaced with a different solution, such as SaaS or an existing off-the-shelf product. Here, the implementation cost (and the general impact) may be high, but the running cost is considered to be lower (or zero, if the application is completely retired), as less maintenance will be required. The new software is intended to perform better and offer enhanced features.

In the following table, we can see a summary of the characteristics of each of the approaches:

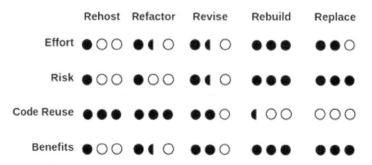

Table 9.1 – The characteristics of the five Rs

As you can see in the preceding table, getting the most benefits means a trade-off of taking on the most effort and risk.

In the following table, some considerations of the benefits of each of these approaches can be seen:

Rehost	Refactor	Revise	Rebuild	Replace
- Refreshed infrastructure	- Refreshed infrastructure - Autoscaling, self healing - Automation and CI /CD	- Refreshed infrastructure - Autoscaling, self healing - Automation and CI /CD - Uses PAAS / Cloud provided services	- Refreshed infrastructure - Autoscaling, self healing - Automation and CI /CD - Uses PAAS / Cloud provided services - Modernized code - Redesigned architecture (Microservices)	- Brand new product (or retired) - Cheaper to maintain (in case of SAAS) - Standard / Modern architecture (depends on the product adopted)

Table 9.2 – The benefits of the five Rs

Once again, most of the benefits come with the last two or three approaches.

However, it's worth noticing that the last two (**Rebuild** and **Replace**) fit into a much bigger discussion, often considered in the world of software development: that of *build versus buy*. Indeed, **Rebuild** is related to the *build* approach: you design the architecture and develop the software tailored to your own needs. It may be harder to manage this, but it guarantees maximum control. Most of this book, after all, is related to this approach.

Buy (which is related to **Replace**), on the other hand, follows another logic: after a software selection phase, you find an off-the-shelf product (be it on-premises or SaaS) and use it instead of your old application. In general, it's easier, as it requires limited to no customization. Very often, maintenance will also be very limited, as you will have a partner or software provider taking care of it. Conversely, the new software will give you less control and some of your requirements and use cases may need to be adapted to it.

As said, an alternative to *buy* in the **Replace** strategy is simply to ditch the software altogether. This may be because of changing requirements, or simply because the features are provided elsewhere.

The five Rs approach is to be considered in a wider picture of application modernization and is often targeted at big chunks of an enterprise architecture, targeting tens or hundreds of applications.

I would like to relate this approach to something more targeted to a single application, which is the **strangler pattern**. The two approaches (five Rs and strangler) are orthogonal and can also be used together, by using the strangler pattern as part of revising (**Revise**) or rebuilding (**Rebuild**) an application. Let's look into this in more detail.

The strangler pattern

As outlined in the previous section, the five Rs model is a programmatic approach to identify what to do with each application in an enterprise portfolio, with changes ranging from slightly adapting the existing application to a complete refactoring or replacement.

The strangler pattern tackles the same issue but from another perspective. Once an application to be modernized has been identified, it gives specific strategies to do so, targeting a path ranging from small improvements to a progressive coexistence between old and new, up to the complete replacement of the old technologies.

This approach was originally mentioned by Martin Fowler in a famous paper and relates to the *strangler fig*, which is a type of tree that progressively encloses (and sometimes completely replaces) an existing tree.

The metaphor here is easy to understand: new application architectures (such as microservices) start growing alongside existing ones, progressively strangling, and ultimately replacing, them. In order to this, it's essential to have control of the ingress points of our users into our application (old and new) and use them as a *routing layer*. Better yet, there should be an ingress point capable of controlling each specific feature. This is easy if every feature is accessed via an API call (SOAP or REST), as the routing layer can then simply be a network appliance with routing capabilities (a load balancer) that decides where to direct each call and each user. If you are lucky enough, the existing API calls are already mediated by an API manager, which can be used for the same purposes.

In most real applications, however, this can be hard to find, and most likely some of the calls are internal to the platform (so it is not easy to position a network load balancer in the middle). It can also happen that such calls are done directly in the code (via method calls) or using protocols that are not easily redirected over the network (such as Java RMI).

In such cases, a small intervention will be needed by writing a piece of code that adapts such calls from the existing infrastructure to standard over-the-network APIs (such as REST or SOAP), on both the client and server sides.

An alternative technique is to implement the routing functionality in the client layers. A common way to do so is to use feature flags, which have hidden settings and are changeable on the fly by the application administrators who set the feature that must be called by every piece of the UI or the client application.

However, while this approach can be more fine-grained than redirecting at the network level, it may end up being more complex and invasive as it also changes the frontend or client side of the application.

Once you have a mechanism to split and redirect each call, the strangler pattern can finally start to happen. The first step is to identify the first feature – one as isolated and self-contained as possible – to be reimplemented with a new stack and architecture.

The best option is to start with simple but not trivial functionality, in order to keep the difficulty low but still allow you to test the new tools and framework on something meaningful. In order to exactly identify the boundary of each feature, we can refer to the concept of bounded context in DDD, as we saw in *Chapter 4, Best Practices for Design and Development*.

In order to finance the project, it's a common practice to piggyback the modernization activities together with the implementation of new features, so it is possible that the new piece we are developing is not completely *isofunctional* with the old one, but contains some additional new functionalities.

Once such a piece of software is ready, we start testing it by sending some traffic toward it. To do so, we can use whatever routing layer is available, be it a network load balancer or a piece of custom code, as we have seen before. For such a goal, advanced routing techniques, such as canary or A/B testing, can be used (more on this in *Chapter 13, Exploring the Software Life Cycle*).

If something goes wrong, a rollback will always be possible, as the old functionalities will still be present in the existing implementation. If the rollout is successful and the new part of the application works properly, it's time to extend and iterate the application.

More features and pieces of the application are implemented in the new way, and deprecated from the old implementation in an approach that can be parallelized but needs to be strictly governed, in order to easily understand and document which functionality is implemented where and potentially switch back in case of any issue.

Eventually, all the features of the platform will now be implemented in the new stack, which will most likely be based on microservices or something similar.

After a grace period, the old implementation can be discarded and our modernization project will finally be completed, delivering the maximum benefit it has to offer (even more so as we no longer need to keep running and maintaining the old part).

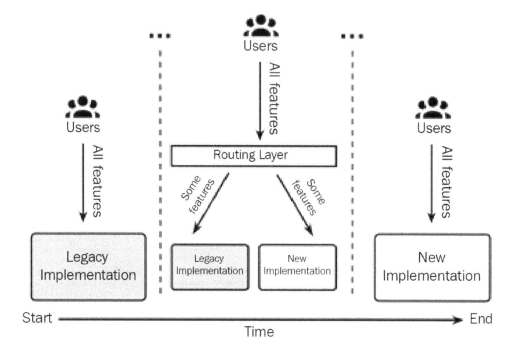

Figure 9.7 – The strangler pattern approach

The preceding diagram that you see is simplified for the sake of space. There will be more than one phase between the start (where all the features are running in the legacy implementation) and the end (where all the features have been reimplemented as microservices, or in any other modern way).

In each intermediate phase (not fully represented in the diagram, but indicated by the *dotted* lines), the legacy implementation starts receiving less traffic (as less of its features are used), while the new implementation grows in the number of functionalities implemented. Moreover, the new implementation is represented as a whole block, but it will most likely be made up of many smaller implementations (microservices), growing around and progressively strangling and replacing the older application.

Note that the strangler pattern as explained here is a simplification and doesn't take into account the many complexities of modernizing an application. Let's see some of these complexities in the next section.

Important points for application modernization

Whether the modernization is done with the strangler pattern (as seen just now) or a more end-to-end approach covering the whole portfolio (as with the five Rs approach, seen earlier), the approach to modernize an existing application must take care of many, often underestimated, complexities. The following gives some suggestions for dealing with each of them:

- **Testing suit**: This is maybe the most important of them all. While we will see more about testing in *Chapter 13*, *Exploring the Software Life Cycle*, it's easy to understand how a complete testing suite offers the proof needed to ensure that our modernization project is going well and ultimately is complete. In order to ensure that the new implementation is working at least as well as the old one, it's crucial that everything is covered by test suites, possibly automated. If you lack test coverage on the existing implementation, you may have a general feeling that everything is working, but you will likely have some bad surprises on production release. So, if the test coverage on the whole platform is low, it's better to invest in this first before any refactoring project.

- **Data consistency**: While it wasn't underlined in the techniques we have seen, refactoring often impacts the existing data layer by adding new data technologies (such as NoSQL stores) and/or changing the data structure of existing data setups (such as altering database schemas). Hence, it is very important to have a strategy around data too. It is likely that, if we migrated one bounded context at a time, the new implementation would have a dedicated and self-consistent data store.

  However, to do so, we will need to have existing data migrated (by using a data integration technique, as we saw in *Chapter 8*, *Designing Application Integration and Business Automation*). When the new release is ready, it will likely have to exchange data with the older applications. To do so, you can provide an API, completely moving the integration from the data layer to the application layer (this is the best approach), or move the data itself using, as an example, the *change data capture pattern*. As discussed earlier, however, you must be careful of any unwanted data inconsistency in the platform as a whole.

- **Session handling**: This is another very important point, as for a certain amount of time, the implementation will remain on both the old and new applications and users will share their sessions between both. This includes all the required session data and security information as well (such as if the user is logged in). To handle such sessions, the best approach is to externalize the session handling (such as into an external in-memory database) and make both the old and new applications refer to it when it comes to storing and retrieving session information. An alternative is to keep two separate session-handling systems up to date (manually), but as you can imagine, it's more cumbersome to implement and maintain them.

- **Troubleshooting**: This has a big impact. For a certain amount of time, the features are implemented using many different systems, across old and new technologies. So, in case of any issue, it will be definitely harder to understand what has gone wrong where. There is not much we could do to mitigate the impact of an issue here. My suggestion is to maintain up-to-date documentation and governance of the project, in order to make clear to everybody where each feature is implemented at any point in time. A further step is to provide a unique identifier to each call, to understand the path of each call, and correlate the execution on every subsystem that has been affected. Last but not least, you should invest in technical training for all staff members to help them master the newly implemented system, which brings us to the next point.

- **Training**: Other than for the technical staff, to help them support and develop the new technologies of choice, training may be useful for everybody involved in the project, sometimes including the end users. Indeed, while the goal is to keep the system *isofunctional* and improve the existing one, it is still likely that there will be some impact on the end users. It may be that the UI is changed and modernized, the APIs will evolve somehow, or we move from a fat client for desktop to a web and mobile application. Don't forget that most of these suggestions are also applicable to the five Rs methodology, so you may end up completely replacing one piece of the application with something different (such as an off-the-shelf product), which leads us to the final point.

- **Handling endpoints**: As in the previous point, it would be great if we could keep the API as similar as possible to minimize the impact on the final customers and the external systems. However, this is rarely possible. In most real-world projects, the API signature will slightly (or heavily) change, along with the UIs. Hence, it's important to have a communication plan to inform everybody involved about the rollout schedule of the new project, taking into account that this may mean changing something such as remote clients; hence, the end users and external systems must be ready to implement such changes, which may be impactful and expensive. To mitigate the impact, you could also consider keeping the older version available for a short period.

As you have seen, modernizing an application with a microservice or cloud-native architecture is definitely not easy, and many options are available.

However, in my personal experience, it may be really worth it due to the return on investment and the reduction of legacy code and technical debt, ultimately creating a target architecture that is easier and cheaper to operate and provides a better service to our end users. This section completes our chapter on cloud-native architectures.

Summary

In this chapter, we have seen the main concepts pertaining to cloud-native architectures. Starting with the goals and benefits, we have seen the concept of PaaS and Kubernetes, which is currently a widely used engine for PaaS solutions. An interesting excursus involved the twelve-factor applications, and we also discussed how some of those concepts more or less map to Kubernetes concepts.

We then moved on to the well-known issues in cloud-native applications, including fault tolerance, transactionality, and orchestration. Lastly, we touched on the further evolution of microservices architectures, that is, miniservices and serverless.

With these concepts in mind, you should be able to understand the advantages of a cloud-native application and apply the basic concepts in order to design and implement cloud-native architectures.

Then, we moved on to look at a couple of methodologies for application modernization, and when and why these kinds of projects are worth undertaking.

In the next chapter, we will start discussing user interactions. This means exploring the standard technologies for web frontends in Java (such as Jakarta Server Pages and Jakarta Server Faces) and newer approaches, such as client-side web applications (using the React framework in our case).

Further reading

- The JBoss community, *Undertow* (`https://undertow.io/`)

- Martin Fowler, *Patterns of Enterprise Application Architecture* (`https://martinfowler.com`)

- Netflix OSS, *Hystrix* (`https://github.com/Netflix/Hystrix`)

- Eclipse MicroProfile, *Microprofile Fault Tolerance* (`https://github.com/eclipse/microprofile-fault-tolerance`)

- Chris Richardson, *Microservices Architecture* (`https://microservices.io`)

- Richard Watson, *Five Ways to Migrate Applications to the Cloud* (`https://www.pressebox.com/pressrelease/gartner-uk-ltd/Gartner-Identifies-Five-Ways-to-Migrate-Applications-to-the-Cloud/boxid/424552`)

- Martin Fowler, *StranglerFigApplication* (`https://martinfowler.com/bliki/StranglerFigApplication.html`)

- Paul Hammant, *Strangler Applications* (`https://paulhammant.com/2013/07/14/legacy-application-strangulation-case-studies`)

10
Implementing User Interaction

User interaction constitutes a very important layer in software architecture. This layer comprises all the ways, such as web interfaces and mobile applications, that end users can approach and use our applications.

For this reason, user interaction needs to be implemented with very high attention to detail. A badly designed, poorly performing user interface will compromise the overall user experience, even if the rest of the application is well written and performs really well. And indeed, the user interface can use a number of different tricks to hide issues (such as performance issues) in other layers of the software architecture (that is, *the backend*).

In this chapter, we are going to explore the most widely used technologies for Java applications, both for cloud-native and traditional applications.

This will include frameworks built using the **Java Enterprise Edition** platform (such as **Jakarta Server Pages** and **Jakarta Server Faces**) and more modern JavaScript frameworks for single-page applications (React, in our case).

In this chapter, you will learn about the following topics:

- User interface architecture – backend versus frontend
- Web user interfaces using Jakarta Server Faces and Jakarta Server Pages
- Introducing single-page applications
- Learning about mobile application development
- Exploring IVR, chatbots, and voice assistants
- Omnichannel strategy in enterprise applications

Let's start by concentrating on the architecture of the user interaction layer, or rather, where to put each component and functionality. We will touch on the architectural aspects of building frontend layers for our applications.

User interface architecture – backend versus frontend

It may seem silly to discuss where a **User Interface** (**UI**) must live. After all, it's almost a given – the UI is the forefront of our software architecture, providing the interaction with end users, and for this reason, it must stay at the front, hence the term *frontend*, which is used as a synonym for UI. And everybody agrees on that, without a doubt.

Except that it's not that easy to draw a line as to where a UI starts and where it ends. And, depending on the particular implementation, a number of different components may provide the functionalities needed to build the experience we want to eventually present to our customers. The UI will be made of, more or less, the following components:

- **Assets, also referred to as static files**: These are the pieces of the web application that must be sent (where relevant) to our clients. They include, usually, HTML files, JavaScript scripts, other graphical artifacts (images, CSS files, and movie clips), and even fully built, self-contained applications (as in the case of mobile apps).
- **Data**: This is the content shown using the assets. This implies having a way to retrieve updates (usually involving web services or similar technology).

- **Behavior**: This refers to how the UI reacts to user inputs and other events. This is a broader area that includes *interactivity* (what changes, and how, when our user does something), *validation* (checking user inputs for formal and substantial consistency), *navigation* (how different views, or pages, must be shown one after the other to implement the features requested by the user), and *security* (how to be sure that each user is properly identified and profiled, able to do only what they are allowed to do, and able to access only the appropriate set of data).

The point is more or less this – different implementations will have different ways of providing assets to end users, different ways of providing (and collecting) data, and different implementations of behavior (such as navigation or validation being implemented on the client side or the server side). In this chapter, we will look at the most common ways to arrange all of those components to provide a good user experience. Our next section will be about the most traditional ways to implement this in Java Enterprise Edition – Jakarta Server Faces and Jakarta Server Pages.

Web user interface using Jakarta Server Pages and Jakarta Server Faces

If you have ever worked with Java Enterprise applications developed from 2000 to 2015, chances are you have seen **Java Server Pages** (**JSP**) and **Java Server Faces** (**JSF**) in action. Now widely considered legacy, these two technologies still appear widely in existing Java deployments and are worth knowing about, at least for historical reasons.

Introducing basic Java web technology – Jakarta Server Pages

Jakarta Server Pages (formerly Java Server Pages) is, in essence, a templating technology, allowing you to mix dynamic content written in Java with static content (usually written in HTML). By using JSP, an application server can build a web page to provide to a client (and visualize in a web browser). We already talked about JSP in *Chapter 6, Exploring Essential Java Architectural Patterns*, when talking about server-side **Model View Controller** (**MVC**). If you remember, JSP plays the role of the *View* in the MVC pattern. In the same context, we mentioned servlets as another core component, taking care of the *Controller* part of MVC.

It's now time to clarify the relationship between servlets and JSP. Assuming that most of you know what a servlet is, I will provide just a very brief description.

A **servlet**, in Java, is a special class that implements an API (the Jakarta Servlet API) and is designed to simplify communication over a client-server protocol. Usually, a servlet is created to model HTTP communication, as a way to provide HTML content to a browser. For this reason, in the hierarchy of classes and interfaces supporting the servlet model, there are a number of methods used to handle conversation over HTTP between a client and a server. Some such methods are adhering to the servlet life cycle (such as `init()`, called when the servlet is loaded, and `destroy()`, called before the servlet is unloaded), while others are called when HTTP actions are performed against the servlet (such as `doPost(...)`, to handle an HTTP `POST` request, and `doGet(...)`, to do the same with a `GET` request).

So far so good! Servlets are specialized components, able to handle HTTP conversations, and for this reason, they are used to complement views (such as JSP files) in providing user experience features (such as form submission and page navigation). As we have discussed, it is common for a servlet to be the *controller*, whereas a JSP file is used as the *view*, although more sophisticated frameworks have been developed to deal with JSP, such as Struts and the Spring MVC.

But there is another point worth noting – JSP files are basically servlets themselves.

A JSP file, indeed, is just a different way to implement a servlet. Each JSP file, at runtime, is translated into a servlet by the application server running our code. As we discussed in *Chapter 7*, *Exploring Middleware and Frameworks*, we need an application server fully or partially compliant with the JEE specification in order to run each JEE API (including the servlet and JSP APIs).

So, now that we know that JSP is a templating technology and that each JSP file is translated to a servlet, which outputs to the client what we have modeled in the template, it is time to see what a JSP file looks like.

A JSP file is somewhat similar to a **PHP** one. It mixes HTML fragments with special scripts implementing functions and business logic. The scripts are enclosed within tags, such as `<% ... %>`, which is called a **scriptlet** and contains arbitrary Java code. A scriptlet is executed when a client requests a page.

Another tag used in JSP is marked using `<%= %>` and is called an **expression**. The Java code is evaluated when the page is loaded, and the result is used to compose the web page. Another commonly used JSP feature is **directives**, delimited by `<%@ ... %>` and used to configure page metadata. A very basic JSP page might look like this:

```
<%@ page contentType="text/html;charset=UTF-8"
   language="java" %>
<%@ page import = "java.util.Date" %>
```

```
<%@ page import = "java.text.SimpleDateFormat" %>
<html>
    <head>
        <title>Simple JSP page</title>
    </head>
    <%
       SimpleDateFormat sdf = new SimpleDateFormat("dd-MM-
          yyyy HH:mm:ss");
       String date = sdf.format(new Date());
    %>
    <body>
        <h1>Hello world! Current time is <%=date%> </h1>
    </body>
</html>
```

Some things worth noticing are listed as follows:

- At the top of the files, directives are used. They define some metadata (the page content type and the language used) and the Java packages to be imported and used on the page.

- There is some HTML code interleaved with the scripts. The code inside the script delimiters is largely recognizable to Java developers, as it simply formats a date.

- In the first scriptlet, we define the date variable. We can then access it in the expression inside the <h1> HTML tag. The expression simply refers to the variable. The engine will then replace the variable value in that spot in the generated HTML page.

In a real-world application, we can imagine some useful ways to use such syntax:

- We could use scriptlets to retrieve useful data by calling external services or using database connections.

- By using Java code in the scriptlets, it is easy to implement iterations (usually to display tabular data) and format data in a preferred way (both in scriptlets and expressions).

- Java code in scriptlets can be used to get and validate data provided by the user (usually using HTML forms). Moreover, user sessions and security can be managed in different ways (usually by leveraging cookies).

It's also worth noticing that JSP directives can be used to include other JSP files. In this way, the logic can be modularized and reused.

Moreover, JSP allows defining custom tag libraries. Such libraries are collections of personalized tags that embed custom logic that is executed when a tag is used. A widely used tag library is the one provided by default by the JSP implementation, which is the **Jakarta Standard Tag Library (JSTL)**. In order to use a tag library (JSTL core, in this case), we need to use a directive to import it like this:

```
<%@ taglib prefix="c"
  uri="http://java.sun.com/jsp/jstl/core" %>
```

The JSTL provides a set of tags that offer the following functionalities:

- `Core` provides basic functionalities such as flow control (loops and conditional blocks) and exception handling.

- `JSTL` is used mostly for string manipulation and variable access.

- `SQL` implements basic database connection handling and data access.

- `XML` is used for XML document manipulation and parsing.

- `Formatting` is a set of functions useful for formatting variables (such as dates and strings), according to character encodings and locale.

So, the date formatting that we did in our previous example can be summarized with the appropriate JSTL tags (included from the `Core` and `Formatting` collections) as follows:

```
. . .
<c:set var = "now" value = "<% = new java.util.Date()%>" />
<fmt:formatDate pattern=" dd-MM-yyyy HH:mm:ss "
  value="${now}" />
    <body>
        <h1>Hello world! Current time is <%=date%> </h1>
    </body>
. . .
```

But even if JSTL tags are supposed to reduce the amount of Java code in JSP files, it's very hard to entirely remove *all* Java code. Java code mixed with presentation code is considered an antipattern to avoid, and it's not the only consideration to be made. In the next section, we will see why JSP is considered a legacy technology and almost every Java project today relies on different options for frontend development.

JSP – the downsides

Now, it's time to look at the bad news. There are a number of reasons why JSP is nowadays widely considered unsuitable for modern applications. I can summarize some of those reasons:

- JSP allows for Java code to be interleaved with HTML code. For this reason, it becomes very easy to mix presentation logic with business logic. The result is often an ugly mess (especially in big applications), as it becomes tempting to have presentation logic slip into Java code (such as conditional formatting and complex loops), with the final result being JSP pages that are both hard to read and maintain.

- For similar reasons, a collaboration between different teams with different skills is very difficult. Frontend teams (such as graphic designers) are supposed to work on the HTML sections of a JSP file, while the same file could be being worked on by the backend team for adding business logic-related functionalities. This leads to a resource contention that is hard to solve, as each team may break the other team's implementations.

- Focusing on frontend development, the development cycle is cumbersome and has a slow turnaround. Frontend developers and graphic designers are used to editing an HTML file, refreshing the browser, and immediately seeing the result. With JSP, this is just not possible; usually, the project has to be rebuilt into an artifact (such as a .war file) and redeployed to an application server. There can be solutions to this particular issue (such as exploded deployments, where a .war file is deployed as an extracted folder), but they are usually implemented differently depending on the application server and may have some downsides (such as not covering all the different kinds of modifications to the file or incurring out-of-memory exceptions if performed too many times).

- This leads to another very important point – JSP files require an application server that is fully or partially implementing the JEE specification. **Apache Tomcat** is a common choice here. Indeed, instead of a basic web server serving static content (which is what is used with modern client-side frameworks, as we will see in a couple of sections), you will need a **Java Virtual Machine** (**JVM**) and an application server (such as Tomcat) running on top. This will mean slightly more powerful machines are needed, and fine-tuning and security testing must be performed more thoroughly (simply because Java application servers are more complex than static files serving web servers, not because Java is less secure per se). Furthermore, frontend developers will need to use this server (maybe on their local workstation) for development purposes (and that may not be the simplest thing to manage).

- JSP also lacks a simple way of sharing components between different pages and applications. Also, the JSP tags are usually a bit cumbersome to use, especially in complex applications.

- Moreover, the performance in JSP applications is overall worse when compared to single-page applications. Modern JavaScript frameworks for single-page applications are indeed designed to keep the data exchange with the server at a minimum – after you download the HTML files and assets for the first time, and only then, the data is exchanged. This is not so easy to achieve with just JSP, which, in general, is designed to render the server-side page and download it as a whole.

- Finally, the intrinsic nature of JSP makes things more complex from an architectural point of view. Since, in a JSP file, you can use Java code, which entails calling backend services and doing SQL queries, the flow of calls may become complex and convoluted. *Are you supposed to have connections from the frontend directly to the database? What about services exposed by the backend?* Instead of having a thin, simple frontend layer used mostly for visualization and interactivity, you will have business logic and data manipulation sprawling all over the frontend layer. Not the best situation from an architectural standpoint.

So, now we have seen the basics of JSP, which is a complete template engine that's useful for defining HTML websites and providing some dynamic content written in Java. We've also understood what the limitations of the technology are. It is now worth noticing that JEE provides a more complex and complete framework for building web apps, which is Jakarta Server Faces.

Jakarta Server Faces – a complex JEE web technology

Jakarta Server Faces (**JSF**) is a much more complete (and complex) framework compared to JSP. It implements the MVC pattern, and it is much more prescriptive and opinionated. This means that concepts such as variable binding, page navigation, security, and session handling are core concepts of the framework. It also provides a component-centric view, meaning that it provides reusable components, which include complex view functionalities such as tables, forms, inputs, and validation.

As per the *view* component of the MVC pattern, JSF used to rely on JSP templating. In more recent implementations, this has been switched by default to **Facelets**, which is an XML-based templating technology.

The *controller* part of JSF is implemented by a special servlet, **FacesServlet**, which takes care of things such as resources initialization, life cycle, and request processing.

Finally, the *model* part of JSF is implemented using so-called *managed beans*, which are simply Java classes with a set of properties, getters, and setters. Managed beans are used to bind to pages and components in pages, containing values to be displayed, validating the user input, and handling events. Managed beans can be configured to live within different scopes, including a session (attached to an HTTP session), a request (the same, but with an HTTP request), and an application (living as long as the entire web application does).

There are a number of different JSF implementations, with the most famous being the **Mojarra JSF** (backed by Oracle). Other projects extend such implementation, also providing a suite of reusable components. The most famous ones are **RichFaces** (backed by Red Hat and discontinued for many years), **IceFaces**, and **PrimeFaces**.

This is an overview of the internal architecture and basics of JSF. Without going into too much detail, let's analyze, as we have done for JSP, the downsides of JSF.

JSF – the downsides

Let's start by saying that, whereas JSP is currently considered legacy but sometimes still used here and there for basic tasks (for simple internal web interfaces such as administration panels), JSF is today avoided wherever possible. The reason for this is that, on top of the JSP's downsides, JSF adds some more. This is what I can say about it:

- **JSF is very hard to learn**: While basic tasks are easy to perform, JSF does a lot of things behind the scenes, such as managing the life cycle of pages and building stateful sessions, that are very hard to master. For this reason, it's common to use it in the wrong way or take advantage of only a small subset of all the features provided, making it overkill and difficult to manage for most web applications.

- **JSF is difficult to test**: Unit tests can be written for some components (such as managed beans) but it is very hard to automate all the tests, especially on the *view* side (Facelets), mostly because it's almost impossible to remove logic from that layer (and, as we know, having logic in the *view* layer is a terrible idea).

- **JSF is hard to troubleshoot**: Since, as we said, JSF manages many things behind the scenes (above all, the binding of variables handled by the browser with the values contained in the managed beans), it's really hard to understand the cause of things going wrong (such as variables not being updated and performance issues).

- **JSF lacks a proper implementation for some small but very useful features**: The first things here that come to mind are AJAX communication (where some values on a web page are updated without the need of a full-page reload) and friendly URLs (when the URL of a page can be customized, which is helpful because it makes it easily readable and favored by search engines). For both of those features, there are some workarounds, but they are incomplete, not standard, and in general have been added late to the framework. Those are just two examples, but there are many; it's all down to a general inflexibility of the framework.

The preceding points are enough to understand why, as of today, almost everybody agrees with JSF being a legacy technology that must not be adopted in new projects.

For this reason, it is not worth providing code samples of JSF.

In this section, we have looked at the two major web technologies built using the JEE framework. As we have seen, these technologies, even if they are still widely used, have some big limitations (especially JSF), mostly coming from their *monolithic* approach, meaning that they are tightly coupled with backend implementation, and their limited flexibility.

In the next section, we will look at the widely used alternative to JEE native web technology – the single-page application.

Introducing single-page applications

Single-Page Applications (or **SPAs**) is a broad term that came about to simply describe the behavior of some solutions meant to create web UIs in a lighter, more modern way. The first characteristic of the SPA is the one that it relates to the name. An SPA, in general, bundles all the assets necessary to start user interaction into a single HTML document and sends it to the client.

All the following interactions between the client and the server, including loading data, sending data back, and loading other assets (as images or CSS files), are performed within the page using JavaScript. For this reason, SPAs minimize the communication between the client and server (improving performances), avoid the full-page refresh, and allow for a simpler architectural model.

Indeed, a basic, static file-serving web server (such as **Apache HTTPD** or **NGINX**) is all you need on the frontend (no Java application server is needed). Moreover, the interaction between client and server is almost exclusively limited to web service calls (usually JSON over REST), hence mixing backend and frontend logic (such as doing SQL queries from the frontend layer) is highly discouraged. The most significant downside that I see with SPAs is that there is no standardization of them. Unlike JEE technologies, each framework here provides its own different approach.

For this reason, there are a number of different, well-written, but incompatible implementations of frameworks for building SPAs. Most (if not all) heavily rely on JavaScript and are independent of what's used in the backend (provided that the backend can expose a compatible services layer, such as JSON over REST). In this context, we take for granted that a Java backend (JEE or something more modern, exposing a REST service) is provided.

But it is not uncommon for simpler projects to go for a full-stack approach (using JavaScript also on the backend, usually running on a server such as **Node.js**) or using different backend technology (such as Python or PHP). For the sake of brevity in this chapter, we will explore just one SPA framework, **React**, which is backed by Facebook and widely used for building web applications, from small websites to large and popular platforms such as social networks. But it is worth noting that there are a number of similarly powerful alternatives (such as Angular, Vue, and Svelte), and since no standard is provided, there is no guarantee of the life cycle of any such technology, nor is it possible to easily move code written in one implementation to another. In order to start playing with SPAs, a preamble on the JavaScript ecosystem will be needed.

Basics of the JavaScript ecosystem

I suppose that most of the readers of this book are beginners or experienced Java developers and junior architects, with little or no exposure to JavaScript. Of course, JavaScript is a huge and interesting world that cannot be completely described in just a few paragraphs, so the goal of this section is just to give you the basics, enough for the next couple of sections, which will focus on React.

JavaScript was born in 1995, mostly for programming inside a web browser. Originally designed for Netscape, it was of course then implemented as part of most major web browsers. Even if the name looks very similar, JavaScript doesn't share that much with the Java language, being interpreted (whereas Java is compiled into bytecode) and dynamically typed (so it checks for type safety at runtime, while Java is statically typed, checking for type safety at build time). And there are a number of other differences, including the object model, APIs, and dependency management. JavaScript has been standardized into a technical specification called **ECMAScript**.

Another important topic is Node.js. While, as mentioned, JavaScript was initially executed by engines embedded into web browsers, Node.js is a standalone engine, able to execute JavaScript code outside of a web browser. Node.js is used for server-side development, whereas JavaScript is used for developing server-side logic by implementing web services and integrating with other components such as databases.

The reason I'm mentioning Node.js is not for its use as a backend server (or at least, this is not relevant in this particular context) but because it has evolved as a complete toolbox for JavaScript development. Indeed, it includes **Node Package Manager** (**npm**), which is a utility for dependency management in JavaScript (conceptually similar to Maven in the Java world).

Moreover, it's very often used as a local server for JavaScript development, being very lightweight and supporting the hot reloading of updates. Last but not least, a lot of client-side SPA frameworks (including React) distribute utilities for Node.js, such as command-line interfaces, useful for creating the skeleton of a new application, packaging it for distribution, and so on. Now that we have seen the basics of current JavaScript development, is time to have a look at the framework that we have selected for this chapter, React, in the next section.

Introducing the React framework

React (also known as **ReactJS**) is a JavaScript framework for building SPAs. A very interesting feature of React is that as well as being used to build web applications to be accessed using a web browser, React can be used (through the React Native project) to build native applications to be executed on mobile platforms (Android and iOS) and desktop (Windows and macOS).

React is very simple in its approach, which is based on the concept of components (more about that soon). Moreover, it's also very efficient because it uses the concept of the **virtual DOM**.

Many JavaScript frameworks create web pages and interactions by directly accessing and modifying the **Document Object Model** (**DOM**). The DOM is basically the standard object representing the HTML document rendered by the browser, in the form of a tree starting with an HTML tag.

React uses this alternative approach of building a custom object (called the virtual DOM) that is a partial representation of the DOM, modeling the desired state of the DOM itself (hence the appearance and behavior of the web application). React applications act on this representation. It is then the framework that compares the DOM to the virtual DOM and makes only the necessary updates to the DOM, changing it in an effective and efficient way.

JavaScript syntax extension

JavaScript Syntax Extension (**JSX**) is a technology that's widely used with React. It looks similar to HTML and offers the ability to mix JavaScript code with HTML tags. With this in mind, it can be seen as a template technology, not so different from JSP, which we saw a couple of sections ago. It's also worth noting that, just as JSP transforms everything we write to Java code (and, specifically, to a servlet that uses Java code to output HTML), JSX does exactly the same, transforming JSX code into JavaScript code, producing the right HTML.

It's worth noting that in the JSX world, mixing JavaScript and HTML is not considered an antipattern but is instead encouraged (and often done). This is because even if you implement complex logic with JavaScript, such logic mostly entails frontend-related behaviors (such as when to show one component and optional formatting), so you are less likely to pollute frontend code with things that don't belong in the frontend (such as business logic).

The reason JSX is so popular is that can be used to define React components with a very compact and understandable syntax. This is what a basic React component might look like without using JSX:

```
React.createElement('h1', {className: 'welcomeBanner'},
   `Welcome to our payment system, ${user.fullName} ! `);
```

And this is how to implement the same component with JSX:

```
<h1 className="welcomeBanner">
   Welcome to our payment system, {user.fullName} !
</h1>
```

So, the advantage in terms of readability and effectiveness is evident. And it would become more evident with more complex cases, such as with tags that include other tags (such as HTML lists or other nested tags).

Readability and ease of use are not the only qualities of JSX. It's important to note that JSX will also prevent, by default, injection attacks. An **injection attack** is when, using various techniques, a malicious user injects into your page custom code (such as JavaScript code or arbitrary HTML content). JSX, by default, sanitizes the output, hence neutralizing such attacks with no efforts on the development side. Moreover, JSX is a pretty complete language that can embed conditions and loops, call other functions, and so on. It's a very powerful tool for building UIs in React.

Since we have mentioned React components, it's important now to explain what they are and how they work.

Introducing React components

The **component** is a core concept of React. It is basically a small, embeddable piece of UI that includes structure, appearance, and behavior logic that can be reused.

From a technical point of view, React components are JavaScript functions or classes. Components take, by default, a `props` argument, which is basically an object encapsulating the (optional) properties to be passed to the component. So, this is a component modeled using `function` (and JSX):

```
function HelloWorld(props) {
    return <h1> Welcome to our payment system,
        {props.fullName}</h1>;
}
```

And this is the same component, using a class:

```
class HelloWorld extends React.Component {
    render() {
        return <h1> Welcome to our payment system, {this.props.
            fullName}</h1>;
    }
}
```

As you see, the component, whether defined as a function or a class, basically wraps around a JSX template representing the HTML code to be rendered. Whatever way you define it, you can then use it as a tag, in this case `<HelloWorld/>`, which will be replaced with what is evaluated by executing the component logic. It's worth knowing that to pass properties, you can simply use tag attributes, which will be passed as part of the `props` object. So, in our case, to pass `fullName`, it's enough to use the component as `<HelloWorld fullName="Giuseppe Bonocore"/>`.

You may have noticed that when defined as a class, we are adding our presentation using the `render` method. By default, it is also possible to use the `constructor` method, which takes `props` as a parameter. Such a method can be used to initialize the component. If you need to manage a state in the component (such as when saving local variables), you can do so by accessing the `this.state` object.

Such an object can be, of course, modified too, for which it is worth using the `this.`
`setState` method, which will notify React that something in the state of the component
has changed (and that maybe something in the view must be updated). This can be
particularly useful when associated with UI events, such as the click of a button. The
following code snippet represents a component with a button. Every time a user clicks on
the button, a counter is incremented and saved into the local state of the component:

```
class Counter extends React.Component {
    constructor(props) {
        super(props);
        this.state = {counter: 0};
    }
    render(props) {
        return (
        <div>
            <h2>You have clicked {this.state.counter} times
                !</h2>
            <button onClick={() => this.setState({ counter:
                this.state.counter + 1 })}>
                    Click Me!
            </button>
        </div>
        );
    }
}
```

From here, of course, more complex combinations of event handling and internal
state management can be designed. Last but not least, there are a number of other
callbacks associated with the life cycle steps of the React component, such as
`componentDidMount`, called after the component is rendered on the web page (but
there are many other similar life cycle hooks).

React app structure

Now, we have some basic information on how to create a component and where to place our presentation markup and business logic. *But how should we start to create a basic React application and apply the concepts we have just learned about?*

The most common and easy way is to use the npm utility, which, as we saw a couple of sections ago, comes with the Node.js server. In order to download and install the Node.js server, you can refer to the official website at `https://nodejs.org/it/download/`.

Once you have a working setup of Node, it is enough to run the following command:

```
npx create-react-app myAppName
```

You'll need to change myAppName as needed, of course. Node.js (and npm) will then download all the necessary dependencies and create the folder structure and scaffolding for a basic React application.

Such a structure might look as follows (some files are omitted):

```
myAppName /
        README.md
node_modules/
package.json
        public/
index.html
src/
index.js
App.js
```

The most important files are as follows:

- `README.md` is the autogenerated *readme* file associated with your project and is used for documentation purposes.
- `node_modules` contains the JavaScript dependencies.
- `Package.json` contains the project metadata, including the dependencies needed.
- `public/index.html` is the page template.
- `src/index.js` is the JavaScript file executed as the first file (the entry point).
- `src/App.js` is a de facto standard generated by the `create app` utility. It is basically a macro component that includes all the components and references in an `index.js` file.

So, that is the standard empty folder structure. In order to add our custom components, as per our previous example, we can create a `components` subfolder in `src`.

Each file containing a component will be a `.js` file named using the name of the component (*with a capital initial letter*). For our previous component example, that would be `Counter.js`.

The file should declare the imported dependencies (at least React):

```
import React from "react";
```

Then, make this component available to other components (the last line in the file):

```
export default Counter;
```

In order to use such components in our app, we will need to import them into our `App.js` file as follows:

```
import Counter from "./components/Counter";
```

We can then use them as a tag (`<Counter/>`) in our JSX content.

Finally, in order to test our React application, you can execute this command from inside the project folder:

```
npm start
```

This will execute the `node.js` server and launch your browser to the right page (`http://localhost:3000/`) to see your application running. Another important aspect of a React application is how to interact with the backend APIs. We will look at this in the next section.

Interacting with REST APIs

What we have looked at so far is basically presentation and behavior. A very important feature to consider, in order to implement a real application, is making requests to a backend. A common way to do that is to call REST APIs.

The standard way to call a REST API from a React application is by using the `axios` library.

To install the `axios` dependency in React, you can use the `npm` command:

```
npm install axios
```

You will then need to import the library into the component that is going to make REST requests:

```
import axios from 'axios';
```

And you can then use `axios` to make the usual REST calls (GET, POST, and so on). This is a quick snippet of a REST `get` call reading from an API and saving data to the local state:

```
axios.get(`http://localhost:8080/rest/payments/find/1`)
    .then(response => response.data)
        .then((data) => {
            this.setState({ data: data })
            console.log(this.state.data)
        })
```

Of course, this example can be extended to make use of other REST verbs.

React – where to go from here

The goal of the previous sections was to give you a taste of what it's like to program a web interface using a client-side JavaScript framework. React is one of the most popular choices at the time of writing, so I think it's a good investment to learn at least the basics of it. However, what we have just learned is far from being complete. Here are a few more topics that I suggest exploring in more depth:

- Forms and event handling, in order to implement rich user interaction, including validation and file uploads

- Advanced visualizations, such as lists, tables, and conditional formatting

- Packaging and deploying to production, with considerations about file size optimization, progressive web apps, and best practices

- React Native, or how to target alternative platforms to web browsers, such as Android, iPhone, and desktop apps

- Routing (provided by the React Router dependency), which provides a way to implement navigation between different views

In some cases, React provides a solution, while other times third-party plugins are required. There are a number of resources online; I suggest starting with the official React website and the other resources listed in the *Further reading* section.

Let's quickly recap the evolution of frontend development in Java over time:

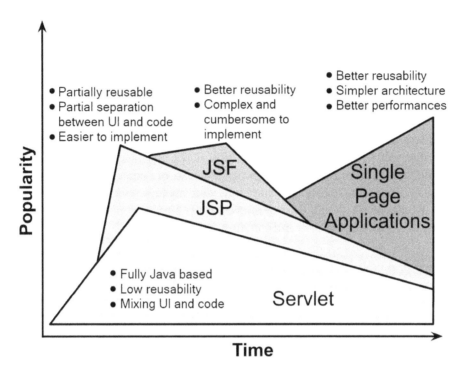

Figure 10.1 – The evolution of frontend development in Java

As we can see, **Servlet** was the first approach and is still somewhat used, regardless of its limitations, for basic use cases and as a supporting technology for more complex frameworks. The same is true for **JSP** (introduced shortly after **Servlet**), which provides some advantages (such as having a markup language that allows for development without directly using Java code).

The usage of JSP has slowed down over time, but it's still used for some use cases. **JSF** started getting traction after **JSP** but stopped gaining popularity soon after, and it is now almost completely abandoned and basically only used in legacy applications. SPAs (based on frameworks such as React) have since emerged and are now very popular and widely used.

With this section, we have completed our overview of web frameworks. In the next section, we are going to take a look at mobile application development.

Learning about mobile application development

Mobile application development shares a lot of concepts (and challenges) with web application development. However, there are also some core differences. In this section, we are going to analyze some core concepts to keep in mind when designing the architecture of a mobile application.

In this context, we are mostly referring to mobile applications as a further channel to access the functionalities offered by a more complex ecosystem that is also accessible in other ways (via a web frontend, at least). Also, most of the considerations made in this section should be seen from an enterprise perspective. So, of course, if you are working in a different environment (such as in a start-up), your mileage may vary. First of all, let's start by looking at why we should consider developing a mobile application as a way to enable interaction with our features and functionalities.

The importance of mobile applications

In today's world, it's trivial to point out that a mobile application is often our first point of contact with many services, such as banking, shopping, and entertainment.

There are around 7 billion mobile users in the world, and it's more and more common for people to possess one or more mobile devices (smartphones and tablets) rather than a laptop or desktop PC.

Mobile devices offer a regulated environment through application permission settings and app stores that, while being a bit more restrictive, highly improve stability and performance standards. That's often a key reason for choosing app interaction over web interaction – it often offers a more standardized user experience and simpler access.

Then, of course, there is convenience – it's way easier to complete a task, whether it's purchasing something or just accessing some information, using a device that you have in your pocket compared to having to use a laptop or desktop PC.

Last but not least, mobile devices are equipped with sensors and functionalities that are key for offering an integrated experience. I can search for a restaurant and immediately call them to book a table or ask for directions while driving there.

Given all that, it's nowadays the default position to think about a *mobile-first* user experience when developing user interactions. But there are a number of challenges to think about.

The challenges of mobile application development

When it comes to the development of mobile applications, the first issue that comes to mind is **fragmentation**. In web applications, modern browsers have almost eliminated incompatibilities between devices, and indeed today you will get the same user experience whether you are using Firefox, Chrome, or Edge on a Macintosh, Windows, or Linux machine. And modern frameworks (such as React, which we saw earlier) make it almost effortless to create such a unified experience.

This is unfortunately not true for mobile devices:

- First of all, you have the form factor to consider. Different models of smartphones have different screen sizes and ratios. They can be used in landscape or portrait mode. Tablet devices add even more variants to the mix.

- The hardware resources may be limited. The rendering of complex animations or heavily interactive features may slow down some low-end devices.

- Connectivity can be unstable. You have to manage what happens to your application when the bandwidth is low or there is a network interruption.

- Devices often offer additional hardware, such as sensors, cameras, and GPS. However, you have to manage what happens when you have permissions to access such devices and what happens when they are denied.

Last but not least, there are, at the time of writing, at least two ecosystems to consider, Google and Apple, which have different distribution channels, different supporting services (such as notifications and updates), and different programming languages and frameworks. This last point we are going to discuss in more detail in the next section.

Mobile application development options

Since the inception of mobile application development, a common topic has been how to manage different platforms (or, at least, Google and Apple) and whether there is a way to partially reuse the effort spent on development for other platforms (such as the web).

The first viable option for reducing fragmentation and leveraging web development efforts is to completely ditch mobile application development and instead go for mobile-optimized web applications. This is a smart option, as with modern web development frameworks and languages (such as HTML5 and CSS3), it's easy to target mobile devices.

These technologies, other than making it easy to create responsive designs that fit nicely in a mobile-optimized layout, create a standard for accessing the most common mobile features, such as position tracking (via GPS), cameras, and microphones.

The most important benefit of this approach is that we can manage a single code base. Even if we want to differentiate between the mobile and web versions of our user interface (it's our choice – we could even just have one single version), at least we can have a single version across all mobile devices, regardless of the underlying technology. The second benefit is that we can keep our publishing process outside app stores, so we are not subject to the timing and regulations that are typically enforced by such distribution channels.

However, there are of course some limitations, with the most significant one being performance. Mobile web applications generally perform worse than their native counterparts. That is particularly true for heavily interactive experiences, such as games or very visual user interfaces. Moreover, mobile web applications have more limited options (if any) to run in an offline or limited-connectivity scenario.

Moreover, mobile web applications are usually less *ergonomic* to access, meaning that the user needs to access the browser and load the application. Even if it's possible to use shortcuts, it's still a more uncomfortable experience than directly finding an app icon in an application list (which is also better from a branding perspective). Last but not least, mobile web applications do not benefit from the visibility that can come from app stores.

A possible alternative to mobile web applications is hybrid applications. In this approach, a mobile web application is enclosed into a native *shell*, which basically is just a slimmed-down, full-screen browser used to act as a bridge between a mobile web application and a device. In such a setup, our application can be published to app stores and can access more native features of the host device. Moreover, it is possible to implement unique code bases, or at least that's the case with the two main technologies of Apple and Google. The downside is that performance and access to native hardware devices will still be limited compared to a fully native application. A notable framework to develop hybrid apps is React Native, which we mentioned previously.

The last option is, obviously, to develop a fully native application. To do that, you will have to use the languages and tools provided for your specific target platform.

Such languages are commonly Swift or Objective-C for Apple devices, and Java or Kotlin for Android ones. The vendors also distribute development environments and tools for building and distributing the applications.

In this way, you will have full control over the device's capabilities and can exploit all the available resources, which can be crucial for some applications. The obvious downside is that you will have to manage two completely different development lines, which means having different skills on the team, dedicated build pipelines, and in general, a duplicated effort.

Regardless of your development choice, you will still have to face a challenge in testing, as checking the application's behavior for all the available platforms properly can be very expensive. In order to partially solve this challenge, it is possible to rely on simulators, which emulate the major mobile devices for testing purposes. Another viable alternative is to use specialized services. There are a number of companies offering a range of mobile devices for testing purposes, which can be rented (as cloud resources) and remotely controlled in order to execute test suites.

In this section, we have looked at web and mobile applications, which are the two most common channels for enterprise services nowadays. But there are some other options that are being used more and more. We are going to learn about them in the next section.

Exploring IVR, chatbots, and voice assistants

Providing more channels for customer interaction is often a very smart investment. It means reaching more people, having a high customer satisfaction rate, and a reduced need for manual interaction (such as assistance provided by a human operator), which can be expensive and also less effective. These goals are important to achieve, and in this section, we'll look at some ways to do so.

Interactive voice response

Interactive Voice Response (**IVR**) is one kind of technology that helps us achieve the aforementioned goals. It provides a way for a human user to interact with services over a phone call. I think that pretty much every one of us has first-hand experience of interacting with an IVR system, as they are pretty common in helpdesk hotlines. The system offers a number of options to choose from. The user can then choose one of the options using a **Dual-Tone Multi-Frequency** (**DTMF**) tone (a tone generated by pressing a button on the phone's number pad) or via voice recognition (which can be harder and more expensive to implement than the DTMF method, as it requires speech-to-text capabilities).

Every option can lead to another set of options. At some point, the customer gets to the desired information, provided by text-to-speech or a recorded message. Another option is to have the call ultimately dispatched to a human operator. While still requiring a human, the presence of the IVR system will most likely filter the most common requests, reducing the number of human operators required, and can provide the operator with data collected from the automatic interactions, such as the user's identity or the problem to resolve.

Algorithmically, an IVR system basically involves *tree traversing*. The customer starts at the root node. At each interaction, the customer is provided with a set of options (the child nodes). The customer can then pick one of the nodes or step back a level (but no further than the root node, of course). At some point, the customer will reach a leaf (the desired information or a human operator).

IVR systems, as we have seen, involve a lot of different technologies, starting with integrating phone calls (both inbound and outbound) and spanning media handling (recorded voice playback), speech-to-text, and text-to-speech. In other words, they are rarely implemented from scratch. In almost every case, in order to implement an IVR system, it is common to rely on packaged solutions. Asterisk, which is a piece of open source PBX software, is used as a common choice for implementing these kinds of systems. Nowadays, however, SaaS solutions are commonly used, requiring just configuration tweaks in order to implement the desired behavior. And since the interaction is so standardized with packaged solutions, and in terms of branding you're limited to the provided recorded voices, a custom IVR implementation is not worth the effort.

Chatbots

Chatbots are basically the same concept transposed to text chats. They achieve the same goal (providing a customized user experience while reducing pressure on human operators), but they don't require text-to-speech, speech-to-text, or recorded voice messages.

The interaction can still be modeled as a tree by providing multiple options to the customer. However, it is common for most chatbot platforms to provide freeform input to customers and try to interpret what the customer is looking for, by parsing the messages and doing what is called **Natural Language Processing** (**NLP**). This process can be complex, involving looking for keywords to analyze the customer's request or even decoding the meaning of the customer's entire message.

Chatbots are less *invasive* than IVR, as they don't require integration with phone infrastructures. There are a number of frameworks available for implementing such solutions, and they are often identified as a perfect use case for the serverless deployment model (see *Chapter 9, Designing Cloud-Native Architectures*). However, as with IVR technology, it is unusual to implement such solutions from scratch nowadays, and it's more common to rely on packaged applications or SaaS solutions.

Voice assistants

Voice assistants are one of the most modern takes on the same issue. Conceptually, voice assistants are kind of a mix between the user experience provided by an IVR system and the one provided by a chatbot. From a user's perspective, voice assistants are consumed from a proprietary hardware and software stack, implemented by what is commonly called a *smart speaker*. The most widespread implementations at the time of writing are Google Home and Amazon Alexa.

This topic is particularly hot, as currently voice assistant applications are still in their infancy, and implementing one is a really unique feature. However, to do so, you will require specific skills, and each vendor relies on proprietary SDKs to build their platforms, which are usually hosted and powered by the cloud provider behind them (AWS and Google Cloud Platform).

Omnichannel strategy in enterprise applications

In this chapter, we explored a number of different options for user interaction, from web applications (which are the most common channel for user interaction), through mobile applications, to some alternative channels, such as IVR, chatbots, and voice assistants.

This opens up a big consideration as to which is the best strategy to go for. Indeed, it is pretty common for enterprise applications to provide many, if not all, of these channels at once. And as a user, we want to interact with applications and get the same information and the same user experience regardless of the channel used.

This poses some serious challenges, from user identity to state management.

There are a number of ways to face such challenges.

The most important thing is to provide a unified backend for all channels. To do so, it is common to use the same services (for example, for identifying a user or searching for saved information) and wrap things using a mediation layer (also known as a *backend for frontend*) in order to optimize the inputs and outputs for a specific device (such as a phone call, a mobile application, or a web user interface). In this way, we can make sure that we provide the same results regardless of what channel is used for interaction.

In doing so, we will provide what is called **multichannel** functionality – the same features are available on different channels and devices (of course, with minor modifications due to the limitations of each device). But there is a further step that can be taken for a more complete user experience, and this is called **omnichannel** functionality. With an omnichannel experience, the user can switch channels during a complex transaction and continue an operation started on a different kind of device with limited or no impact on the user experience and the final result.

The classical example is a mortgage application. A user can call an IVR system asking for information to start a mortgage application. This mortgage application can then be continued using a web application, where the customer can more comfortably provide personal information. After an asynchronous approval process, the customer can then be notified on a mobile app of the mortgage application outcome and complete the process in the mobile app itself.

In order to implement an omnichannel approach, our enterprise application must be capable of storing the state and details of multi-step transactions (such as the mortgage application in our example) in a so-called *state machine*, commonly implemented as a business workflow (as seen in *Chapter 8*, *Designing Application Integration and Business Automation*). It will then be necessary to implement some services to interact with the workflow from the desired channel (as previously mentioned, using a mediator or backend-for-frontend pattern).

A common strategy is also to *codify* some checks (possibly by using a business rule) in order to identify which step can be implemented by which specific channel (and device), as due to the specificity of each channel, it may be impossible (or at least not advisable) to perform certain steps on certain devices. A typical example is IVR. It is usually difficult to properly identify a customer over a phone call. It is possible to check the phone number and to ask for a PIN, but this may be not enough for some operations that are better suited to a mobile device (where we can ask for biometric authentication) or a web application (where we can enforce **Two-Factor Authentication** (**2FA**)).

With this section, we have completed our overview of the most common interaction channels for our users.

Let's summarize what we have learned in this chapter.

Summary

In this chapter, we have explored the core server-side web technologies provided by the JEE platform (JSP and JSF). We explored the pros and cons of these technologies and the main ideas behind them, including interaction with other JEE technologies and standards.

We then moved on to client-side frameworks for building SPAs. We saw how simple and powerful the React framework is and how it can be used to implement componentized interfaces.

We also studied the basics of mobile application development, which is now essential to provide a complete customer experience and can leverage some of the concepts of web application development.

Moreover, we had a look at other interaction channels, such as phone calls (using IVR systems), text chats (using chatbots), and voice assistants. Lastly, we looked at some considerations on how to harmonize all those technologies into a multichannel and omnichannel user experience.

In the next chapter, we are going to focus on the data layer. This will include coverage of relational databases as well as alternatives, such as key-value stores and NoSQL. This will represent another fundamental layer of application architecture.

Further reading

- The Eclipse Foundation – the JSP specification (`https://projects.eclipse.org/projects/ee4j.jsp`)

- *The Problems with JSP, Jason Hunter* (`http://servlets.com/soapbox/problems-jsp.html`)

- Jakarta EE – the JSF specification (`https://jakarta.ee/specifications/faces/`)

- *Why You Should Avoid JSF, Jens Schauder* (`https://dzone.com/articles/why-you-should-avoid-jsf`)

- Meta Platforms, Inc. – the official React website (`https://reactjs.org/`)

- W3Schools – *React Tutorial* (`https://www.w3schools.com/react/`)

11
Dealing with Data

You should know that no matter what your application does, you will end up dealing with persistence sooner or later. Whether it's a payment, a post on social media, or anything else, information has no value if it's not stored, retrieved, aggregated, modified, and so on.

For this reason, data is very much a point of concern when designing an application. The wrong modeling (as we saw in *Chapter 4*, *Best Practices for Design and Development*, when talking about **Domain-Driven Development**) can lead to a weak application, which will be hard to develop and maintain.

In this chapter, we are taking data modeling a step further and discussing the ways your objects and values can be stored (also known as *data at rest*, as opposed to *data in motion*, where objects are still being actively manipulated by your application code).

In this chapter, we will cover the following topics:

- Exploring relational databases
- Introducing key/value stores
- Exploring NoSQL repositories
- Looking at filesystem storage
- Modern approaches – a multi-tier storage strategy

As we have seen with many topics in this book so far, data persistence has also evolved a lot. Similar to what happened to the software development models and the **Java Enterprise Edition** (**JEE**) framework, when we deal with data, we also have a lot of different options to implement in several use cases.

However, just as we have seen elsewhere (namely, in JEE applications versus cloud-native alternatives), the old ways have not been abandoned (because they are still relevant in some cases); instead, they are being complemented by more modern approaches that are suited for other use cases. And this is exactly what happened with the first technology that we are going to discuss – relational databases.

Exploring relational databases

Relational databases are hardly a new idea. The idea was first introduced by Edgar F. Codd in 1970. Omitting the mathematical concepts behind it (for brevity), it says that data in a relational database is, as everybody knows, arranged into *tables* (we had a quick look at this in *Chapter 7*, *Exploring Middleware and Frameworks*, in the *Persistence* section).

Roughly speaking, each table can be seen as one of the objects in our business model, with the columns mapping to the object fields and the rows (also known as *records*) representing the different object instances.

In the following sections, we are going to review the basics of relational databases, starting with keys and relationships, the concept of transactionality, and stored procedures.

Keys and relationships

Depending on the database technology, it's a common idea to have a way to identify each row. This is commonly done by identifying a field (or a set of fields) that is unique to each record. This is the concept of a **primary key**. Primary keys can be considered a constraint, meaning that they represent some rules with which the data inserted into the table must comply. Those rules need to be maintained for the table (and its records) to stay in a valid state (in this case, by having each record associated with a unique ID). However, other constraints are usually implemented in a relational database. Depending on the specific technology of the database system, these constraints may be really complex validation formulas.

Another core concept of the database world is the concept of **relations**. This is, as you can imagine, a way to model links between different objects (similar to what happens in the world of **Object-Oriented Programming** (**OOP**), where an object can contain references to other objects). The relations can fall into one of the following three cardinalities:

- A **one-to-one** relationship represents a mapping of each record to one, and only one, record from another table. This is usually referring to a relationship in which each row points to a row containing further information, such as a user record pointing to a row representing the user's living address in another table.

- A **one-to-many** relationship is where we model a relation in which each record maps to a set of records in another table. In this case, the relation between the two tables is unbalanced. One record in a table refers to a set of related records in another table, while the reverse is not valid (each record maps to one and only one record in the source table). A practical example is a user versus payment relationship. Each user is associated with one or more payments, while each payment is linked to only one user.

- A **many-to-many** relationship is the last option. Basically, in this case, multiple rows from a table can relate to multiple rows in the related tables, and vice versa. An example of this kind of relationship is movies and actors. A record in a movie table will link to more than one row in the actor table (implementing the relation of actors starring in a movie). And the reverse is true – a row in the actor table will link to many records in the movie table, as each actor will most likely be part of more than one movie.

Here is a diagram of the three types of relationship cardinalities:

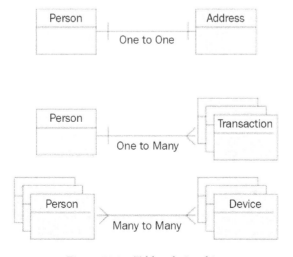

Figure 11.1 – Table relationships

As you can see in the preceding diagram, there is a graphical representation of three examples of relationships:

- **One to One**, as in a person with address – each person can have just one primary home address.

- **One to Many**, as in a person with transactions – each person can be associated with more than one payment transaction.

- **Many to Many**, as in people with devices – each person can have many devices, and a device can be used by more than one person.

These relationships are nothing new; the same is true for Java objects, with which you can model the same kinds of relationship:

- A class can be linked one-to-one with another one, by having a field of it.

- A class can be linked in a one-to-many scenario by having a field containing a list (or a set) of objects of the target class type.

- A class can implement a many-to-many scenario by extending the previous scenario and having the target class type with a field containing a list (or set) of objects of the source class type (hence linking back and forth).

All of those models can then be propagated into SQL databases, and this is indeed done by JPA, which we introduced in *Chapter 7, Exploring Middleware and Frameworks*.

It used to be common (and it still happens in some cases) to define the domain model of an application, starting with the design of the database that will store the data. It's quite a simplistic approach since it cannot easily model every aspect of object-oriented applications (such as inheritance, interfaces, and many other constructs), but it works for some simple scenarios.

Transactionality

One of the more interesting (and widely used) capabilities of a relational database is related to **transactionality**. Transactionality refers to a set of characteristics of relational databases that are the basis for maintaining data integrity (especially in the case of failures). These characteristics are united under the **ACID** acronym (which stands for **Atomicity**, **Consistency**, **Isolation**, and **Durability**):

- **Atomicity**: Each transaction (which is typically a set of different operations, such as the creation, modification, or deletion of records in one or more tables) is treated as a single unit; it will be successful as a whole, or it will fail completely (leaving all the tables as they were before the transaction started).

- **Consistency**: Each transaction can only change the database into a valid state by maintaining all the existing constraints (such as primary keys).

- **Isolation**: The concurrent transactions must be executed correctly with no interference from other transactions. This basically means that the final effect of a number of transactions executed in parallel should be the same as the same transactions being executed sequentially.

- **Durability**: This simply refers to the guarantee that a persisted transaction will be maintained (and can be retrieved) after a failure of the database system. In other words, the database should persist the data into non-volatile storage (a disk or similar technology).

> **Tip**
>
> Consider that the concept of transactionality is usually not very well suited to heavily distributed environments, such as microservices and cloud-native architecture. We will discuss this more in *Chapter 9, Designing Cloud-Native Architectures*.

Last but not least, many different technologies allow us to execute custom code directly on the database.

Stored procedures

Many widely used databases can run complex programs. There is no standard for this kind of feature, even if the languages that are often used are similar to extensions of SQL, including conditions, loops, and similar statements. Occasionally, some general-purpose languages (such as **Java** and **.NET**) are available on some database systems.

The reason for storing code in a database is mainly data locality. By executing code in a database, the system has complete control over execution and transactional behavior (such as locking mechanisms); hence, you may end up getting very good performance. This may be particularly useful if you are doing batch operations and calculations on a large amount of data. But if you ask me, the advantages stop here and are not very worthwhile anyway.

When using stored procedures on a database system, you will observe small performance improvements, but the overall solution will be ugly from an architectural point of view and hard to maintain. Putting business logic in the data layer is never a good idea from a design point of view, and using special, procedural languages (such as the ones often available on such platforms) can only make things worse. Moreover, such languages are almost always impossible to port from one database system to another, hence strongly coupling your application with a specific technology and making it hard to change technology if needed.

> **Tip**
> Unless it's really needed, I advise avoiding stored procedures at all costs.

Now that we have seen a summary of the basic features, let's see the commonly used implementations of relational databases.

Commonly used implementations of relation databases

Let's quickly discuss some commonly used products that provide the relational database features we have seen so far:

- We cannot talk about relational databases without mentioning **Oracle** (`https://www.oracle.com/database/`). The name of this vendor has become synonymous with databases. They provide many variants, including versions with clustering and embedded caching. This database is considered a de facto standard in many enterprises, and most commercially available software packages are compatible with Oracle databases. Oracle databases support Java and PL/SQL (a proprietary language) as ways to define stored procedures.

- **Microsoft SQL Server** (`https://www.microsoft.com/sql-server/`) is another widely used database server. It became popular for its complete features and proximity with the Microsoft ecosystem, as many widespread Microsoft applications use it. It also offers extensions for running .NET languages as part of stored procedures. It's worth noting that for a couple of years, SQL Server has also been supported on Linux servers, widening the use cases for SQL Server, especially in cloud environments.

- **MySQL** (`https://www.mysql.com/`) is another widely used database technology. It's one of the first examples of an open source database and provides advanced features comparable to commercial databases. After the MySQL project was acquired by Oracle, a couple of forks have been created in order to keep the project autonomous. The most important fork currently available is called **MariaDB**.

- **PostgreSQL** (`https://www.postgresql.org/`) is another open source relational database and has been available for a very long time (it was released shortly after the first release of MySQL). In contrast with MySQL, however, it's still independent, meaning that it hasn't been acquired by a major software vendor. For this reason and because of the completeness of its features, it is still a widely used option in many setups. Also, it's worth noting that many different third-party vendors provide commercial support and extensions to cover some specific use cases (such as clustering and monitoring).

- **H2** (`https://www.h2database.com/`) is an open source database written in Java. We played with this technology in *Chapter 7*, *Exploring Middleware and Frameworks*. It's very interesting to use because, being written in Java and released as a `.jar` file, it's easy to use it in an *in-memory* setup as part of the development process of Java applications.

  This includes scenarios such as embedding the database as part of a development pipeline or a Maven task, when it can be programmatically destroyed, created, and launched any time you want. This makes it particularly useful in testing scenarios. Despite more complex setups being available (such as client servers), H2 is usually considered unsuitable for production usage. The most common use case, other than testing and development, is to ship it embedded with applications in order to provide a demo mode when an application is first started, suggesting that a different database should be set up and used before going into production.

- **SQLite** (`https://www.sqlite.org/`) is another type of embeddable database. In contrast with H2, it's written in the **C** language and does not offer any setup other than embedded. Moreover, SQLite lacks some features (for example, it doesn't support some advanced features of SQL). However, due to its robustness and exceptional performance, SQLite is widely used in production environments. The most widespread use case is to embed it as part of a client application. Many web browsers (such as Firefox and Chrome) and desktop applications (such as Adobe Photoshop) are known to use SQLite to store information. It's also widely used in Android applications.

Now that we have seen a brief selection of commonly used databases, let's have a look at the use cases where it's beneficial to use a relational database and when other options would be better.

Advantages and disadvantages of relational databases

Transactionality is the key feature of relational databases and is one of the advantages of using the technology. While other storage technologies can be configured to offer features similar to ACID transactions, if you need to reliably store structured data consistently, it's likely that a relational database is your best bet, both from a performance and a functionality standpoint. Moreover, through the SQL language, databases offer an expressive way to retrieve, combine, and manipulate data, which is critical for many use cases.

Of course, there are downsides too. A database needs a rigid structure to be defined upfront for tables, relations, and constraints (that's pretty much essential and inherent to the technology). Later changes are of course possible, but they can have a lot of side effects (typically in terms of performance and potential constraint violations), and for this reason, they are impactful and expensive. On the other hand, we will see that alternative technologies (such as NoSQL storage) can implement changes in the data structure more easily.

For this reason, a relational database may not be suitable in cases where we don't exactly know the shape of the data objects we are going to store. Another potential issue is that, given the complexity and rigidity of the technology, you may end up with performance and functional issues, which are not always easy to troubleshoot.

A typical example relates to complex queries. A relational database typically uses indexes to achieve better performance (each specific implementation may use different techniques, but the core concepts are often the same). Indexes must be maintained over time, with operations such as defragmentation and other similar ones (depending on each specific database implementation). If we fail to properly perform such maintenances, this may end up impacting heavily on the performance. And even if our indexes are working correctly, complex queries may still perform poorly.

This is because, in most practical implementations, you will need to combine and filter data from many different tables (an operation generally known as a join). These operations may be interpreted in many different ways by databases that will try to optimize the query times but will not guarantee good results in every case (especially when many tables and rows are involved).

Moreover, when doing complex queries, you may end up not correctly using the indexes, and small changes in a working query may put you in the same situation. For this reason, my suggestion is, in complex application environments, to make sure to always double-check your queries in advance with the database administrators, who are likely to have tools and experience for identifying potential issues before they slip into production environments.

As we have seen in this section, relational databases, while not being the most modern option, are still a very widespread and useful technology for storing data, especially when you have requirements regarding data integrity and structure. However, this comes at the cost of needing to define the data structure upfront and in having some discipline in the maintenance and usage of the database.

You should also consider that, sometimes, relational databases may simply be overkill for simple use cases, where you just need simple queries and maybe not even persistence. We are going to discuss this scenario in the next section.

Introducing key/value stores

There are scenarios in which you simply need temporary storage and are going to access it in a simple way, such as by a known **unique key**, which will be associated with your object. This scenario is the best for key/value stores. Within this concept, you can find a lot of different implementations, which usually share some common features. The basic one is the **access model** – almost every key/value store provides APIs for retrieving data by using a key. This is basically the same mechanism as **hash tables** in Java, which guarantee maximum performance. Data retrieved in this way can be serialized in many different ways. The most basic way, for simple values, is strings, but **Protobuf** is another common choice (see *Chapter 8*, *Designing Application Integration and Business Automation*, where we discussed this and other serialization technologies).

A key/value store may not offer persistent storage options, as that is not the typical use case. Data is simply kept in memory to be optimized for performance. Modern implementations, however, compromise by serializing data on disk or in an external store (such as a relational database). This is commonly done asynchronously to reduce the impact on access and save times.

Whether the technology you are using is providing persistent storage or not, there are other features for enhancing the reliability of a system. The most common one is based on data replication. Basically, you will have more than one system (also called *nodes*) running in a clustered way (meaning that they are talking to each other). Such nodes may be running on the same machine or, better yet, in different locations (to increase the reliability even more).

Then, the technology running your key/value store may be configured to propagate each change (adding, removing, or modifying data) into a number of different nodes (optionally, all of them). In this way, in case of the failure of a node, your data will still be present in one or more other nodes. This replication can be done synchronously (reducing the possibility of data loss but increasing the latency of each write operation) or asynchronously (the other way around).

In the upcoming sections, we are going to see some common scenarios relating to caching data and the life cycle of records stored in the key/value store. Let's start looking at some techniques to implement data caching.

Data caching techniques

A typical use case for key/value stores is **caching**. You can use a cache as a centralized location to store disposable data that's quickly accessible from your applications. Such data is typically considered disposable because it can be retrieved in other ways (such as from a relational database) if the key/value store is unavailable or doesn't have the data.

So, in an average case (sometimes referred to as a *cache hit*), you will have better performance and will avoid going into other storage (such as relational databases), which may be slow, overloaded, or expensive to access. In a worst-case scenario (sometimes referred to as a *cache miss*), you will still have other ways to access your data.

Some common scenarios are as follows:

- **Cache aside**: The key/value store is considered part of the application, which will decide programmatically which data should be stored on it, which data will go into persistent storage (such as a database), and how to keep the two in sync. This is, of course, the scenario providing the maximum flexibility, but it may be complex to manage.

- **Read-through** and **write-through**: The synchronization between the key/value store and the persistent storage is done by the key/value store itself. This can be only for read operations (read-through), only for write operations (write-through), or for both. What happens from a practical point of view is that the application interacts with the key/value store only. Each change in the store is then propagated to the persistent storage.

- **Read-behind** and **write-behind**: Basically, this is the same as read-through and write-through, but the sync with the persistent storage is not completed immediately (it's asynchronous). Of course, some inconsistency may happen, especially if you have other applications accessing the persistent storage directly, which may see incorrect or old data.

- **Write-around**: In this scenario, your application reads from the key/value store (by using a read-through or read-behind approach) and directly writes on the persistence store, or maybe other applications perform the write on the persistence store. Of course, this scenario can be dangerous, as your application may end up reading incorrect things on the key/value store.

This scenario can be managed by notifying the key/value store about any change occurring in the persistent storage. This can be done by the application writing data, or it can be done directly by the persistent storage (if it is a feature provided by the technology) using a pattern known as **change data capture**. The key/value store may then decide to update the changed data or simply delete it from the cached view (forcing a retrieve from the persistent store when your application will look again for the same key).

Another common topic when talking about key/value stores is the life cycle of the data.

Data life cycle

Since they use memory heavily, with huge datasets you may want to avoid having everything in memory, especially if the access patterns are identifiable (for example, you can foresee with reasonable accuracy which data will be accessed by your application). Common patterns for deciding what to keep in memory and what to delete are as follows:

- **Least recently used**: The system keeps track of the time of last access for each record and ditches the records that haven't been accessed for a set amount of time.

- **Tenure**: A variant of the previous scenario that simply uses the creation time instead of the last access time.

- **Least frequently used**: The system keeps a count of how many times a record is accessed and then, when it needs to free up some memory, it will delete the least accessed records.

- **Most recently used**: The opposite of least recently used, this deletes the most recently accessed records. This can be particularly useful in some scenarios, such as when it's unlikely that the same key will be accessed twice in a short amount of time.

Key/value stores lack a standard language, such as SQL. It's also for this reason that key/value stores are a big family, including many different products and libraries, often offering more features than just key/value management. In the next section, we are going to see a few of the most famous implementations of key/value stores.

Commonly used implementations of key/value stores

As previously mentioned, it's not easy to build a list of key/value store technology implementations. As we will see in the next few sections, this way of operating a database is considered to be a subcategory of a bigger family of storage systems, called NoSQL databases, offering more options and alternatives than just key/value storage. However, for the purpose of this section, let's see a list of what is commonly used in terms of key/value stores:

- **Redis** is likely the most famous key/value store currently available. It's open source, and one of the reasons for its success is that, despite offering a lot of advanced features and tunings, it just works well enough in its default setting, making adopting it very easy. It provides client libraries for almost every language, including Java. It offers a lot of advanced features, such as clustering, transactions, and embedded scripting (using the **Lua language**). It can operate on in-memory only, or persist the data on the filesystem using a configurable approach in order to balance performance impact and reliability.

- **Oracle Coherence** is a widely used commercial key/value storage. It's particularly used in conjunction with other Oracle products, in particular with the database. It offers a wide range of features, including a complete set of APIs and a custom query language. Since 2020, a community edition of Coherence is available as open source software.

- **Memcached** is a simple key/value store that is light and easy to operate. However, it lacks some features, such as persistence. Moreover, it provides only the cache-aside use case, so other scenarios must be implemented manually.

- **Infinispan** is an open source key/value store that provides features such as persistence, events, querying, and caching. It's worth noting that Infinispan can be used both in an embedded and a client/server setup. In the embedded setup, Infinispan is part of the **WildFly JEE application server**, providing caching services to Java Enterprise applications.

Now that we have seen some widespread key/value stores, let's see when they are a good fit and when they are not.

The pros and cons of key/value stores

The most important advantage of key/value stores is the performance. The access time can be incredibly fast, especially when used without any persistent storage (in-memory only). This makes them particularly suitable for low-latency applications. Another advantage is simplicity, both from an architectural and a usage point of view.

Architecturally speaking, if your use case doesn't require clustering and other complex settings, a key/value store can be as simple as a single application exposing an API to retrieve and store records. From a usage point of view, most use cases can be implemented with primitives as simple as `get`, `put`, and `delete`. However, some of these points can become limitations of key/value stores, especially when you have different requirements. If your application needs to be reliable (as in losing as little data as possible when there's a failure), you may end up with complex multi-node setups and persistence techniques. This may, in turn, mean that in some cases, you can have inconsistency in data that may need to be managed from an application point of view.

Another common issue is that, usually, data is not structured in key/value stores. This means that it is only possible to retrieve data searching by key (or at least, that's the most appropriate scenario). While some implementations allow it, it can be hard, performance-intensive, or in some cases impossible to retrieve data with complex queries on the object values, in contrast with what you can do with SQL in relational databases.

In this section, we have covered the basics of data caching and key/value stores. Such techniques are increasingly used in enterprise environments, for both their positive impact on performances and their scalability, which fit well with cloud-native architectures. Topics such as data caching techniques and the life cycles of objects are common considerations to be made when adopting key/value stores.

Key/value stores are considered to be part of a broader family of storage technologies that are alternatives to relational databases, called NoSQL. In the next section, we will go into more detail about this technology.

Exploring NoSQL repositories

NoSQL is an umbrella term comprising a number of very different data storage technologies. The term was coined mostly for marketing purposes in order to distinguish them from relational databases. Some NoSQL databases even support SQL-like query languages. NoSQL databases claim to outdo relational databases in terms of performance. However, this assurance only exists because of some compromises, namely the lack of some features, usually in terms of transactionality and reliability. But to discuss these limitations, it is worth having an overview of the CAP theorem.

The CAP theorem

The **CAP theorem** was theorized by Eric Brewer in 1998 and formally proven valid in 2002 by Seth Gilbert and Nancy Lynch. It refers to a distributed data store, regardless of the underlying technology, so it's also applicable to relational databases when instantiated in a multi-server setup (so, running in two or more different processes, communicating through a network, for clustering and high-availability purposes). The theorem focuses on the concept of a *network split*, when the system becomes partitioned into two (or more) subsets that are unable to communicate with each other due to connectivity loss.

The CAP theorem describes three core characteristics of distributed data stores:

- **Consistency** refers to keeping the stored data complete, updated, and formally correct.

- **Availability** refers to providing access to all the functionalities of the data store, especially the reads and writes of the data itself.

- **Partition tolerance** refers to the system functioning correctly, even in a case of network failure between servers.

The CAP theorem states that, when a partition occurs, you can only preserve consistency or availability. While a mathematical explanation is available (and beyond the scope of this book), the underlying idea can be understood easily:

- If a system preserves availability, it may be that two conflicting operations (such as two writes with two different values) arrive in two different partitions of the system (such as two servers, unable to communicate between each other). With availability in mind, both servers will accept the operation, and the end result will be data being inconsistent.

- If a system preserves consistency, in case of a network split, it cannot accept operations that will change the status of the data (to avoid the risk of conflicts damaging the data consistency); hence, it will sacrifice availability.

However, it's worth noticing that this theorem, while being the basis for understanding the distributed data store limits, must be considered and contextualized in each particular scenario. In many enterprise contexts, it is possible to make the event of a network split extremely unlikely (for example, by providing multiple network connections between each server).

Moreover, it's common to have mechanisms to elect a primary partition when there's a network split. This basically means that if you are able to define which part of the cluster is primary (typically, the one with the greater number of survival nodes, and this is why it's usually recommended to have an odd number of nodes), this partition can keep working as usual, while the remaining partition can shut down or switch to a degraded mode (such as read-only). So, basically, it's crucial to understand the basics of the CAP theorem, but it's also important to understand that there are a number of ways to work around the consequences.

This is exactly the reasoning behind NoSQL databases. These databases shift their point of view, *stretching* a bit over the CAP capabilities. This means that, while traditional relational databases focus on consistency and availability, they are often unreliable to operate in a heavily distributed fashion. Conversely, NoSQL databases can operate better in horizontally distributed architectures, favoring scalability, throughput, and performance at the expense of availability (as we saw, becoming read-only when there are network partitions) or consistency (not providing ACID transaction capabilities).

And this brings us to another common point of NoSQL stores – the **eventual consistency**.

Indeed, most NoSQL stores, while not providing full transactionality (compared to relational databases) can still offer some data integrity by using the pattern of eventual consistency. Digging into the details and impacts of this pattern would require a lot of time. For the sake of this section, it's sufficient to consider that a system implementing eventual consistency may have some periods of time in which data is not coherent (in particular, enquiring for the same data on two different nodes can lead to two different results).

With that said, it's usually possible to tune a NoSQL store in order to preserve consistency and provide full transactionality as a traditional relational database does. But in my personal experience, the impacts in terms of reduced performance and availability are not a worthwhile compromise. In other words, if you are looking for transactionality and data consistency, it's usually better to rely on relational databases.

With that said, let's have an overview of the different NoSQL database categories.

NoSQL database categories

As we discussed in the previous sections, NoSQL is an umbrella term. There are a number of different categories of NoSQL stores:

- **Key/value stores**: This is the easiest one, as we have already discussed the characteristics of this technology. As should be clear by now, key/value stores share some core characteristics with NoSQL databases – they are generally designed to be horizontally scalable, to focus on performance over transactionality, and to lack full SQL compliance.

- **Document stores**: This is one of the most widespread categories of NoSQL databases. The core concept of a document store is that instead of rows, it stores documents, serialized into various formats (commonly JSON and XML). This often gives the flexibility of storing documents with a different set of fields or, in other words, it avoids defining a strict schema in advance for the data we are going to store. Documents then can be searched by their contents. Some notable examples of document stores include MongoDB, Couchbase, and Elasticsearch.

- **Graph databases**: This category of stores is modeled around the concept of a graph. It provides storage and querying capabilities optimized around graph concepts, such as nodes and vertex. In this way, concepts such as roads, links, and social relationships can be modeled, stored, and retrieved easily and efficiently. A famous implementation of a graph database is **Neo4j**.

- **Wide-column databases**: These stores are similar to relational databases, except that in a table, each row can have a different set of fields in terms of the number, name, and type of each one. Two known implementations of wide-column databases are Apache Cassandra and Apache Accumulo.

Of course, as you can imagine, there is a lot more to say about NoSQL databases. I hope the pointers I gave in this section will help you quickly understand the major features of NoSQL databases, and I hope one of the examples I've provided will be useful for your software architecture. In the next section, we are going to have a look at filesystem storage.

Looking at filesystem storage

Filesystems are a bit of a borderline concept when it comes to data storage systems. To be clear, filesystem storage is a barely structured system providing APIs, schemas, and advanced features, like the other storage systems that we have seen so far. However, it is still a very relevant layer in many applications, and there are some new storage infrastructures that provide advanced features, so I think it's worth having a quick overview of some core concepts.

Filesystem storage should not be an alien concept to most of us. It is a persistent storage system backed by specific hardware (spinning or solid-state disks). There are many different filesystems, which can be considered the protocol used to abstract the read and write operations from and to such specific hardware. Other than creating, updating, and deleting files, and the arrangement of these files into folders, filesystems can provide other advanced features, such as journaling (to reduce the risk of data corruption) and locking (in order to provide exclusive access to files).

Some common filesystems are the **New Technology File System** (**NTFS**) (used in Windows environments) and the **Extended File System** (**ext**) (used in Linux environments). However, these filesystems are designed for working on a single machine. A more important concept relates to the filesystems that allow interactions between different systems. One such widespread implementation is networked filesystems, which is a family of filesystem protocols providing access to files and directories over a network. The most notable example here is NFS, which is a protocol that provides multi-server access to a shared filesystem. The **File Transfer Protocol** (**FTP**) and the **SSH File Transfer Protocol** (**SFTP**) are other famous examples, and even if they are outdated, they are still widely used.

A recent addition to the family of network storage systems is **Amazon S3**. While it's technically an object filesystem, it's a way to interact with Amazon facilities using APIs in order to store and retrieve files. It started as a proprietary implementation for providing filesystem services on AWS infrastructure over the internet; since then, S3 has become a standard, and there are a lot of other implementations, both open source and commercial, aiming to provide S3-compliant storage on-premises and in the cloud.

The advantages and disadvantages of filesystems

It's hard to talk about the disadvantages of filesystems because they are an essential requirement in every application, and it will stay like this for a long time. However, it's important to contextualize and think logically about the pros and cons of filesystems to better understand where to use them.

Application interaction over shared filesystems is particularly convenient when it comes to exchanging large amounts of data. In banking systems (especially legacy ones), it's common to exchange large numbers of operations (such as payments) to be performed in batches, in the form of huge `.csv` files. The advantage is that the files can be safely chunked, signed, and efficiently transferred over a network.

On the other hand, filesystems don't usually offer native indexing and full-text search, so these capabilities must be implemented on top. Moreover, filesystems (especially networked filesystems) can perform badly, especially when it comes to concurrent access and the locking of files.

With this section, we have completed our overview of storage systems.

In the next section, we are going to see how, in modern architecture, it is common to use more than one storage solution to address different use cases with the most suitable technology.

Modern approaches – a multi-tier storage strategy

In the final section of the chapter, we'll be exploring a concept that may seem obvious, but it's still worth mentioning. **Modern architecture** tends to use multiple data storage solutions, and I think that this could be a particularly interesting solution.

In the past, it was common to start by defining a persistence strategy (typically on a relational database or on another legacy persistence system) and build the application functionalities around it. This is no longer the case. Cloud-native technologies, through microservices, developed the idea that each microservice should own its own data, and we can extend this concept in that each microservice could choose its own persistent storage technology. This is better suited for the particular characteristics of that business domain and the related use cases. Some services may need to focus on performance, while others will have a strong need for transactionality and data consistency.

However, even if you are dealing with a less innovative architecture, it's still worthwhile evaluating different ideas around data persistence solutions. Here are some discussion points about it:

- Relational databases are your best bet when data is structured upfront and such a structure doesn't change very often. Moreover, if you will need ACID-compliant transactions, relational databases are generally the most performant solution.

- Key/value stores, especially in their in-memory setup, are useful in a number of use cases. The more common scenarios include the storage of user sessions, which will demand high performance (as it's related to web and mobile use cases, where there is heavy user interaction and high expectation in terms of availability) and consistency/reliability is less of an issue (in a worst-case scenario, the user will be logged out and will need to log in again). Another widely used scenario is database offloading – implementing some of the described scenarios (read-through, write-through, and so on) where cached entries will boost the overall performance and reduce the load on the database.

- NoSQL databases can be used for scenarios particularly suited to the specific technology of choice. In particular, if some entities in our architecture have a variable or unstructured representation, they can be suitable for document repositories. Graph databases can be useful for other scenarios in which algorithms on graphs are needed (such as the shortest path calculation).

- As previously mentioned, filesystems are almost always a fundamental infrastructure. They may be needed by some middleware (such as message brokers) for writing journals, and they can be used explicitly by an application as a data exchange area for large amounts of information (especially when dealing with legacy systems).

So, once again, choosing the right data storage technology can be crucial to have a performant and well-written application, and it's a common practice to rely on more than one technology to meet the different needs that different parts of our application will require.

Summary

In this chapter, we have seen an overview of different possibilities on the data layer, ranging from traditional SQL databases to more modern alternatives.

While most of us are already familiar with relational databases, we have had a useful examination of the pros and cons of using this technology. We then broadened our view with alternative, widespread storage technologies, such as key/value stores, NoSQL, and even filesystems.

Eventually, we looked at how the choice of a particular way of storing data may affect both the application design and the performance of our system. Indeed, in modern architecture, we may want to pick the right storage solution for each use case by choosing different solutions where needed.

In the next chapter, we are going to discuss some architectural cross-cutting concerns. Topics such as security, resilience, usability, and observability are crucial to successful application architecture and will be analyzed to see their impacts and best practices.

Further reading

- *Database Systems: Concepts, Languages, & Architectures*, by Paolo Atzeni, Stefano Ceri, Stefano Paraboschi, and Riccardo Torlone

- *Relational Databases 101: Looking at the Whole Picture*, by Scott W. Ambler (`http://www.agiledata.org/essays/relationalDatabases.html`)

- NoSQL database list – Edlich (`https://hostingdata.co.uk/nosql-database/`)

- *Making Sense of NoSQL: A guide for managers and the rest of us*, by Dan McCreary and Ann Kelly

Section 3: Architectural Context

In this section, we will take an overview of orthogonal context related to architecture design and software architecture implementations. Those cross-cutting topics are crucial to the success of a project and have to be thoroughly considered during the different phases of implementation.

In this section, we are going to cover some additional concepts, useful to define the context of software architectures. These include cross-cutting concerns (such as identity management and security), the software development life cycle (such as source code management, building, and deployment), and visibility (including log management, monitoring, and tracing). Our last chapter will be about the Java framework per se, including versioning, the vendor ecosystem, and what's new in the latest release.

This section comprises the following chapters:

- *Chapter 12, Cross-Cutting Concerns*
- *Chapter 13, Exploring the Software Life Cycle*
- *Chapter 14, Monitoring and Tracing Techniques*
- *Chapter 15, What's New in Java?*

12
Cross-Cutting Concerns

Throughout the previous chapters, we have explored many different aspects of Java application development. Starting from the beginning of the development life cycle (including requirements collection and architecture design), we've focused on many different technological aspects, including frameworks and middleware.

At this point, several cross-cutting concerns need to be examined, regardless of the kind of application we are building and the architectural style we choose. In this chapter, we are going to look at a few of these aspects, as follows:

- Identity management
- Security
- Resiliency

The cross-cutting concerns discussed in this chapter provide some very useful information about topics that are crucial for a project's success. Indeed, implementing identity management, security, and resiliency in the right way can be beneficial to the success of our application, both from an architectural point of view (by providing elegant, scalable, and reusable solutions) and a functional point of view (by avoiding reinventing the wheel and approaching these issues in a standardized way).

With that said, let's get started with a classic issue in application development: identity management.

Identity management

Identity management is a broad concept that deals with many different aspects and involves interaction with many different systems.

This concept is indeed related to identifying a user (that is, who is asking for a particular resource or functionality) and checking the associated permissions (whether they are allowed to do so and so, or not). So, it's easy to see how this is a core concept, common in many applications and many components inside the application. If we have different functionalities provided by different components (as in a microservices application), then obviously each of them will need to perform the same kind of checks, to be sure about the user's identity and act accordingly.

However, having an ad hoc identity management infrastructure for each application can be considered an *antipattern*, especially in a complex enterprise environment, since each application (or component) has the same goal of identifying the user and its permissions.

For this reason, a common approach is to define a company-wide identity management strategy and adopt it in all of the applications, including the off-premises and microservices architectures.

Now, to come back to the fundamentals, identity management is basically about two main concepts:

- **Authentication**: This is a way of ensuring, with the maximum possible degree of certainty, that the person asking for access to a resource (or to perform an action) is the person that they claim to be. Here is a diagram of the username and password authentication method:

Figure 12.1 – Authentication

- **Authorization**: This is a way of declaring who can access each resource and perform a specific action, as shown in the following diagram. This may involve authenticated and non-authenticated entities (sometimes referred to as anonymous access).

Figure 12.2 – Authorization

Both authentication and authorization include two main scenarios:

- **Machine to machine**: This is when the entity requesting access is an application, for example, in batch calculations or other processes that do not directly involve the interaction of a human user. This is also called **server to server**.

- **Interactive** or **use**: This is the other scenario, with a human operator interacting directly with the resource, hence requesting authentication and authorization.

Now that we have the hang of some basic concepts, let's learn a bit more about authentication and authorization.

Authentication

As stated, **authentication** is about verifying that the entity performing a request (be it a human or a machine) is who they claim to be. There are many different ways to perform this verification. The main differentiator is what the user presents (and needs to be checked). It falls into one of the following three categories:

- **Something that the user knows**: This refers to secrets, such as passwords, pins, or similar things, like the sequence to unlock a mobile phone.

- **Something that the user has**: This refers to physical devices (such as badges or hardware tokens) or software artifacts (such as certificates and software tokens).

- **Something that the user is**: In this case, authentication is linked to biometric factors (such as a fingerprint or face identification), or similar things like a signature.

There are several things to consider here, as follows:

- The first is that a piece of public information, such as a username, may be associated with the authentication factor. In this case, multiple users can share the same factor (such as a password or a badge) and we can tell them apart by using the username. The unintentional occurrence of this pattern (such as two users choosing the same password by accident) may be harmless, whereas intentional implementations (multiple users using the same badge) can be a security issue.

- You also have to consider that a combination of more than one authentication factor is considered a best practice and is encouraged for stronger security implementations. This is called **multi-factor authentication** (**MFA**). Moreover, in some specific environments (such as banking) this may be mandated by specific regulations. Strong authentication is often one of those specifics and refers to an authentication process leveraging at least two different factors, belonging to different groups (for example, *something that a user knows*, plus *something that a user has*).

- Some authentication factors may be subject to policies. The most common examples are password rules (length, complexity) or expiration policies (forcing a user to change a factor after a certain time where possible).

Of course, an immediate concern that comes to mind is how and where to store the information relevant for implementing authentication – in other words, where to save our usernames and passwords (and/or the other kinds of secrets used for authentication).

The most common technology used for this goal is **LDAP**, which is short for **Lightweight Directory Access Protocol**. LDAP is a protocol for storing user information. An LDAP server can be seen as a standard way to store information about users, including things such as usernames, emails, phone numbers, and, of course, passwords. Being quite an old standard, around since the 1990s, it's widely adopted and compatible with a lot of other technology.

Without going into too much detail, we can look at it as just another datastore, which we can connect to using a connection URL. Then, we can query the datastore by passing specific attributes to search for specific entries.

The authentication operation against an LDAP server is called **Bind**. LDAP can typically encrypt the passwords in various ways. One very famous implementation of an LDAP server (technically, an extension of it, providing more services than just the standard) is **Microsoft Active Directory**.

LDAP is not the only way to store user information (including passwords) but is likely the only widely adopted standard. Indeed, it is common to store user information in relational databases, but this is almost exclusively done in a custom way, meaning that there is no standard naming nor formats for tables and columns storing usernames, passwords, and so on.

One other way to store user information is to use files, but this is an approach that's not scalable nor efficient. It works mostly for a small set of users or testing purposes. A common file format used to store user information is `.htpasswd`, which is simply a flat file storing a username and password, in a definition originally used by the Apache httpd server for authentication purposes.

It is a commonly accepted best practice to store passwords in an encrypted form whenever possible. This is a crucial point. Whatever the user store technology (such as LDAP or a database), it is crucial that the passwords are not stored in cleartext. The reason is simple and quite obvious: if our server gets compromised in some way, the attacker should not be able to access the stored passwords.

I have used the word *encryption* generically. A solution, indeed, can be to encrypt the passwords with a symmetrical algorithm, such as AES. Symmetrical encryption implies that by using a specific secret key, I can make the password unusable. Then, I can again decrypt the password using the same key.

This approach is useful, but we will still need to store the key securely since an attacker with the encrypted password and the key can access the original password as cleartext. Hence, a more secure way is to store the hashed password.

By hashing a password, you transform it into an encrypted string. The great thing, compared to the previous approach, is that we are implementing asymmetrical encryption. There is no way (if we are using a proper algorithm) to reverse the encrypted string to the original one in a reasonable amount of time. In this way, we can store the encrypted passwords without requiring any key. To validate the passwords provided by the clients, we simply apply the same hashing algorithm used for saving it initially and compare the results. Even if an attacker gains access to our user information store, the stolen encrypted passwords will be more or less useless.

> **Important Note**
>
> It's certainly better to encrypt a password rather than store it in cleartext;
> even the encrypted ones are not 100% secure. Indeed, even if it is impossible,
> in theory, to reconstruct the original password from a hashed value, some
> techniques attempt to do so. In particular, it is possible to try to run a brute-
> force attack, which basically tries a lot of passwords (from a dictionary, or
> simply random strings), hashes them, and compares the output with a known
> actual hash. A more efficient alternative is to use **rainbow tables**, which
> are basically tables of passwords and their pre-computed hashes. Defenses
> against these kinds of techniques are possible, however, by using longer and
> more complex passwords and using salting, which is a way to add some more
> randomness to hashed passwords.

Authorization

User **authorization** is complementary to authentication. Once we are sure that a user is who they claim to be (using authentication), we have to understand what they are allowed to do. This means which resources and which operations they are permitted to use.

The most basic form of authorization is no authorization. In simple systems, you can allow an authenticated user to do everything.

A better approach, in real-world applications, is to grant granular permissions, differentiated for different kinds of users. This is basically the concept of roles.

A **role** can be considered a link between a set of users and a set of permissions. It is usually mapped to a job function or a department and is defined by a list of permissions, in terms of resources that can be accessed and functionalities that can be used. Each user can be associated with a role, and with this, they inherit the permissions associated with that role.

This kind of authorization methodology is called **Role-Based Access Control** (**RBAC**). Based on the kind of RBAC implementation, each user can be assigned to more than one role, with different kinds of compositions. Normally, policies are additive, meaning that a user belonging to more than one role gets all the permissions from both roles. However, this may be subject to slight changes, especially if the permissions conflict, up to the point that there may be implementations denying the possibility of having more than one role associated with each user.

Another aspect of RBAC implementations concerns role inheritance. Some RBAC implementations employ the concept of a hierarchy of roles, meaning that a role can inherit the set of permissions associated with its parent role. This allows for a modular system. In the Java Enterprise world, **JAAS** (short for **Java Authentication and Authorization Service**) is the implementation standard for authentication and authorization. It can be regarded as a reference implementation of an RBAC-based security system.

An alternative to RBAC is **Policy-Based Access Control** (**PBAC**). In this approach, the permission is calculated against a set of attributes, using Boolean logic, in the form of an `if then` statement, where more than one attribute can be combined with `AND`, `OR`, and other logic operators. The attributes can be simply related to the user (such as checking whether a user belongs to a particular group), or to other conditions (such as the time of the day, the source IP, and the geographical location).

`SELinux`, which is a security module underlying some **Linux** OS variants (including **Android**) is a common implementation of PBAC.

Identity and Access Management

Identity and Access Management (**IAM**) is a term usually associated with systems that provide authentication, authorization, and other identity security services to client applications. The function of an IAM system is to implement such features in a unified way, so each application can directly use it and benefit from an adequate level of security. Other than what we have seen here in terms of authentication and authorization, an IAM system also provides the following:

- **Decoupling the user store**: This means that usernames, passwords, and other information can be stored in the technology of choice (such as LDAP or a database), and the client application does not need to know the implementation details. An IAM can also usually unify multiple storage systems in a unique view. And of course, if the user storage system needs to change (such as being moved from LDAP to a database), or we have to add a new one, we don't need to make any changes to the client applications.

- **Federating other authentication systems (such as more IAM systems)**: This can be particularly useful in shared systems where access is required from more than one organization. Most of us have experienced something like this when accessing a service through a third-party login using **Google** or **Facebook**.

- **Single sign-on** (**SSO**): This means that we only need to log in (and log out) once, and then we can directly access the set of applications configured in the IAM.

There are many different ways (and standards) to implement such features, depending on each specific IAM product used. Such standards often boil down to some key concepts:

- **Provisioning and connecting each application managed by the IAM**: This usually means configuring each application to point to the IAM. In the Java world, a common way to achieve this is to configure a servlet filter to intercept all requests. Other alternatives are agent software or reverse proxies that implement the same functionality of intercepting all the requests coming to our application.

- **Checking each request coming to each application**: In case a request needs to be authenticated (because it is trying to access a protected resource or perform a limited action), check whether the client is already authenticated. If not, redirect to an authentication system (such as a login form).

- **Identifying the user**: Once the client provides a valid authentication credential (such as a username and password), it must be provided with a unique identifier, which is regarded as the *ID card* of the user, used to recognize it across different requests (and potentially log in to other applications in an SSO scenario). To do so, the client is often provided with a session token, which may then be stored by the client application (as in a cookie) and usually has a limited lifespan.

A standard way to implement this kind of scenario is the **OAuth protocol**.

However, IAM is not the only security aspect that we need to take care of in a cloud-native architecture. Indeed, the topic of security in an application (especially in a cloud-native one) includes many more considerations. We are going to discuss some of them in the next section.

Security

Security is a very complex aspect, as well as a foundational and crucial one. Unless security is your main focus (which is unlikely if you are in charge of defining the whole architecture of a cloud-native application), chances are that you will have some experts to work with. Nevertheless, it's important to take care of some simple security implications right from the outset of software implementation (including requirement collection, design, and development), to avoid going through a security check after you have completed architecture and development, only to realize that you have to make a lot of changes to implement security (thereby incurring costs and delays).

This approach is often referred to as **shift-left security**, and it's a common practice in **DevOps** teams.

Intrinsic software security

The first aspect to take care of is **intrinsic software security**. Indeed, software code can be subject to security vulnerabilities, often due to bugs or poor software testing.

The main scenario is software behaving unexpectedly as a result of a malformed or maliciously crafted input. Some common security issues of this kind are the following:

- **SQL injection**: A malicious parameter is passed to the application and is attached to a SQL string. The application then performs a special SQL operation that is different from the expected operation and can allow the attacker access to unauthorized data (or even to damage existing data).

- **Unsafe memory handling**: A purposely wrong parameter is passed to the application and is copied to a special portion of memory, which the server interprets as executable code. Hence, unauthorized instructions can be executed. A well-known instance of this kind of bug is the *buffer overflow*.

- **Cross-site scripting**: This is a specific security issue in web applications where an attacker can inject client-server code (such as JavaScript) that is then executed and the attacker can use it to steal data or perform unauthorized operations.

There are several techniques for avoiding or mitigating these issues:

- **Input sanitizing**: Every input should be checked for special characters and anything unnecessary. Checking the format and the length is also important.

- **Running as a user with limited permissions on the local machine (the fewer permissions, the better)**: If there's an unexpected security exception, the impact may be limited.

- **Sandboxing**: In this case, the application will run within a limited and constrained environment. It is kind of an extension of the previous approach. There are various techniques for doing this, depending on the specific application technology. The JVM itself is kind of a sandbox. Containers are another way to implement sandboxing.

The preceding topics are a quick list of common issues (and advice to mitigate them) with regard to software development. However, these approaches, while crucial, are not exhaustive, and it's important to take a look at the overall security of our applications and systems, which will involve some other considerations.

Overall application security

Good overall security starts with the way we write our application but doesn't end there. There are several other security techniques that may involve different IT departments, such as network administrators. Let's look at some of them here:

- **Network firewalls**: They are an integral piece of the enterprise security strategy and are very often completely transparent to developers and architects (at least until you find that some of the connections you want to make are failing due to a missing network rule). The primary duty of firewalls is to block all the network connections unless they are explicitly allowed. This includes rules on ports, protocols, IP addresses, and so on.

  Nowadays, however, firewalls are way more sophisticated than they used to be. They are now capable of inspecting the application-level protocols and are often not only deployed at the forefront of the infrastructure but also between each component, to monitor and limit unauthorized accesses.

  For the same reason, some orchestrator tools (such as **Kubernetes**, but also the public cloud providers) offer the possibility to implement the so-called *network policies*, which are essentially **Access Control Lists** (**ACLs**) acting as a network firewall, hence not allowing (or dropping) unwanted network connections. Firewalls can be hardware appliances (with major vendors including **Cisco** and **Check Point**, among others), or even software distributions (such as **PFSense** and **Zeroshell**).

- **Intrusion Protection Systems** (**IPSes**) (similar to **Intrusion Detection Systems**, with a slight difference in the implementation): These are an extension to firewalls. An IPS, like a firewall, is capable of inspecting network connections. But instead of just identifying authorized and unauthorized routes, an IPS is also capable of inspecting the packages to identify signatures (recurrent patterns) of well-known attacks (such as SQL injections or similar behaviors).

  Moreover, an IPS can inspect other aspects of an application beyond just its network connections. Typically, an IPS can access application logs or even inspect the application behavior at runtime, with the same goal of identifying and blocking malevolent behavior. In this context, IPSes are similar to antivirus software running on workstations. Two common IPS implementations are **Snort** and **Suricata**.

- **Source code inspection**: This is focused on analyzing the code for well-known bugs. While this is a general-purpose technique, it can be focused on security issues. In most cases, this kind of analysis is integrated into the software delivery cycle as a standard step for each release. This kind of test is also named **static software analysis** because it refers to inspecting the software when it is not being executed (hence, looking at the source code).

A technique similar to the previous point is checking the versions of dependencies in an application. This may refer to libraries, such as Maven dependencies. Such modules are checked against databases for known vulnerabilities linked to the specific version. This is part of following the general recommendation of keeping the software constantly patched and upgraded.

All of the aspects seen so far are relevant best practices that can be partially or completely adopted in your project. However, there are contexts where security checks and considerations must be applied in a standardized and well-defined way, which we will see next.

Security standards and regulations

Security is a core concept in applications, especially in some specific industries, such as financial services, defense, healthcare, and the public sector. But it's really a cross-concept that cannot be ignored in any context. For this reason, there are sets of regulations, sometimes mandated by law or industry standards (for example, banking associations), that mandate and standardize some security practices. These include the following:

- **Payment Card Industry Data Security Standard** (**PCI DSS**): This is a very widespread standard for implementing and maintaining IT systems that provide credit card payments. The goal is to reduce fraud and establish the maximum level of trust and safety for credit card users. PCI DSS mandates a set of rules not only on the system itself (such as access control and network security) but also in the way IT staff should handle such systems (by defining roles and responsibilities).

- **Common Criteria** (**CC**): This is an international standard (under the denomination ISO/IEC 15408) that certifies a set of tests for checking the security of an IT system. Such certification is conducted by authorized entities, and the certified systems are registered on an official list.

- **Open Web Application Security Project** (**OWASP**): This approach is a bit different from what we have seen so far. Instead of being a centralized testing institution providing a certification, OWASP is an open and distributed initiative that provides a set of tools, tests, and best practices for application security (especially focused on web application security). OWASP also shares and maintains a list of well-known security issues. The association distributes the **Dependency-Check** tool (`https://owasp.org/www-project-dependency-check`), which helps in identifying vulnerable software dependencies, and the Dependency-Track tool monitors and checks dependency usage.

As we explained, security is a crucial topic that must be considered important in all project phases (from design to implementation to testing) and across all different teams (from developers to testers to sysadmins). This is the reason why we decided to consider it a cross-cutting concern (and why we discussed it in this chapter). To establish and maintain security in our applications, best practices must be taken into account at every step of a development project, including coding. But to maintain a safe system, we should also consider other potential sources of disruption and data loss, and ways to avoid or mitigate them, which we will look at in the next section.

Resiliency

Security is about preventing fraudulent activities, the theft of data, and other improper behavior that could lead to service disruptions. However, our application can go down or provide degraded service for several other reasons. This could be due to a traffic spike causing an overload, a software bug, or a hardware failure.

The core concept (sometimes underestimated) behind the resiliency of a system is the **Service Level Agreement** (**SLA**).

An SLA is an attempt to quantify (and usually enforce with a contract) some core metrics that our service should respect.

Uptime

The most widely used SLA is **uptime**, measuring the availability of the system. It is a basic metric, and it's commonly very meaningful for services providing essential components, such as connectivity or access to storage. However, if we consider more complex systems (such as an entire application, or a set of different applications, as in microservices architectures), it becomes more complex to define. Indeed, our application may still be available, but responding with the wrong content, or simply showing static pages (such as a so-called *courtesy page*, explaining that the system is currently unavailable).

So, the uptime should be defined carefully in complex systems, by restricting it to specific features and defining their expected behaviors (such as the data that these features should provide).

Uptime is usually measured as a percentage over a defined period, such as 99.9% per year.

When considering the uptime, it's useful to define the two possible types of outages:

- **Planned downtime**: This refers to service disruption occurring due to maintenance or other predictable operations, such as deployments. To reduce planned downtime, one technique is to reduce the number of releases. However, this kind of technique may be impractical for modern systems because it will reduce agility and increase time to market. So, an alternative approach is to implement rolling releases or similar techniques to continue to provide services (eventually in a degraded mode) while performing releases or other maintenance activities.

- **Unplanned downtime**: This is, of course, linked to unpredictable events, such as system crashes or hardware failures. As we will see in this section, there are several techniques available for increasing uptime, especially in cloud-native architectures.

  With regard to unplanned downtime, there are several further metrics (I would say *sub-metrics*) that measure certain specific aspects that are useful for further monitoring of the service levels of a system:

  - **Mean time between failures**: This measures the average time between two services outages (as said before, an outage can be defined in many ways, ranging from being completely down to services answering incorrectly). A system with a short mean time between failures, even if still respecting the overall uptime SLA, should be considered unstable and probably fixed or strengthened.

  - **Mean time to recovery**: This measures the average time to restore a system to operation following a period of downtime. This includes any kind of workaround or manual fix to resolve an issue. These kinds of fixes are considered temporary. A system with a high mean time to recovery might need some supporting tools or better training for the team operating it.

  - **Mean time to repair**: This is similar to the previous metric but measures the complete resolution of an issue definitively. The difference between this metric and the previous one is subtle and subjective. A high mean time to repair can signify a poorly designed system or the lack of good troubleshooting tools.

Uptime is not the only SLA to consider when monitoring a system with regard to resiliency. Several other metrics can be measured, such as the response time of an API (which can be measured with something such as *90% of the calls should respond in under 1 millisecond*), the error rate (the percentage of successful calls per day), or other related metrics.

But as we said, there are several techniques to achieve these SLAs and increase system reliability.

Increasing system resiliency

The most commonly used (sometimes overused) technique for increasing system resiliency is **clustering**. A **cluster** is a set of components working concurrently in a mirrored way. In a cluster, there is a way to constantly share the system status between two or more instances. In this way, we can keep the services running in case downtime (planned or unplanned) occurs in one of the systems belonging to the cluster itself.

Moreover, a cluster may involve a redundant implementation of every subsystem (including network, storage, and so on) to further improve resiliency in the event of the failure of supporting infrastructure.

Clusters are usually complex to set up and there is a price to pay for the increase in reliability, usually a performance impact due to the replication of the state. Moreover, depending on the specific application, there are several restrictions for implementing a cluster, such as network latency and the number of servers (nodes).

We discussed a related topic in *Chapter 11, Dealing with Data*, when talking about **NoSQL** repositories and the **CAP** theorem. Since the data inside a cluster can be considered distributed storage, it must obey the CAP theorem.

A cluster often relies on a networking device, such as a network load balancer, that points to every node of the cluster and re-establishes full system operativity in case of a failure, by redirecting all the requests to a node that is still alive.

To communicate with each other, the cluster nodes use specific protocols, commonly known as a **heartbeat** protocol, which usually involves the exchange of special messages over the network or filesystem. A widely used library for implementing heartbeat and leader election in Java is **JGroups**.

One common issue with clusters is the **split-brain** problem. In a split brain situation, the cluster is divided into two or more subsets, which are unable to reach each other via the heartbeat. This usually occurs because of a network interruption between the two system subsets, caused by a physical disconnection, a misconfiguration, or something else. In this situation, one part of the cluster is unaware if the other part is still up and running (but cannot be seen using the heartbeat) or is down. To maintain data consistency (as seen in the CAP theorem in *Chapter 11, Dealing with Data*) in the case of split-brain, the cluster may decide to stop operating (or at least stop writing functionalities) to avoid processing conflicting operations in two parts of the cluster that are not communicating with each other.

To address these scenarios, clusters may invoke the concept of a quorum. A **quorum** is the number of nodes in a cluster required for the cluster to operate properly. A quorum is commonly fixed to *half of the cluster nodes + 1*.

While the details may vary with the type of specific cluster implementation, a quorum is usually necessary to elect a cluster leader. The leader may be the only member of the cluster running, or, more commonly, having other duties related to cluster coordination (such as declaring a cluster fully functional or not). To properly handle split-brain situations, a cluster is usually composed of an odd number of nodes, so if there's a split between two subsets, one of the two will be in the majority and continue to operate, while the other will be the minority and will shut down (or at least deny write operations).

An alternative to the use of a quorum is the technique of **witnesses**. A cluster may be implemented with an even number of nodes, and then have a special node (usually dislocated in the cloud or a remote location) that acts as a witness.

If there's a cluster split, the witness node can reach every subset of the cluster and decide which one should continue to operate.

As we have said, clustering can be expensive and has lots of requirements. Moreover, in modern architectures (such as **microservices**), there are alternative approaches for operating in the case of a failure in distributed setups. One common consideration is about the eventual consistency, discussed in the previous chapter, under the *Exploring NoSQL repositories* section.

For all these reasons, there are other approaches to improving system availability, which can be used as an alternative or a complement to clustering.

Further techniques for improving reliability

An alternative approach to clustering used to improve reliability is **High Availability (HA)**. An HA system is similar to a clustered system. The main difference is that in normal conditions, not all nodes are serving requests. Conversely, in this kind of setup, there is usually one (or a limited number of) primary nodes running and serving requests, and one or more failover nodes, which are ready to take over in the case of a failure of the primary node.

The time for restoring the system may vary depending on the implementation of the systems and the amount of data to restore. The system taking over can already be up and running (and more or less aligned with the primary). In this scenario, it's called a **Hot Standby**. An alternative scenario is when the failover servers are usually shut down and lack data. In this case, the system is called **Cold Standby** and may take some time to become fully operational.

An extreme case of Cold Standby is **Disaster Recovery** (**DR**). This kind of system, often mandated by law, is dislocated in a remote geographical location, and aligned periodically. How remote it should be and how often it is aligned are parameters that will vary depending on how critical the system is and how much budget is available. A DR system, as the name implies, is useful when recovering from the complete disruption of a data center (due to things such as a fire, an earthquake, or flooding). Those events, even if unlikely, are crucial to consider because being unprepared means losing a lot of money or being unable to re-establish a system.

DR is also linked to the concept of **Backup and Restore**. Constantly backing up data (and configurations) is crucial to re-establishing system operation in the case of a disaster or unforeseen data loss (think about a human error or a bug). Backed-up data should also be periodically tested for restore to check the completeness of data, especially if (as is advised) the data is encrypted.

Whether you are planning to use clustering, HA, or DR, two special metrics are commonly used to measure the effectiveness and the goals of this kind of configuration:

- **Recovery time objective** (**RTO**): This is the time needed for a failover node to take over after the primary node fails. This time can be 0 (or very limited) in the case of clustering, as every node is already up and running, or can be very high in the case of DR (which may be acceptable as the occurrence of a disaster is usually very unlikely).

- **Recovery point objective** (**RPO**): This is the amount of data that it is acceptable to lose. This may be measured in terms of time (such as the number of minutes, hours, or days since the last sync), the number of records, or something similar. An RPO can be 0 (or very limited) in a clustered system, while it can be reasonably high (such as 24 hours) in the case of DR.

A last important topic is the **physical location** of the application. Indeed, all of the approaches that we have seen so far (clustering, HA, and DR, with all the related metrics and measurements, such as RPO and RTO) can be implemented in various physical setups, greatly varying the final effect (and the implementation costs).

One core concept is the **data center**. Indeed, each node (or portion) of a cluster (or of an HA or DR setup) can be running on a physically different data center in a specific geographical location.

Running servers in different data centers usually provides the maximum level of resiliency, especially if the data centers are far away from each other. On the other hand, the connection between applications running in different data centers can be expensive and subject to high latency. Cloud providers often call the different data centers **availability zones**, further grouping them by geographical area, to provide information about the distance between them and the users.

However, even if an application is running in just one data center, there are techniques to improve resilience to failures and disasters. A good option is running the application copies in different rooms in a data center. The rooms of a data center, even if belonging to the same building, can be greatly independent of each other. These data centers may apply specific techniques to enforce such independence (such as dedicated power lines, different networking equipment, and specific air conditioning systems). However, it's easy to understand that major disasters such as earthquakes, floods, and fires will be disruptive for all the rooms in the same way. However, hosting in separate rooms is usually cheaper than in separate data centers, and rooms have quite good connectivity with each other.

A lower degree of isolation can be obtained by running different copies of our application (different nodes of a cluster) on different racks. A **rack** is a grouping of servers, often all running in the same room (or close to each other, at least). In this sense, two applications running on different racks may be unaffected by minor issues impacting just one rack, such as a local network hardware failure or power adapter disruption, as these physical devices are commonly specific to each rack.

Of course, a blackout or a defect in the air conditioning system will almost certainly impact all the instances of our cluster, even if running on different racks. For all of these reasons, different racks are cheaper than the other implementations seen so far, but can be pretty poor in offering resilience to major disasters, and are completely unsuitable for implementing proper DR.

A closer alternative to different racks is running our application in the same rack but on different machines. Here, the only redundancy available is against local hardware failures, such as a disk, memory, or CPU malfunctioning. Every other physical issue, including minor ones (such as a power adapter failing), will almost certainly impact the cluster availability.

Last but not least, it is possible to have the instances of a cluster running on the same physical machine thanks to containers or server virtualization.

Needless to say, this technique, while very easy and cheap to implement, will not provide any protection against hardware failures. The only available reliability improvement is against software crashes, as the different nodes will be isolated to some degree.

All of the approaches that we have seen so far offer ways to improve the overall application availability and were available long before cloud-native applications. However, some modern evolutions of such techniques (such as the **saga pattern**, seen in *Chapter 9, Designing Cloud-Native Architectures*) happen to better suit modern applications (such as microservices-based ones).

A topic that is worth highlighting is **reliability**. In the past, reliability was treated exclusively at the infrastructure level, with highly available hardware and redundant network connections. However, nowadays, it is more common to design application architectures that are aware of where they run, or of how many instances are running concurrently. In other words, applications that take reliability and high availability into consideration have become common. In this way, it is possible to implement mixed approaches that provide degraded functionalities if failure is detected in other nodes of the cluster. So, our application will still be partially available (for example in read-only mode), thereby reducing the total outage.

Another technique is to apply **tiering** to functionalities (for example, to different microservices). To do so, it's possible to label each specific functionality according to the severity and the SLA needed. Hence, some functionalities can be deployed on highly resilient, expensive, and geographically distributed systems, while other functionalities can be considered disposable (or less important) and then deployed on cheaper infrastructure, taking into account that they will be impacted by outages in some situations (but this is accepted, as the functionalities provided are not considered core).

All of these last considerations are to say that even if you will never have the job of completely designing the availability options of a full data center (or of more than one data center) in your role as a software architect, you will still benefit from knowing the basics of application availability so that you can design applications properly (especially the cloud-native, microservices-based ones), thereby greatly improving the overall availability of the system.

With this section, we have completed our overview of cross-cutting concerns in software architectures.

Summary

In this chapter, we have seen an overview of the different cross-cutting concerns that affect software architecture. This also included some solutions and supporting systems and tools.

We have learned the different ways of managing identity inside our application (especially when it involves several different components, such as in a microservice architecture).

We had an overview of the security considerations to be made when designing and implementing an application (such as intrinsic software security and overall software security), which are crucial in a shift-left approach, which is the way security is managed in DevOps scenarios.

Last but not least, we had a complete overview of application resiliency, discussing what a cluster is, what the implications of using clustering are, and what other alternatives (such as HA and DR) can be implemented.

In the next chapter, we are going to explore the tooling supporting the software life cycle, with a particular focus on continuous integration and continuous delivery.

Further reading

- Neil Daswani, Christoph Kern, and Anita Kesavan: *Foundations of Security: What Every Programmer Needs to Know*

- The Keycloak community: *The Keycloak OpenSource IDM* (`https://www.keycloak.org`)

- Evan Marcus and Hal Stern: *Blueprints for High Availability: Timely, Practical, Reliable*

13
Exploring the Software Life Cycle

In previous chapters, we explored many different aspects of Java application development. Starting from the **Software Development Life Cycle** (**SDLC**), which includes requirements collection and architecture design, we focused on many different technological aspects, including frameworks and middleware.

At this point, several cross-cutting concerns need to be looked at, regardless of the kind of application we are building and the architectural style we choose.

In this chapter, we are going to explore such aspects. We will start with things such as the maintenance of source code (including versioning and branching strategies), ranging through to some core topics related to deploying, **Continuous Integration/Continuous Delivery** (**CI/CD**), and other concepts related to the SDLC in its entirety.

In this chapter, these are the topics we are going to cover:

- Source Code Management
- Testing
- Deploying
- Continuous integration/continuous delivery (and deployment)

- Releasing
- Maintenance

I'm almost sure that in your professional life, you will already have had the opportunity to become familiar with these topics. Indeed, these are often taken for granted.

Nevertheless, by the end of this chapter, you will have a complete view of the entire process, which will be really useful in structuring and maintaining a functional and efficient software toolchain.

Now, let's start with **Source Code Management** (**SCM**).

Technical requirements

You can find the source code used in this chapter here: `https://github.com/PacktPublishing/Hands-On-Software-Architecture-with-Java/tree/master/Chapter13`.

Source Code Management

SCM is a pretty basic concept and should be considered, of course, mandatory in any software project (including very small ones). Nowadays, SCM is synonymous with Git (more on that soon); however, many alternatives have been used over the years, including **Concurrent Versions System** (**CVS**) and Apache **Subversion** (**SVN**).

The basic function of SCM is backing up, sharing, and versioning source code. However, there are many nuances to these features. So, let's have a closer look at Git.

Introducing Git

Git was created by Linus Torvalds, the creator of the Linux OS, as a tool for supporting the development of the OS itself.

Apart from the history of the project, Git has many interesting characteristics that make it a de facto standard:

- It is heavily decentralized. With Git, every developer can work with a local repository, benefitting the versioning of files, branching, and more features, even in the absence of a remote server (such as in a disconnected environment). This also makes it really scalable from a performance point of view.

- With Git, every version is associated with a cryptographical hash ID. In this way, the history of files can be easily reconstructed, and it makes it hard to tamper with them.
- Git relies on well-known and frequently used protocols, such as HTTP, SSH, and FTP. This makes it easy to use in existing environments.

Git encompasses a lot of different commands and features; however, the basics for using it are as follows:

- `git init [local folder]`: This is the command used to initialize a new repository locally.
- `git clone [repository]`: This creates a local copy of an existing repository. In the case of an authenticated repository, there are many different ways of authenticating, including passing the username and password as part of the URL (using something such as git clone: `https://username:password@remote`). However, better options (such as using tokens) are advisable.
- `git add [files to be added]`: This adds a set of files to a staging area (which is basically an intermediate step before committing to a repository).
- `git commit -m [commit message]`: This commits the files from the staging area to a repository.
- `git branch [branch name]`: This creates a new branch. A **branch** (which is a concept common to many SCM systems) is a way of storing a set of implementations that can potentially have an impact on the rest of the system (such as a new major version) in an isolated area. Such developments can then be merged with the main developments.
- `git tag [tag name]`: This creates a new tag. A **tag** is similar to a branch, but it's basically immutable. It is commonly used to mark a specified important event in the code (such as a release) to make it easier to identify the situation of the code in that particular moment, and potentially rebuild it.
- `git push [remote] [branch]`: This pushes the local changes to a remote repository on the specified branch.

These Git commands and, in particular, the concept of branching and tagging are very powerful tools. How they are used has become more and more structured over time, creating some specified workflows. In the next sections, we'll see some ideas on this.

Git Flow

Despite the name, **Git Flow** (and other similar techniques) is not a prerequisite of Git, and, in theory, could also be implemented with SCM tools, which are different from Git. However, it is common to implement this kind of technique together with Git.

Git Flow is an articulated way of managing and storing developments, creating releases, and, in general, structuring the way the code is handled.

Git Flow is based on a number of branches coexisting constantly and can be implemented manually or by using some tools created to support such a way of working.

The core line where the code is stored is called the **Main** branch. The developers are not supposed to work directly on this branch. Instead, a **develop (Dev)** branch is created from it to store the work in progress. In order to work on a feature, each developer copies the **Dev** branch into a purposely created **Feature** branch, which is created to contain a specific feature. When a feature is completed, it's merged back into the **Dev** branch. In theory, since just a few features are developed, the merge operation should not be too difficult (since not much code has changed). The following diagram illustrates this:

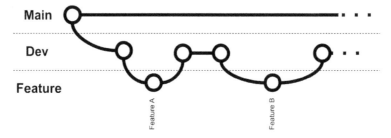

Figure 13.1 – Feature development in Git Flow

As we can see, the simplest situation is when features are developed one after the other, hence the feature we just developed has to be merged back. However, in real situations, it's common to have more than one feature developed in parallel, so the merge back into the **Dev** branch can be slightly more difficult.

When enough features are developed (and have been merged into the **Dev** branch), a new branch is created from the **Dev** branch, called **Release**. The **Release** branch should have some kind of a *feature freeze*, meaning that all the code committed into this branch must only have the goal of releasing and not adding any new features. This means that while tests are going on against the code in the **Release** branch, developers are supposed to commit bug fixes (if any) in this branch.

Other files needed for the release (such as documentation and scripts) can be added there. When the release is ready, the code in the branch will be tagged (that is, *freezed* to a specific version). Then, the **Release** branch is merged back into the **Main** and **Dev** branches so that the developments for the upcoming versions can begin, as shown in the following diagram:

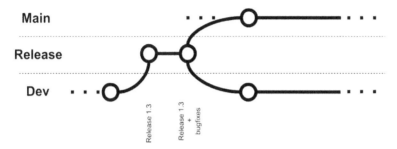

Figure 13.2 – Release management in Git Flow

As we can see, when working on a release, all the code is supposed to be modified in the **Release** branch itself for fixing the issues that prevent this particular release from going into production. Once everything is ready and the production release is successful, the code in the **Release** branch (including the *freezed* code for that release plus the bug fixes, if any) is merged back into the **Main** and **Dev** branches.

If an issue happens in production, an ad hoc **Hotfix** branch is created from the **Main** branch for the purpose of production fixes and merged back as soon as possible, as shown here:

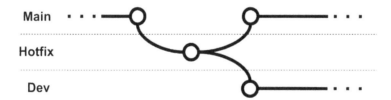

Figure 13.3 – Hotfix development in Git Flow

As seen in the diagram, in the case of a hotfix, the code should come from the **Main** branch and the fixes must be done in the **Hotfix** branch. The changes must then be merged back to both the **Main** and **Dev** branches.

Git Flow seems a bit difficult, and indeed it requires a lot of different branches and merge operations. But it's also considered not very well suited for modern application development techniques, such as CI/CD and DevOps. For such situations, trunk-based development is considered a better choice.

Trunk-based development

Trunk-based development is much simpler than Git Flow. Basically, every developer works on the same branch (the main branch, usually). They are allowed to create branches for local developments, but it's advised to make them as short-lived as possible, and merge them back to the main branch as soon as possible (at least daily). This needs to be done so that the developments are consistent, the tests should pass, and the changes should not break anything else in the project.

With this in mind, trunk-based development is often seen as a perfect pair with CI/CD (more on this later in the chapter). It is possible (and common) to have automated processes constantly checking for the integrity of the main branch (such as after every merge), and in the case of tests failing, changes could be reverted; or, someone in the team (usually the developer of such changes) should focus on fixing the issues. The main branch can be released in production at any time (carrying all the latest developments). Before each release, the code is tagged for traceability and reproducibility of the release.

Trunk-based development, other than being easier to implement and maintain, requires less effort for change management, as merges are smaller and they happen quite often.

On the other hand, it requires great experience and dedication from every project contributor as it increases the possibility of bugs or other bad code slipping into the main trunk (and, theoretically, into production). A related topic to branching strategies is the versioning standard.

Semantic versioning

As we said, during each release, the source code is *frozen* (usually with a tag) and uniquely identified for maintenance purposes.

However, there are many different theories on what the best way for versioning releases is, as in choosing a unique identifier.

A method that I find particularly elegant and effective is semantic versioning.

Semantic versioning is commonly used in open source projects and basically associates each release with three numbers, in the form of *x.y.z* (for example, *1.2.3*). Changes to each of these numbers have a precise meaning:

- The first number (represented by *x* in our example) is called the **major version**. An increase in the major version implies major updates, including new features, re-architecture, technology changes, and, most importantly, potentially breaking changes (including changes in the APIs exposed).

- The second number (represented by *y* in our example) is called the **minor version**. An increase in the minor version implies new functionalities, which can also be non-trivial but are supposed to be backward compatible, so avoid changing the APIs exposed.

- The third number (represented by *z* in our example) is called the **patch version**. An increase in this version just implies bug fixes. No new features should be included (unless very trivial) and, of course, no breaking changes in the APIs exposed.

An increase in the major version implies that minor and patch versions are reset to *0*. So, if we make big changes for version *1.2.3* (by breaking the APIs), the next release should be *2.0.0*.

Similarly, an increase in the minor version resets the patch version to *0*, so for version *1.2.3*, if there are new features that are backward compatible, we go to version *1.3.0*. Needless to say, each version can go to double figures with no impact on the other versions. Hence, an increase in the minor version of the software in version *1.9.3* means going to version *1.10.0*.

After the three numbers compose the version, it is possible to add an optional label. Common labels are **RELEASE** (identifying a version released in production), **ALPHA** (identifying a preliminary version, not intended for production), and **Release Candidate** (**RC**) (this is something almost ready for production, but likely needs some more testing).

It is also a common convention to set the major version to *0* in order to identify the first project draft (such as a prototype not intended to be stable).

In this section, we have learned some interesting concepts. Starting with Git, which is the de facto standard, and SCM, we learned about a couple of different branching strategies, and we had a look at a standard way for versioning releases. Now, our next step is to make some considerations about testing.

Testing

In *Chapter 4*, *Best Practices for Design and Development*, we had a look at **Test-Driven Development** (**TDD**), quickly touching on the concept of **unit testing**. Now is the right time to make some deeper considerations around the concept of testing and return to some topics that we have taken for granted so far.

Unit testing

Unit testing is the most basic technique for software quality assurance and, as we have seen, the tool behind TDD.

Unit testing aims to provide testing (usually automated) for the smallest unit of identifiable software. In the Java world, this means testing at a class and method level. The tests involve calling the method with a defined set of inputs and checking (with assertions) that the output complies with the expectation (including expected failures).

The reasoning behind it is that each method is tested individually, so the tests can be simple and pervasive. This also allows bugs to be identified early and in the exact spot where they are introduced (at least in the exact method). The limitation of this approach is that it doesn't easily detect bugs caused by corner cases or interaction between complex systems, or with external systems.

In the Java world, unit testing means JUnit, which is a very famous library widely used for implementing unit tests and more. Let's learn more about it.

JUnit

JUnit is the de facto standard for unit testing in Java. The current version at the time of writing is **version 5**. JUnit provides some standards and facilities for defining unit tests and integrating them into common toolchains, such as Maven and Gradle. JUnit is also easy to run from a common IDE, such as IntelliJ.

The Maven standard defines that the test classes must be placed in the `src/test/java` folder, whereas the application code is supposed to stay in the `src/main/java` folder.

In this way, the test classes can mirror the same package structure as the application files, and in the release phase, the test classes can then be discarded and not be part of the release artifacts.

JUnit automatically considers (and runs) tests contained in classes whose name starts or ends with `Test`.

Each test method is identified by the `@Test` annotation. It's possible to annotate some method for setting up resources before tests, with annotations such as `@BeforeAll` and `@BeforeEach`. At the same time, it's possible to clean up things after tests, using `@AfterAll` and `@AfterEach`. Test execution can be controlled by using `@Order`.

Moreover, JUnit provides a set of facility methods, such as `AssertEquals`, `AssertTrue`, and `AssertFalse`, which can be used to check for the expected results.

JUnit execution is commonly integrated as a step into a build chain (acting as a part of a Maven build or of a more complex pipeline). You can constantly have a view of what is working and what is failing, often with a visual representation with green and red lights for building reports.

Now, we have a simple class such as the following:

```
package it.test;
 public class HelloWorld {
     private String who;
     public HelloWorld() {
        this.who="default";
    }
     public String getWho() {
        return who;
    }
     public void setWho(String who) {
        this.who = who;
    }
     public String doIt()
     {
        return "Hello "+ this.who;
    }
  }
```

The preceding class basically has a field with a getter and setter, and a method to do the classic *hello world* (with a string concatenation). The unit test class associated with the preceding class is as follows:

```
package it.test;
import org.junit.jupiter.api.Assertions;
import org.junit.jupiter.api.BeforeEach;
import org.junit.jupiter.api.Test;
import io.quarkus.test.junit.QuarkusTest;
public class HelloWorldTest {
    HelloWorld hello;
    @BeforeEach
    public void buildHello()
    {
        this.hello= new HelloWorld();
    }
...
```

A few considerations about the preceding test are as follows:

- The `it.test` package, where the test resides, is the same as the package where the implementation is. As said, this is possible because the implementations stay in the `src/main/java` folder, while the tests stay in the `src/test/java` folder. During the testing phase, you can consider the preceding two folders as the source folders, while, when building the artifact, you can ditch the test folder. This allows us to access `protected` fields and methods on the class to be tested.

- The class name ends with `Test`. This will suggest to the JUnit framework that the class includes some tests.

- The `buildHello` method is annotated with `@BeforeEach`, hence it's executed before each test method. In this case, of course, the implementation is straightforward for the example purpose, but in the real world, there are a lot of meaningful things to be done there, such as initializing fake data and connecting to external systems.

  You can also use `@BeforeAll`, which is executed once before all tests. Also, it's worth noticing that `@AfterEach` and `@AfterAll` are available for the teardown of resources that need to be safely closed (such as database connections) or for cleaning up the necessary data IDs.

- Each test method is annotated with `@Test` and does some assertions on expected output by using the `Assertions.assertEquals` utility method. Other methods, such as `assertTrue` and `assertFalse`, are available as well. As it's easy to spot, simple things such as setters are tested here, which are usually probably not so vulnerable to bugs:

```java
. . .
    @Test
    public void testConstructor()
    {
        Assertions.assertEquals(this.hello.getWho(),
            "default");
    }
    @Test
    public void testGetterSetter()
    {
        String name="Giuseppe";
        this.hello.setWho(name);
        Assertions.assertEquals(this.hello.getWho(),
```

```
            name);
    }
    @Test
    public void testDoIt()
    {
            String name="Giuseppe";
            String expected="Hello "+name;
            this.hello.setWho(name);
            Assertions.assertEquals(this.hello.doIt(),
              expected);
    }
}
```

The previous code can be made more readable by using a `static` import on `Assertions`, and then directly using the methods provided by the class.

When running these tests, you can easily see a recap of test execution. By way of an example, by running the `mvn clean test` command, you should see something similar to this screenshot:

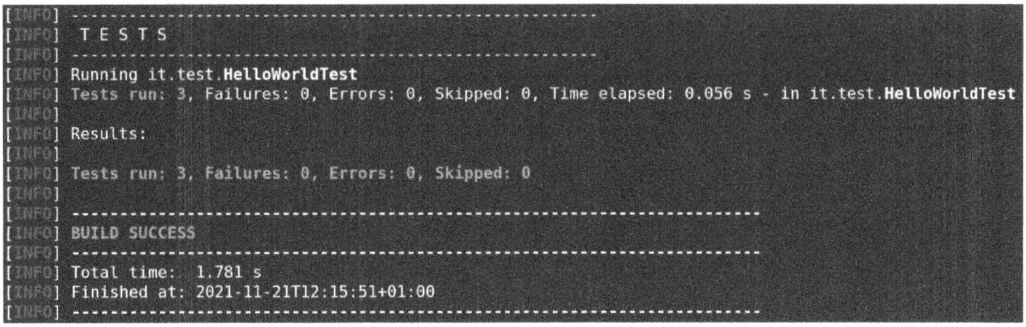

```
[INFO] ------------------------------------------------------
[INFO]  T E S T S
[INFO] ------------------------------------------------------
[INFO] Running it.test.HelloWorldTest
[INFO] Tests run: 3, Failures: 0, Errors: 0, Skipped: 0, Time elapsed: 0.056 s - in it.test.HelloWorldTest
[INFO]
[INFO] Results:
[INFO]
[INFO] Tests run: 3, Failures: 0, Errors: 0, Skipped: 0
[INFO]
[INFO] ------------------------------------------------------
[INFO] BUILD SUCCESS
[INFO] ------------------------------------------------------
[INFO] Total time:  1.781 s
[INFO] Finished at: 2021-11-21T12:15:51+01:00
[INFO] ------------------------------------------------------
```

Figure 13.4 – Test execution in the command line

As you can see, the build succeeds, and there is a recap of the executed tests (that were successful). If a test fails, by default, the build fails. If we know that there is a test intentionally failing (because, as an example, the method is not yet implemented, as it happens in the TDD methodology), we can skip that particular test (by annotating it with `@Disable`) or skip the testing phase completely (which is usually not advised).

In the case of a big project, usually, the testing results are then saved and archived as part of a build process. This may simply mean saving the console output of the build with the test recap (as seen in the preceding screenshot) or using more sophisticated techniques. By using a widespread Maven plugin (**Surefire**), it's easy to save test results as `.xml` or `.html` files, although more complete commercial test suites are able to do similar things.

But this was just about unit testing. To complete our view, it's useful to understand that more ways of testing are possible (and advised). Let's have a look at them in the following sections.

Beyond unit testing

Unit testing, indeed, can be seen as the (basic and essential) lowest step in the testing world. Indeed, as we have already said, unit testing is unable to catch some bugs that depend on more complex interactions between classes. To do so, more testing techniques are usually implemented, such as integration, end-to-end, performance, and **User Acceptance Testing** (**UAT**).

Integration testing

Integration testing is the immediate next step after unit testing. While unit testing tests the most atomic modules of software, such as methods and classes, integration testing focuses on the interaction of such modules with each other (but not on the entire system). So, the classes are put together and call each other to check (and realize) more complex testing scenarios. Each test involves more than one method call, usually from different classes.

There is no fixed rule for defining the granularity of each integration test, even if someone completely ditches this testing technique in favor of end-to-end testing (more on this in the next section). My personal suggestion is to at least add integration testing for the more complex functionalities by trying to involve at least two or three classes simulating the core features or at least the ones most impacted by changes (and by issues).

While there are libraries that can be implemented specifically for integration testing (`arquillian` comes to mind), JUnit is perfectly usable (and widely used) for integration testing, too, by using the same facility (such as assertion and setup methods) as seen in the previous section. Of course, instead of building tests for testing each method and class, more complex interactions are supposed to be implemented by chaining method calls and plugging different classes together.

End-to-end testing

End-to-end testing, also known as **system testing**, takes the integration testing ideas a bit further. End-to-end testing involves the testing of each functionality as a whole, including external systems (such as databases), which are commonly dedicated and maintained as part of the testing efforts. The functionality can be defined at many different times, but nowadays usually overlaps with the concept of an API.

End-to-end testing includes calling an API (or triggering a functionality differently) by passing a known set of inputs and then checking the expected outputs. This will include, alongside the API response, also checking the status of external systems that are supposed to be changed (such as things edited in databases or external systems that are supposed to be contacted, such as sending emails).

It is implied that the system is then tested *from the outside*, as opposed to the other testing techniques seen so far, which are more focused on the source code (and then more looking at the project *from the inside*).

End-to-end testing provides a good idea of system behavior *as a whole*, and it's usually less stable than unit and integration testing because a small change in any of the methods can cause failures in many end-to-end tests depending on that specific method (and go undetected or have a smaller impact on unit and integration tests). However, it is also coarser-grained, so it can be a bit more difficult to understand where and why things are breaking.

There are a lot of tools for end-to-end testing, both free and commercial. Such tools are usually more of a kind of standalone platform, as opposed to the tools seen so far (such as JUnit), which are more libraries and frameworks. Moreover, end-to-end testing is basically language-independent, hence, Java projects don't usually need any specific testing tool, as the entry points for testing are APIs or user interfaces. So, any tool capable of interacting at that level can be used.

Commonly used solutions include LoadRunner, a commercial solution originally built by HP and now part of Micro Focus, which is the standard in some projects. Other alternatives are the SmartBear testing suite and other free testing suites, such as JMeter, Cypress, Karate, Gatling, and Selenium. The last two tools are more focused on automating the user interface interactions, which means that there are basic ways to automate the programmatic use of web browsers, simulating a real user accessing a web application, checking all the expected behavior.

Performance testing

Performance testing is a special case of end-to-end testing. Instead of being exclusively focused on the correct implementation of each API (or feature) tested (which is basically taken for granted), performance tests focus on system capacity and the response time under different loads. While the tools can be (and often are) similar to end-to-end testing, the final goal is different. The metrics measured when doing performance testing are the ones described in *Chapter 6, Exploring Essential Java Architectural Patterns*, in the *Designing for large-scale adoption* section, and include throughput, response time, and elapsed time.

Performance testing can include the following scenarios:

- **Load testing**, which is measuring the performance of the system against a defined load, is usually similar to the one expected in production (or an exaggerated case of it, such as doubling the expected number of concurrent users).

- **Spike testing**, which is similar to the previous one (and indeed they are often run together), basically involves sudden changes in the load of traffic to simulate spikes. This test aims to check the scalability of the system and the time needed for recovery following a sudden traffic increase. In other words, it's often allowed to have a slight slowdown after an unexpected increase in traffic (because the system is expected to adapt to such traffic, such as using an autoscaling technique), but it's worth measuring how long it takes for the system to recover following such a slowdown.

- **Stress testing**, which takes the previous test types to the extreme, aims to benchmark the system by measuring the maximum traffic that can be correctly handled by the system.

During a performance test, regardless of the tools and objectives, it's strongly advised to observe the system *as a whole*, including OS parameters (such as memory, CPU, and network) and external systems, such as databases, in order to check for bottlenecks and understand how the system can be fine-tuned to perform better.

User acceptance testing

UAT is a crucial step of the testing process, and I'd say of the whole software development process. Technically speaking, it is quite similar to end-to-end testing, by focusing on testing functionalities as a whole. There are, however, some crucial differences. The first one is that the test is supposed to be governed by a functional analyst, business people, or the project sponsor.

This, of course, doesn't mean that these people should be running the tests themselves, but that the test structure (including the acceptance criteria and the priority of the features tested) should be set by those teams, and this is usually done by focusing on the point of view of the end user (hence the name of this phase).

It's accepted that part of this test is done manually, with users directly navigating the application feature as a final user is supposed to. A more deterministic approach is to also run the UAT using automated tools, similar to the ones used in end-to-end testing. Even in this case, it's common to still perform a small part of this phase manually, by doing what is called a **smoke test**, which is less structured and aimed at giving a general idea of how the application behaves.

Whether being done manually or automated, there is a core difference between UAT and the other tests seen so far, and this difference is that the tests need to be designed around business capabilities. Indeed, each test case is supposed to be related to a specific requirement in order to prove that this requirement is currently implemented in that particular software release. We already discussed this in *Chapter 2, Software Requirements – Collecting, Documenting, Managing*, where we saw how each software requirement is supposed to be testable.

This is where that loop closes. The successful execution of UAT is the gateway for the production release (hence the word *acceptance*); if all the tests succeed, it is, of course, safe and accepted to release in production. In case of any failure, a choice needs to be made (usually discussed with an extended team).

If minor issues occur, this may mean that the release will go into production anyway, with several known issues. Of course, if this is not the case and the issues are too many (or related to critical features), then the issues need to be fixed and the production release may be canceled or delayed. UAT is basically the higher rank of tests, but it's important to understand that it's crucial to have a strategy around every other testing technique seen so far; otherwise, it's likely to have software that is not completely tested and prone to errors. However, some other considerations around testing are worth noticing.

Further testing considerations

In the previous sections, we saw quite a few interesting things on testing, including the different testing phases and techniques. However, there are a number of other considerations that are worth a few words.

Interacting with external systems

As we have seen, in most of the phases of testing (sometimes even in unit testing), external systems may be involved. Databases, mail servers, and web services are common examples.

There are a number of different techniques for dealing with such external systems in a testing phase. The easiest one, better suited for some specific testing phases, such as unit tests, is to simply mock such systems. This means implementing special custom classes that simply simulate the interaction with such systems, instead of just providing fake values.

A very widespread library for mocking in Java is `Mockito`, which offers a simple but very complete setup to implement methods that react to the requests in a programmable way, simulating the behavior of external systems.

Mocking is handy to use because it's mostly driven by code and requires minimal maintenance. However, as it's easy to understand, it provides limited effectiveness in tests because it tests just a small part of the interaction, often neglecting some aspects such as the connection to the external system (and things that may go bad there), and in general, doesn't test against real (or close to real) systems.

A step further is to effectively use an external system in tests, but a simplified one. The most common example is using H2 (the embeddable database we saw in *Chapter 7, Exploring Middleware and Frameworks*) in place of a full-fledged database system. The reason behind it is that it may be harder (and more expensive) to use compared to the real system, while such simplified tools are usually easier to automate and use in a testing environment.

However, as we discussed in *Chapter 9, Designing Cloud-Native Architectures*, when talking about 12-factor applications, using external services (backing services, as defined in that context) different from the production ones should be considered a source of potential instability as, of course, the behavior may be different from real systems. So, especially in phases such as end-to-end testing and UAT, it's strongly advised to use external systems that are as close as possible to the production ones. This leads us to the next consideration on ephemeral testing.

Ephemeral testing

Ephemeral testing is a technique for creating complete test environments when needed. This basically means that the set of components needed for testing, including the application and the external systems, is created on-demand before each test runs, populated with the data and the configuration needed for the test execution.

Such environments can then be disposed of after each test runs, avoiding wasting computational resources when not needed. This paradigm is particularly suited for IaaS and PaaS environments (as seen in *Chapter 9, Designing Cloud-Native Architectures*) because such infrastructures will facilitate the scripting and automation around environment creation and disposal. IaaS and PaaS are also suited to recreating not only the application components themselves but also the external services (such as databases), and so are a good way to overcome the limitations that we have described in the previous section, and in particular with mocks.

Testcontainers (`www.testcontainers.org`) is an open source framework very well suited for this kind of scenario. It supports JUnit (as well as other testing frameworks) and provides throwaway containerized instances of testing utilities (such as common databases, Selenium browsers, and more).

But having all the right components is not the only consideration to be made in order to have a meaningful and complete testing strategy.

Code coverage, test coverage, and maintenance

A crucial topic of testing is coverage. **Code coverage** basically implies that every line of code, including the ones reached after `if` conditions, loops, and so on, is hit by at least a test case.

These kinds of metrics are not easy to measure manually and indeed are commonly calculated by relying on external tools. The most commonly used technique by such tools is **bytecode instrumentation**, which uses special features of the JVM to check code execution per line, as a result of tests running. Common libraries used for calculating code coverage, such as Cobertura and JaCoCo, use a similar approach.

Code coverage is the baseline of test completeness: a certain threshold must be defined, and a lower coverage should be considered as incomplete testing, especially when creating new functions and modules. But code coverage doesn't ensure that all the features are tested, nor that the data used for tests is complete and variable enough. Hence, a further concept—**test coverage**—must be introduced.

Test coverage is, in my opinion, a bit less *scientific* to calculate. Whether code coverage is exactly measurable (even if it requires tools) as the percentage of lines of code executed during tests versus the total lines of code, test coverage revolves around many different points of view. Some common ones are as follows:

* **Features coverage**, as in the number of features tested versus the total features of the application.

- **Requirements coverage**, as in the number of requirements effectively tested versus the total requirements that the software implements.

- **Device coverage**, particularly meaningful in web and mobile applications, is related to the number of different configurations (different mobile devices, multiple OS versions, multiple browsers versions, and so on) that our application is tested against.

- **Data coverage**, related to the different inputs and configuration that our application is tested against. This is, of course, very difficult to test against, as the combination can really be limitless. On the other hand, having a good variety of inputs to test ensures better protection against unexpected behaviors.

Both code coverage and test coverage should be constantly measured and possibly improved. That implies covering the code added with new features (if any), and checking against the bugs found in the current software releases in order to understand whether there is a way to improve test coverage to check for such bugs in the future. Customer reports are particularly useful in this sense.

Most of us are familiar with issues in the applications we use (especially in mobile applications) when, following an error message, there is the opportunity to send details of the error to the application team. By doing so, the application team has the opportunity to check for the particular conditions (inputs used, device used, and software version) at the time of that particular error, and can potentially extend the test suite (and, hence, the test coverage) to check for similar situations and avoid this family of errors in the future.

Last but not least, it's important to understand when to run what kind of tests.

Since running a whole test suite (including acceptance tests run by human operators) could be expensive and time-intensive, it's a common choice to have different test suites run in different situations. If we are fixing a minor bug or adding a small functionality, then we can probably take the risk of not testing the whole application, but just a subset.

However, with test automation becoming more and more pervasive and disposable test environments made possible by the cloud, the advice is to test as much as possible, especially for performance. In this way, it will become easier to understand whether the release we are testing introduces any performance issues. This concept, taken to the extreme, is called **continuous testing**, and basically implies running the complete suite of tests, in an automated way, after every code or configuration change (even the smallest one).

In this section, we have seen a complete overview of the different testing techniques and phases. Starting with unit testing, we also explored integration, end-to-end, and performance tests.

In the next section, we are going to talk about a step that is contiguous to (and mutually dependent on) testing: deployment.

Deploying

Software deployment is a very broad term, and can extensively be used to refer to the whole software life cycle, from development to release into production. However, in this particular context, I am referring to the deployment phase as the one in which the software is compiled and opportunely packaged, and the right configurations are applied. The software is then supposedly run and made ready for users to access (which is part of the process of releasing; more on this in a couple of sections). While we already mentioned some of these topics in this book, I would like to highlight a couple of them, useful for the purpose of this chapter.

Building the code

The phase of building the code, intended as compilation and packaging into a deployable artifact (`.jar`, `.war`, and `.ear` in the case of Java), is done by utilities shipped with the JDK (in particular, the `javac` tool).

However, this process often includes at the very least the management of dependencies, but in the real world, many other steps can be involved, such as code formatting, the parsing of resources or configuration files, and the execution of unit tests (as seen before).

A widely used technology to perform all of those steps, mentioned and used many times in this book, is Apache Maven.

While I assume that most of you already know and have used Maven, I think it is relevant to highlight some features that are useful to consider from an architect's perspective:

- Maven uses a standard way (the `pom.xml` file) to define instructions about building the software. Indeed, the `pom` file collects the list of dependencies (including the one needed just for testing purposes and the one needed just at development time). It can also specify the steps needed for the compilation and packaging of the software and provide some configurations for each step.

- It provides an extensible system, based on plugins. Hence, you can find (or implement) different plugins, to run tests, create documentation, generate code, and other steps that could be useful at deployment time.

- It can define a hierarchy between different projects and provide a unique way to identify each software artifact (also called a Maven artifact) by setting what is called the **Group, Artifact, Version (GAV)** coordinate standard, a triplet made by `GroupId` (basically a namespace or package for the project), `ArtifactId` (the identifying name of the project), and the version.

As said, Maven is basically a standard technology for building in the Java world, though it's not the only one. **Ant** is another option that used to be widely used some years ago and is more similar to scripting and less flexible. It has been progressively abandoned for its verbosity and a number of shortcomings in dependency management.

Gradle is a modern alternative to Maven, mostly widespread in the context of Android application development. However, the concept of code building raises the need of archiving and managing the dependencies, in the form of built artifacts (which, as we have seen, are referenced uniquely into the `pom.xml` configuration files). Let's have a quick overview of this concept.

Managing artifacts

Software artifacts are essential, both as part of the building process of bigger components (as dependencies) and to be directly deployed (as an example, to production environments). While it's possible to directly manage such contents (which, in the Java world, are basically `.jar`, `.ear`, and `.war` files) in filesystem folders, it's way better to do so in optimized systems. Such systems are called **artifact repositories**.

An artifact repository often provides many advanced features, including the following:

- **Web interfaces**, to simplify the search of the artifacts and the management of them.
- **Role-based access control and authenticated access**, providing differentiated access to different artifacts. A common example is that some artifacts can be changed only by certain groups, while others can be accessed in read-only mode, and maybe others have no access at all.
- Other **security features**, such as inspecting the code for known vulnerabilities (as we have seen in *Chapter 12*, *Cross-Cutting Concerns*, when talking about security).
- **Versioning of the dependencies**, including the cleanup of older versions under configurable policies.
- **Mirroring and hierarchy**, by providing the possibility of querying other artifact repositories over the internet to look up dependencies not available locally, and then mirroring it in order to avoid downloads when not necessary. A very famous repository available over the internet is **Maven Central**.

Two very famous implementations of Maven artifact repositories are JFrog Artifactory and Sonatype Nexus.

It's worth noting that the container technology (as discussed in *Chapter 9, Designing Cloud-Native Architectures*) is often seen as an extension (but not a replacement) of Java artifacts. Indeed, a container image contains a complete application component, including dependencies at an OS level, where JVM and other middleware are needed. Moreover, a container image is immutable and can be uniquely identified (by using a version and a signature), so it is really similar to a Java artifact and raises similar needs in terms of management.

For this reason, the ideas exposed about Java artifact repositories can be extended to container repositories. It's not incidental that both the mentioned technologies (Artifactory and Nexus) have extensions used for handling containers.

Popular container repositories available online include Docker Hub and Quay.io.

Completing the deployment

As we have looked at code compiling and the management of artifacts, it is now time to complete the deployment. This may include different steps, which ultimately aim to install the artifacts in the right places, perform some configurations, and execute the software. The details of these steps may vary heavily depending on the technology used:

- In the traditional Java world, there are custom ways to use Java application servers (such as WildFly, as seen in *Chapter 7, Exploring Middleware and Frameworks*) depending on the application server used. Common ones are copying the application artifact in a specific folder or invoking a command-line utility that triggers the deployment process. More steps for configuring things (such as a connection to databases) may be needed and usually involve changes in configuration files or commands issued.

- When using fat `.jar` applications (such as Spring Boot or Quarkus) in a non-containerized environment, the deployment process usually involves just copying the fat `.jar` in a specified location and running it using a command. This may be done with shell scripts, which can then address other steps (where relevant), such as the cleanup of the previous versions and the changes in configuration files.

- In containerized environments, deployment basically involves copying the container (Kubernetes and Docker are usually configured to access remote container repositories) and executing a number of commands to make it run. In Kubernetes, this is almost entirely done by using `kubectl`.

As we have seen, different technologies require slightly different ways to complete deployment and effectively distribute and run the software packages. If you consider that more steps can be required, including the configuration of external systems such as databases and IaaS or PaaS systems, it's often a good idea to orchestrate those steps in a unified way. This is one of the characteristics of CI/CD.

Continuous integration/continuous delivery (and deployment)

CI/CD is the process of automating most of the steps seen so far (and sometimes some more) in order to straighten the process and have a complete overview of it. Since the process includes many steps executed in a mostly sequential way, the tool providing it is commonly called a **pipeline**.

A typical CI pipeline includes these steps, usually executed sequentially:

1. Building the code (as seen in previous sections).

2. Testing, usually limited to static code testing, unit testing, and some limited integration testing. Since deployment has not occurred yet, end-to-end testing is not possible in this phase.

CD includes a few further steps, focused on deployment (also, in this case, usually executed in a sequence):

1. Versioning of software artifacts in repositories.

2. Deployment of artifacts from repositories to testing environments, including all the configuration needed. This may be done in ephemeral environments (as seen in previous sections).

3. End-to-end testing in such environments.

4. Deployment in other non-production environments, with the goal of user acceptance and/or performance testing.

Here is a simplified sample pipeline:

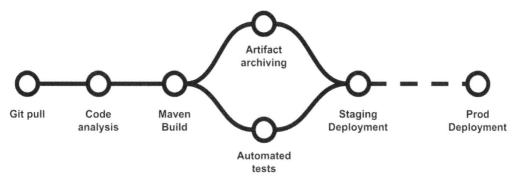

Figure 13.5 – A sample pipeline

In the preceding diagram, you can see a simplified pipeline example. There are a few steps (**Automated tests** and **Artifact archiving**) that are performed in parallel. Moreover, between **Staging Deployment** and **Prod Deployment**, some actions may happen, such as UAT and manual approval.

CD is considered a further extension of the pipeline, and basically includes the deployment of the environments in production environments. This may happen in a completely automated way (with no manual intervention), or it may require manual approval by a release manager (in more traditional environments).

It's of course intended that a failure in any step usually means that the pipeline stops and the release fails. It's also worth noticing that deployment in production does not necessarily mean that the software is released and available to users, as we will see soon. But first, it is worth having a look at widespread pipeline implementations.

Common CI/CD software implementations

It's almost impossible to talk about CI/CD without mentioning Jenkins.

Jenkins is a complete automation server, which is basically a synonym of CI/CD pipelines.

It's written in Java and deployed in a servlet container (usually Tomcat). It's possible to define pipelines in Jenkins, using a domain-specific language, which describes each step with the required parameters. Through a system of plugins, steps in Jenkins can do many different things, including compilation using Maven, performing SSH commands, and executing test suites.

Jenkins can then display the pipeline execution results, archive them, and optionally send notifications (as an example, in case of a build failure).

While still being widely used, Jenkins is nowadays famous for having a monolithic architecture (although some steps can be delegated to agents) and for being resource-intensive.

Attempts to create alternative pipeline software, with a more modern design and better performances in a cloud environment, are currently underway. The most famous ones are Jenkins X and Tekton.

Both of these software types, while created using different languages and frameworks, share the concept of implementing each step in a container, thereby improving horizontal scaling and reusability.

Other famous implementations of CI/CD capabilities include Travis, GitLab, and, more recently, some cloud alternatives such as GitHub Actions.

As we have said, regardless of the implementation, the CI/CD process can automate steps up to the production deployment. However, in order to make the software available to final users, the process requires some final steps.

Releasing

Releasing is usually the final step of a complete CI/CD pipeline. The process can be performed, in simple environments, together with deployment in production.

However, nowadays, it's common to split deployment and releasing into two different steps, and this allows more sophisticated (and often safer) ways of releasing software versions to end users. To do so, the most basic ingredient is to have different versions of the software available in production at the same time and to route users to each version by following different criteria (which is done by operating at a network level, routing each request to the desired target version). Let's look at some scenarios opened by this kind of technique:

- **Blue-green deployment**: Two versions of production environments (including database and other external systems) are released in production. This includes the version we want to release (identified as blue or green) and the previous version (identified by the color left, so either green or blue). The candidate release can then be tested in a real environment. If everything works as expected, the network traffic is then routed to the new version. A rollback can be easily performed at any time. In the next release, the same is done, replacing the previous version with the next one and changing the color-coding. The following diagram illustrates this:

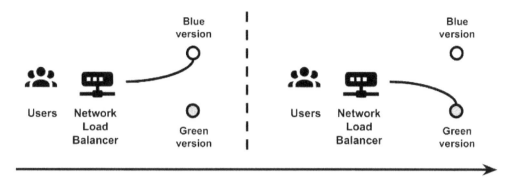

Figure 13.6 – Blue-green deployment

- **Rolling releases**: This implies that our application components are provided on a set of identical instances (hosted on VMs or containers). When a release occurs, the **New version** is installed on a new instance and traffic starts to be sent to such new instances, too. Then, an **Old instance** is shut down (optionally draining all the pending connections). This process keeps going until all the instances are running on the new version, as shown here:

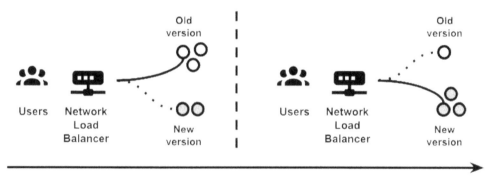

Figure 13.7 – Rolling releases

In the preceding diagram, each *circle* represents an instance (such as a VM or an application instance). The two steps are represented as a sample, but you can imagine that the **New version** starts from an instance, goes to two, and so on, and the **Old version** is progressively shut down, one instance at a time.

While this can be a subtle difference, compared to the blue-green deployment, a rolling release can be seen as a kind of technical trick aiming to reduce downtime in releases, but it provides fewer guarantees if a rollback is needed. Moreover, since old and new instances coexist for a certain amount of time, issues may occur, both on the application and on the external systems (such as databases).

- **Canary releases**: Basically, this is a variant of blue-green deployment. The new release is still provided alongside the previous version, but instead of switching traffic as a whole, this is done progressively, such as routing 1% of the users, then 2%, and so on, until all the users are running on the new release. This technique allows smoother releases and makes it easier to identify issues, if present (and optionally roll back), without impacting the whole customer base, as illustrated here:

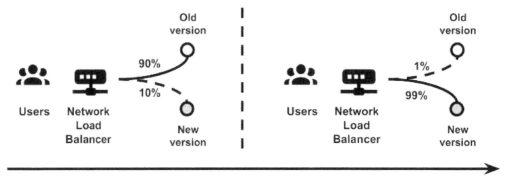

Figure 13.8 – Canary releases

In the preceding diagram, we can see just two sample phases, where we start routing **90%** of the traffic toward the **Old version**, and **10%** to the **New version**, and another one representing **1%** toward the **Old version** and **99%** toward the **New version**. Of course, in a real situation, you can imagine a constant flow going from 1% to 100% and vice versa. The name *canary* refers to the canary used by miners to identify gas leakages. In a similar way, bugs are identified sooner, by a small percentage of users, and it's possible to stop the release before impacting more users.

- **A/B testing**: Technically, this is identical to the blue-green deployment. The most important difference is that, in this case, we are evaluating two alternative versions. The two versions should not be considered as a previous one and a next one, but instead two slightly different variants of the same software that the business wants to test against real users. The following diagram shows this:

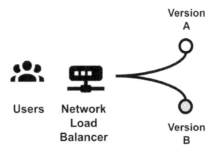

Figure 13.9 – A/B testing

Such variants are usually small changes in the user interface or other features, and the business aims to measure how those changes perform in the real world. The most common example is with online shops, where the business is willing to check whether changes in the purchase process (colors, position of the buttons, and the number of steps) enhance commercial performance.

In A/B testing, the less performant version is usually discarded. It's also worth noticing that the users routed to each version can be chosen randomly, split by percentage (such as 50% for each version), or even selected by specific criteria (where available), such as the geographical location or the age of the user.

- **Traffic shadowing**: This is a bit less common than the other alternatives. It implies that a new release of the software is released in production, where users keep using the older version while the new version gets a copy of all the production traffic:

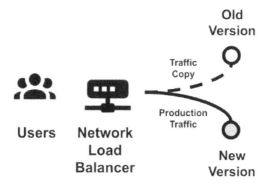

Figure 13.10 – Traffic shadowing

It can be really useful for load testing in the case of major releases, but it may not work in some specific scenarios, and it's necessary to understand the impacts in any specific use case. As an example, if notifications (such as via email) are sent, we should ensure that they are not sent by both the old and the new systems, to minimize the impact on users.

It's worth noticing that these kinds of release techniques (especially the simpler ones) are nothing new, and were also possible before modern cloud and microservice architectures. However, to use them in traditional environments, you may need to coordinate with external teams (such as the ones administering the network), while modern architectures (such as the ones based on public clouds, IaaS or PaaS) provide way more flexibility, allowing the creation of new instances on the fly, and changing network configurations favored by software-defined networking. In the next section, we are going to complete our view of the software life cycle by looking at some considerations regarding software maintenance.

Maintenance

Software maintenance is usually a great part of software life. In my professional experience, it's not uncommon for a project to be fully active (with a lot of new features and developments happening) for a couple of years, followed by many years of maintenance, which is focused on fixing bugs and keeping the product alive (without releasing any new features).

It goes without saying that the maintenance period can become more expensive than the building of the project. Moreover, but this is a consideration purely from an economic perspective, enterprises often find it easier to access the budget for building new applications (which is seen as money generating, or at least associated with business initiatives) than for maintaining and modernizing older ones (which is seen as IT for IT, which means that this is a project with no business impact, hence, purely a cost).

With that said, maintenance activities on existing applications can be roughly categorized into one of the following buckets:

- **Bug fixes** address defects in the software, which may be an existing issue in code or code behaving badly due to an external unforeseeable situation.

- **Requests For Enhancement** (**RFEs**), which are targeted around covering new use cases that were not originally planned.

Of course, both of these types of activities can then have an impact and an effort associated, which may help prioritize the implementation.

The software maintenance process is then further categorized into four types, independent of the previous two categories:

- **Corrective maintenance**, which must be done as a reaction to issues occurring and problems being reported. Think about a classic bug fixing activity as a consequence of the application behaving incorrectly.

- **Adapting maintenance**, which aims at keeping the software correctly working in changing environments. Think about the application needing to be adapted to a newer version of the JVM, or to the release of a new browser.

- **Perfecting maintenance**, which aims at making our application just better, in terms of performance, resource usage, maintainability, or other enhancements (such as a better user experience). This is often neglected, as it's a proactive activity and nothing in the short term usually happens if it's not done. However, it may have a big positive impact (also in terms of savings) because it can, of course, prevent issues in the future. Moreover, avoiding perfecting maintenance may mean growing the so-called *technical debt*, which means that more and more tasks (such as fine-tuning, refactoring, and enhancing test coverage) will pile up, becoming less and less manageable (and more expensive to tackle).

- **Preventive maintenance**, which is really similar to the previous one, but revolves around fixing issues that have been identified (such as known bugs) before they become actual problems. As per the previous point, it risks being neglected and can cause technical debt if not handled properly.

Other than having a direct cost (because somebody needs to do it), all the maintenance categories may have several impacts. Indeed, such activities often involve software releases or configuration changes, and this may impact the application availability (and have a negative influence on the **Service Level Agreement** (**SLA**) established). There may even be legal agreements relegating maintenance activities to well-defined timeframes or allowing it only to solve high-severity issues.

Moreover, software maintenance can have an indirect impact because the activities (both enhancements and bug fixing) can change the application behavior and, in some cases, even leave the API exposed, hence, forcing the users to adapt to such changes or the developers to plan for implementing the retro compatibility.

With that said, application maintenance should never be neglected. Instead, it should be planned (and financed) from the very beginning of the project and be constantly adjusted over time. Indeed, a project that is not correctly maintained can incur security issues or customer dissatisfaction, or also simply lose attractivity on the market. This will risk nullifying all the efforts made to design and implement the project.

This was the last topic of this chapter. Let's now have a look at a summary of what we have learned.

Summary

In this chapter, we have had an overview of many crucial phases of the software life cycle.

Starting with SCM, we had a quick discussion of Git, which is the de facto standard over SCM. This allowed us to understand development models, with a focus on trunk-based development, which is common in CI/CD and DevOps-based projects.

We also briefly discussed semantic versioning, which is a way to identify what changes to expect based on the release version numbering.

We then entered the testing phase, starting with unit testing (and the inevitable description of JUnit, a standard for testing in Java). From unit testing, we moved on to other testing techniques, including integration, end-to-end, and UAT.

Deploying was the next step. We discussed the steps needed to compile software and run it, including an overview of Apache Maven and artifact management. The next topic was CI/CD and pipelines with some consideration around automating most of the steps seen in this chapter.

Then, we focused on releasing, which is the process of making the deployed software available to final users, and we saw many different options to do so, including blue-green and canary releases. Last but not least, we had an overview of maintenance activities (both bug fixes and enhancements) and why they are crucial for the overall success of our project.

In the next chapter, we are going to discuss monitoring and tracing, which are some core concepts for ensuring that our software is performing well in production, and for constantly understanding and governing what's happening.

Further reading

- Richard E. Silverman: *Git Pocket Guide*

- Konrad Gadzinowski: *Trunk-based Development vs. Git Flow* (`https://www.toptal.com/software/trunk-based-development-git-flow`)

- Hardik Shah: *Why Test Coverage is an Important Part of Software Testing?* (`https://www.simform.com/blog/test-coverage/`)

- Himanshu Sheth: *Code Coverage vs Test Coverage – Which Is Better?* (`https://dzone.com/articles/code-coverage-vs-test-coverage-which-is-better`)

- Sten Pittet: *Continuous integration vs. continuous delivery vs. continuous deployment* (`https://www.atlassian.com/continuous-delivery/principles/continuous-integration-vs-delivery-vs-deployment`)

- Michael T. Nygard: *Release It!: Design and Deploy Production-Ready Software*

- Martin Fowler: *BlueGreenDeployment* (`https://martinfowler.com/bliki/BlueGreenDeployment.html`)

14
Monitoring and Tracing Techniques

There is a risk, as developers and architects, of overlooking what happens to our applications and services after production release. We may be tempted to think that it's just a problem for sysadmins and whoever oversees service operations. As is easy to understand, this is the wrong point of view.

Understanding how our application behaves in production gives us a lot of insight into what is and is not working—from both a code and an architecture perspective. As we learned in the previous chapter, maintenance of our application is crucial for the success of each software project, and looking closely at how the application is going in production is the perfect way to understand whether there is something that can be improved.

Moreover, in modern DevOps teams, as we learned in *Chapter 5*, *Exploring the Most Common Development Models*, the separation of concerns must be overcome, and the development and architectural teams are responsible for operating services as well. In this chapter, we will have an overview of the common topics regarding the visibility of what happens to our application during production.

We will look at the following topics in this chapter:

- Log management

- Collecting application metrics

- Defining application health checks

- Application Performance Management

- Service monitoring

The idea is not only to have an overview of these topics and what they are useful for, but also to understand the impact that a correct design and implementation may have on these topics. With that said, let's start discussing log management.

Technical requirements

You can find the source code used in this chapter here: `https://github.com/ PacktPublishing/Hands-On-Software-Architecture-with-Java/tree/ master/Chapter14`.

Log management

Log management has been taken for granted so far in the book. Let's just take a quick glance over some basic concepts related to producing logs in our Java applications.

Logging in Java has had a troubled history. At the very beginning, no standard was provided as part of the Java platform. When a standard (such as **Java Util Logging** (JUL)) was added to the platform (in release **1.4**), other alternative frameworks were available, such as **Apache Commons Logging** and **Log4J**.

At the time of writing, Log4j has been deprecated and replaced by **Log4j2** and **logback**.

Even though the JUL standard has been a part of the platform for many years now, the usage of alternative frameworks such as logback and Log4j2 is still very widespread, due to their features and performance. Regardless of which implementation we choose, there are some common concepts to consider.

Common concepts in logging frameworks

As described previously, no matter what kind of preferred log implementation is used in your projects, there are two main concepts that are common to every one of them:

- **Levels**: Each framework defines its own levels. Logging levels are used to configure the verbosity of the logs at runtime, with a hierarchy defining which level of logs must be produced. Here, the hierarchy means that if a certain level of logging is enabled, all the log entries from that level and above are reported in the logs. The INFO level is commonly present, and defines the *average* level of verbosity, reporting log entries that record basic information useful to understand what's going on, but that can be discarded if you want to reduce the amount of logging or if you are familiar and confident enough with your app's behavior.

  Above the INFO level, there are WARNING and other similar levels (such as ERROR and FATAL) that are used for reporting unusual or incorrect behavior. For this reason, these levels are almost always kept active. Finally, below INFO there are other levels, such as DEBUG and TRACE, which are used for getting details on what is happening in our application and are usually reported only for a limited amount of time. They are also used to gather data on whether there is something wrong in our application that needs troubleshooting, as well as collect data in non-production environments for development purposes. These levels of logging are discouraged in production as they will produce a lot of entries and may impact performance, which leads to our next point.

- **Appenders** define the technology used for reporting log entries. As with the log levels, the appenders are also different in each logging implementation. CONSOLE and FILE are two common ones used to report log entries in the console or in a file. Other alternatives may include appenders to a database and to other external systems (such as logging to a socket).

  As described in the previous point, appenders may impact the overall performance. Writing to a file may be slower than writing to the console. For this reason, appenders often offer asynchronous alternatives that buffer the log entries in memory before writing them to the target system. However, this of course increases the risk of losing data should our application crash before the entries are forwarded to the relevant system (such as the file or the database).

These are some very basic concepts of logging in Java (similar concepts can be found in other languages). But there are also some recommendations about logging that I can provide, from personal experience:

- It's a good practice to avoid string concatenation (such as `"My log is " + variable + " ! "`). Other than being ugly, it can have a performance impact since string concatenation operations happen even if the log is not used (because it's related to a disabled level). Most logging frameworks offer alternatives based on placeholders (such as `"My log is {} !", variable`). Most recent Java versions automatically mitigate string concatenation by replacing it at compilation time with more efficient alternatives (such as `StringBuilder`), but it's still a good idea to avoid it.

- Consider differentiating log destinations by content type. You may want to have different appenders (almost every framework allows it) to log different information. So, business information (related to how our application is performing from a business perspective) such as user information or the products used can go to a specific database table (and maybe then can be aggregated and reported), while logs containing technical information (useful for troubleshooting or checking the application's health) may go in a file to be easily accessed by SysOps.

  This can also be done by log severity by sending the `INFO` level on a certain appender, and other levels on other appenders. This may also allow for different *quality of service* for logs: you could log business information that is logged on an asynchronous appender (because you may lose some data in the event of an application issue—this is not a problem), while technical logs should go on synchronous appenders (because you cannot afford to lose anything if you intend to understand the issues behind a misbehaving application).

- **Log rotation** is an essential concept, but it's still sometimes forgotten, especially in older applications. Log rotation can be implemented using the logging framework itself or by external scripts and utilities. It's basically related to file appenders, and defines the way logs are archived by renaming them, moving them, and optionally compressing them. A log rotation policy allows the current logs to be small enough (for easy reading and searching) and makes it easier to find information from previous dates and save space on disk. This will help SysOps, who sometimes have to deal with misconfigured applications that fill the disk because of a misconfigured log rotation policy. Hopefully, this should be less common nowadays.

- Every message should provide meaningful information. As trivial as it may sound, it's very easy to write just basic things in logs, assuming that whoever reads the log will have enough context. *I highly recommend not doing this!* Log messages could be read by people not knowing much about the application (such as first-line support staff) in emergency situations (such as production troubleshooting). When in doubt, be as clear as possible. This doesn't necessarily mean being verbose, but make sure to provide a decent amount of content.

- Logging levels should be defined in a standard way. Especially in big projects composed of many microservices or applications, it should be well documented what is supposed to be logged as INFO, what should be higher, and what should be lower. In this way, logging levels can be set in a uniform way, expecting the same kind of information across all modules.

- The same is true for log format. Almost every logging library supports defining a pattern, which means setting which information (apart from the log message itself) should be written, including date, time, the log level, and more. It's better if this kind of format is uniform across all the components to be easy to read and parse using tools (such as the very basic `grep` utility). Also, I strongly suggest configuring the logging library to provide information about the class that is generating the log. It's usually a bit expensive from a computational perspective (often negligible) but is worth it for sure.

- You should have a discussion with security and legal advisors as soon as possible (if present) about what can, must, and should not be present in logs. This varies from application to application, but there may be information (such as personal information or credit card data) that is prohibited from being present in logs (or needs to be anonymized), and other information that is required to be present by law (such as audit information). You need to know about this and implement the requirements accordingly.

- As a follow-up from the previous point, most applications have legal (or other kinds of) requirements for log storage and archiving. You may need to store logs for many years, sometimes in an immutable way. Hence, log rotation and specialized hardware and software may be needed.

As a final consideration about logging, we cannot avoid having a chat about log aggregation.

Log aggregation

In *Chapter 9*, *Designing Cloud-Native Architectures*, when discussing **Twelve-Factor Applications**, we saw how logs can be seen as an event stream and must be handled by a supporting platform, capable of capturing the event stream, storing it, and making it usable and searchable across different applications. We even mentioned **Fluentd** as a commonly used solution for this. This is exactly what log aggregation is about. A typical log aggregation architecture features the following:

- An agent for collecting logs from the console (or a file) in the form of event streams. Fluentd is a common choice (even though it has some known limitations in terms of performance and logs that can potentially be lost in corner cases). **Filebeat** and **collectd** are some alternatives to this.

- Persistence for log entries. **Elasticsearch** is practically the standard in this area, providing storing, indexing, and searching capabilities.

- A frontend for navigating and monitoring log entries. Software commonly used for this goal are **Kibana** and **Grafana**. Here is a screenshot of the Kibana UI:

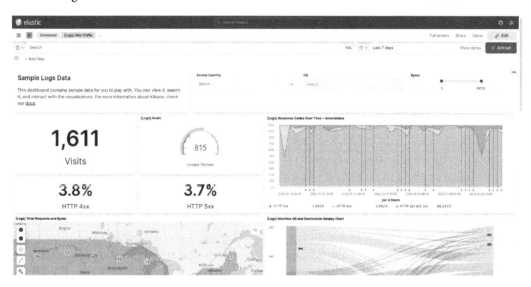

Figure 14.1 – Kibana UI home

Log aggregation should be considered a must in cloud-native architecture, because having a heavily distributed application based on microservices will mean having a lot of systems to monitor and troubleshoot in the event of issues.

With a log aggregation strategy, you will have a centralized way to access logs, hence everything will become a bit easier.

In this section, we had a quick overview of logging in Java, and then we highlighted the characteristics and the advantages of log aggregation. In the next section, we are going to have a look at another key topic about monitoring, which is metrics collection.

Collecting application metrics

Metrics are a way to instrument your source code to provide real-time insights into what's happening. Metrics are also known as **telemetry**. Instead of logs, which represent information pushed into a file, a console, or another appender, metrics are values exposed by the application and are supposed to be pulled by whoever is interested in them.

Moreover, while a log contains what's happening in our application, collected in a sequential way, metrics expose a snapshot of how the application was behaving in that instant, summarized into some well-known values (such as the number of threads, the memory allocated, and so on). It's also possible to define some custom metrics, which can be useful to define figures that are specific to our particular use case (such as the number of payments, transactions, and so on).

There are many widespread frameworks useful for exposing metrics in Java. Micrometer is an open source façade implementation, while other commercial solutions exist, such as **New Relic** and **Datadog**. However, I think that one of the most interesting efforts in this area is one part of the MicroProfile standard. We looked at MicroProfile in *Chapter 7, Exploring Middleware and Frameworks*, when discussing Quarkus as an implementation of it.

I think that a quick example (MicroProfile compliant) will be useful here to better explain what metrics look like. Let's see a simple hello world REST API:

```
@GET
    @Path("/hello")
    @Produces(MediaType.TEXT_PLAIN)
    @Counted(name = "callsNumber", description = "How many
      calls received.")
    @Timed(name = "callsTimer", description = "Time for
      each call", unit = MetricUnits.MILLISECONDS)
    public String hello() throws InterruptedException {
        int rand = (int)(Math.random() * 30);
        Thread.sleep(rand*100);
        return "Hello RESTEasy";
    }
```

As you can see, the hello method is annotated with two metric-related annotations (Counted and Timed), which declare the kind of metrics we want to collect. The annotations also provide some documentation (the name and description of the metric). Now, if we query the application via REST, we can see all the metrics values exposed:

```
# HELP application_it_test_MetricsTest_callsNumber_total
   How many calls received.
# TYPE application_it_test_MetricsTest_callsNumber_total
   counter
application_it_test_MetricsTest_callsNumber_total 4.0
...
# HELP application_it_test_MetricsTest_callsTimer_seconds
   Time for each call
# TYPE application_it_test_MetricsTest_callsTimer_seconds
   summary
application_it_test_MetricsTest_callsTimer_seconds_count
   4.0
...
```

A number of other metrics (such as minimum, maximum, and average) are omitted in the preceding output and calculated automatically by the framework.

These kinds of metrics are exposed under the /metrics/application endpoint (/q/metrics/application, in the case of the Quarkus framework).

The MicroProfile specification also defines the /metrics/vendor (vendor-specific), /metrics/base (a meaningful predefined subset), and /metrics (all the metrics available) endpoints. In these endpoints, you may find a lot of useful insights into the application, such as virtual machine stats and similar things. This is a small subset of what can be retrieved from such endpoints:

```
...
# HELP base_memory_usedHeap_bytes Displays the amount of
   used heap memory in bytes.
# TYPE base_memory_usedHeap_bytes gauge
base_memory_usedHeap_bytes 9.4322688E7
# HELP base_thread_count Displays the current number of
   live threads including both daemon and non-daemon threads
# TYPE base_thread_count gauge
```

```
base_thread_count 33.0
...
```

The metrics exposed in this way can then be collected by external systems, which can store them and provide alerts in the event of something going wrong. A widely used framework to do so is **Prometheus**.

Being a part of the **Cloud-Native Computing Foundation** (**CNCF**) effort, Prometheus is able to collect the metrics from various systems (including OpenMetrics-compliant endpoints, similar to the ones exposed by the example we saw previously), store them in a so-called **Time Series Database** (**TSDB**) (which is basically a database optimized for storing events on a temporal scale), and provide capabilities for querying the metrics and providing alerts. It also offers a built-in graphical interface and integration with Grafana.

But metrics are just one of the aspects of application monitoring. Another similar and important one is health checks.

Defining application health checks

Health checks are a kind of special case for metrics collection. Instead of exposing figures useful for evaluating the trends of application performance, health checks provide simple *on/off* information about the application being healthy or not.

Such information is particularly useful in cloud and PaaS environments (such as Kubernetes) because it can allow self-healing (such as a restart) in the event of an application not working.

OpenMetrics currently defines three kinds of health checks: **live**, **ready**, and **started**. These checks come from concepts in the Kubernetes world:

- By using a live (health) check, Kubernetes knows whether an application is up and running, and restarts it if it's not healthy.

- By using a readiness check, Kubernetes will be aware of whether the application is ready to take requests and will forward connections to it.

- Startup checks identify the successful completion of the startup phase.

Note that **ready** and **started** are very similar but **started** has to do with the first startup of the application (which may be slow), while **ready** may involve a temporary inability to process requests (as an example, a traffic spike or other temporary slowdowns).

Quarkus provides such checks with the `smallrye-health` extension. The probes are exposed, by default, at the `/q/health/live`, `/q/health/ready`, and `/q/health/started` endpoints and the results are formatted as JSON.

In order to implement the checks, Quarkus provides an infrastructure based on annotations. This is how a basic Liveness probe is implemented:

```
@Liveness
public class MyLiveHealthCheck implements HealthCheck {
    @Override
    public HealthCheckResponse call() {
        // do some checks
        return HealthCheckResponse.up("Everything works");
    }
}
```

As you can see, the preceding method is annotated with @Liveness and returns a message using the up method of the HealthCheckResponse object.

Similarly, a Readiness check will look like this:

```
@Readiness
public class MyReadyHealthCheck implements HealthCheck {
    @Override
    public HealthCheckResponse call() {
        // do some checks
        return HealthCheckResponse.up("Ready to take
          calls");
    }
}
```

Also, in this case, the preceding method is annotated (in this case, with @Readiness) and returns a message using the up method of the HealthCheckResponse object.

Finally, a Startup check will look like this:

```
@Startup
public class MyStartedHealthCheck implements HealthCheck {
    @Override
    public HealthCheckResponse call() {
        // do some checks
return HealthCheckResponse.up("Startup completed");
    }
}
```

For startup checks, the preceding method is annotated (with `@Startup`) and returns a message using the `up` method of the `HealthCheckResponse` object.

As you can see, the API is pretty simple. The objects providing the functionalities are singletons by default. Of course, in a real-world application, you may want to do some complex checks such as testing the database connection or something similar.

You can of course return a negative response (such as with the `down()` method) if you detect any failure. Other useful features include the chaining of multiple checks (where the cumulative answer is `up`, only if every check is up) and the ability to include some metadata in the response.

Implementing OpenTracing

Tracing is a crucial monitoring technique when you have a long chain of calls (for example, a microservice calling other microservices, and so on), as you compose your answer by calling a huge number of internal or external services.

Indeed, it's a very common use case in microservices applications: you have a call coming into your application (such as from a REST web service or an operation on a web user interface, which in turn translates into one or more REST calls). This kind of call will then be served by a number of different microservices, ultimately being assembled into a unique answer.

The issue with this is that you may end up losing trace of whatever happened. It becomes very hard to correlate the incoming call with every specific sub-call. And that may be a big problem, in terms of troubleshooting issues and even simply understanding what's happening.

Tracing allows a way to identify the path made by each request by propagating an identifier code used in each subsystem, hence helping to document and reconstruct the tree of calls used to implement our use case, both for troubleshooting and for other purposes, such as audit logging.

OpenTracing is a standard and part of the CNCF family, which implements this kind of functionality. In Quarkus, as an example, this feature is provided by a library that's part of the SmallRye project, which is called `smallrye-opentracing`.

An interesting feature is that OpenTracing also supports computing the time spent on each sub-call.

Let's see a very simple example to understand how tracing works in Quarkus.

We will start with a simple REST resource, as we have seen many other times in this book:

```
@Path("/trace")
public class TracingTest {
    @Inject
    NameGuessService service;

    @GET
    @Produces(MediaType.TEXT_PLAIN)
    public String hello() {
        String name = service.guess();
        return "Hello "+name;
    }
}
```

As you can see, it's a simple REST method listening on the /trace endpoint. It uses a service (NameGuessService) that has been injected.

It's worth noticing that there is no specific code related to tracing: indeed, tracing in REST endpoints is basically automatically provided by the framework. It's enough to have the smallrye-opentracing extension in the project itself.

Now, let's have a look at the NameGuessService class:

```
@ApplicationScoped
@Traced
public class NameGuessService {
    public String guess()
    {
        Random random = new Random();
        String[] names = {"Giuseppe","Stefano",
            "Filippo","Luca","Antonello"};
        return names[random.nextInt(names.length)];
    }
}
```

As you can see, there is nothing special here: there's just a simple mocked service returning a string, which is chosen randomly. The only notable thing is that the method is annotated with @Traced, because the framework needs to know explicitly whether the method must be traced.

Where do we go from here? The most common and useful way to use tracing is with a Jaeger server. Jaeger basically exposes some services that collect and graphically display what's happening in our application. The basic concept is a **span**, which is an end-to-end method call. In our case, one **span** is made out of our REST call, and another one is the sub-call in our injected service.

A quick way to test our service locally is to use a ready-made Jaeger server containerized.

On a laptop with a container engine (such as Docker) installed, it's enough to run the following command:

```
sudo docker run -p 5775:5775/udp -p 6831:6831/udp -p
6832:6832/udp -p 5778:5778 -p 16686:16686 -p 14268:14268
jaegertracing/all-in-one:latest
```

This will run a jaegertracing *all-in-one* image, specifying the ports used.

We can then run our application, hooking it into a Jaeger server:

```
./mvnw compile quarkus:dev -Djvm.args="-
DJAEGER_SERVICE_NAME=testservice -
DJAEGER_SAMPLER_TYPE=const -DJAEGER_SAMPLER_PARAM=1"
```

These parameters are provided as command-line arguments but could also be provided as part of the properties file. In this case, we are specifying how this service is called and which kind of sampling should be done (it's okay to use the default parameters for the purposes of this test).

Now, we can invoke our REST service a couple of times at `http://127.0.0.1:8080/ trace`, just to generate some traffic to display. If we then navigate to the Jaeger UI, available by default at `http://localhost:16686/`, we will see something similar (click on the **Find Traces** button and select the **test-opentracing** service, if necessary):

Figure 14.2 – Jaeger UI home

As you can see, each of the calls made to our service is displayed with the overall time to respond (a couple of milliseconds, in our example).

If we click on one of those calls, we can see the two spans:

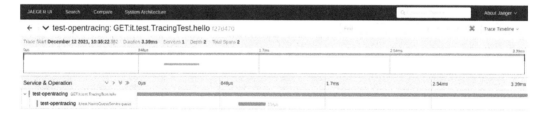

Figure 14.3 – Jaeger UI spans

As you can see, the main span concerns the REST call with a smaller span on the sub-call of the injected service. It's easy to imagine how useful it is to have this kind of information on an application running in production.

As you can see, in this example, we have just one microservice with two methods. However, the same concept can be easily extended to more than one microservice talking to each other.

Tracing and metrics are part of a bigger concept called **Application Performance Management** (**APM**).

Application Performance Management

APM is a broad and very important aspect of running an application in production. It involves a lot of different technologies, and sometimes it has some unknowns around log aggregation, metrics collection, and overall monitoring, among other things.

Each vendor or stack of monitoring technologies has slightly different comprehensions of what APM is about, and somewhat different implementations of it as a result.

I think that it's good to start from the goal: the goal of APM is to have insights into how a set of applications is performing, and what impact the underlying parameters (such as memory usage, database metrics, and more) have on the end user experience (such as user interface responsiveness, response times, and so on).

It is easy to understand that to implement such a useful (and broad) goal, you may need to stack a number of different tools and frameworks.

We have seen some of this in the previous section: you may want to have information coming from logs (to understand what the application is doing), together with metrics (to understand the resource consumption and collect other KPIs such as the number of calls), with health checks (to have a quick view over which service is up), and with tracing (to understand how each specific call is performed, including all the sub-calls).

And there are a number of other tools that you can use. As an example, JVM provides some useful parameters (we saw some when discussing metrics) such as memory and CPU consumption.

Last but not least, for code that is not natively instrumented (such as legacy code that is not providing metrics using frameworks similar to the one seen previously), it is possible to collect some metrics using some more invasive approaches, such as Java agents, which are low-level configurations that act on the JVM to understand how and when each method of our code is called.

With that said, you can imagine how hard it can be to provide a unified, easy-to-read, overall vision of what's happening with our application. You will need to install and maintain a lot of different tools and glue them together in order to display meaningful and uniform information.

For this reason, aside from open source standards and tools, commercial solutions have emerged (such as **Dynatrace**, **Datadog**, and **Splunk**), which allow us to use ready-made stacks to provide such information.

But now that it is clear how important and useful it is to have this kind of information, let's look at some topics to be aware of when talking about APM:

- **It may impact performance**: Many of the approaches seen so far have been designed to have as little impact as possible by using asynchronous and non-blocking techniques. However, especially if we use older approaches such as Java agents, the impact can be significant. And if you think that an APM system might be useful when your application is slow, it's easy to understand that APM must be as lightweight as possible, to avoid putting any further pressure on systems that are already requested.

- **It requires nontrivial maintenance**: The data collected can simply be huge in quantity. Think about every transaction generating a bunch of metrics (timing, error codes, resources consumed), plus a number of lines of logs and tracing information. When all these metrics are multiplied by hundreds or thousands of transactions, it may become difficult to maintain them. Plus, as said, each specific type of information you might want to look for (logs, metrics, and checks) is managed by a different stack, hence we may end up using different servers, storage, and configurations.

- **The information collected may be hard to correlate**: Especially in the event of an issue, you may want to understand whether a specific transaction caused the issue and how the system behaved. While tracing makes it easy to correlate a transaction with each sub-call and subsystem, correlating tracing information with logging information plus metrics and health checks will still be trouble. Moreover, comparing different kinds of data (such as timespans with KPIs and messages) can be hard, especially in user interfaces.

Last but not least, it's crucial to correlate the platform information with the related features implemented. In the next section, we are going to look a bit more into what kind of information is worth collecting, and how to categorize it.

Service monitoring

A very important consideration is what to monitor.

Indeed, it's very important to collect as much data as possible, in terms of metrics and KPIs, as they may reveal interesting trends, and can be very useful if something unpredicted happens. But at the same time, business users are mostly interested in different kinds of metrics and information, such as the number of transactions per second (or per hour, or per day), the amount of money that passes through the platform, the number of concurrent users, and so on.

Hence, there are two different kinds of KPIs to look for, sometimes with a blurred boundary between them:

- **Technical information**: Things such as the memory used, the number of threads, the number of connections, and so on. These things are useful for sizing and scaling systems and trying to forecast whether our system will perform well or some interventions are needed.
- **Business information**: Defining what information is business information heavily depends on the application realm, but usually includes the average transaction time, the number of concurrent users, the number of new users, and so on.

From a technical standpoint, you can use the same frameworks (especially ones for collecting metrics) in order to collect both technical and business information.

But it's very important (and not so easy to do) to try to correlate one kind of metric with another.

In other words, it could be useful to have a map (even simple documentation such as a web page can be enough) that documents where each feature is hosted, and how specific business information is related to a set of technical information.

Let's look at an example: if we have a business KPI about the transaction time of a specific functionality, it is important to understand which servers provide that functionality, and which specific set of microservices (or applications) implements it.

In this way, you can link a business metric (such as the transaction time) to a set of technical metrics (such as the memory used by a number of JVMs, used threads, CPU consumption on the servers that are running such JVMs, and more).

By doing that, you can better correlate a change in performance in that particular feature (in our case, transactions going slower) to a specific subset of technical information (such as an increase in CPU usage on one particular server).

This will help in troubleshooting and quickly fixing production issues (by scaling the resources on impacted systems).

Other than this, business metrics are simply valuable for some users: they may be used for forecasting the economic performance of the platform, the expected growth, and similar parameters. For this reason, it's common to store such information on specific data stores (such as big data or data lakes), where they can be correlated with other information, which is analyzed and further studied.

This completes the topics that were planned for this chapter.

Summary

In this chapter, we have looked at some interesting considerations about monitoring and tracing our applications.

We started by reviewing some basic concepts about logging in Java, and why log aggregation is a good thing to do in microservices and cloud-native applications. We then moved on to the concept of metrics and health checks, and how applications can provide data in real time on the performance and health of our modules.

We then discussed tracing, which is very important when it comes to troubleshooting and managing distributed applications (such as microservices applications). APM was the next topic and is about putting all the information together (such as metrics, health checks, and logs) to create an overview of the application insights.

Last but not least, we saw how service monitoring involves linking business information with the technical KPIs behind it, to support troubleshooting and draw more insights from the collected data.

In the next chapter, we are going to see what's new in the latest version of the Java technology.

Further reading

- Hanif Jetha, *How To Set Up an Elasticsearch, Fluentd and Kibana (EFK) Logging Stack on Kubernetes* (`https://www.digitalocean.com/community/tutorials/how-to-set-up-an-elasticsearch-fluentd-and-kibana-efk-logging-stack-on-kubernetes`)

- Himanshu Shukla, *#Microservices : Observability Patterns* (`https://medium.com/@greekykhs/microservices-observability-patterns-eff92365e2a8`)

- *MicroProfile Metrics* (`https://download.eclipse.org/microprofile/microprofile-metrics-2.3/microprofile-metrics-spec-2.3.html`)

- *The OpenTracing project* (`https://opentracing.io/`)

15
What's New in Java?

Java, as is obvious, has been the *leitmotif* of this book. Even if, in some of the previous chapters, we focused on more general concepts such as architectural design and software life cycle management, the main goal of this book is to provide Java software engineers with a compendium of architectural concepts, ultimately supporting them to become better architects.

With this in mind, we cannot avoid a few words regarding the status of Java technology today, especially regarding the latest releases.

In this chapter, we are going to discuss the following topics:

- Java versioning
- Vendor ecosystem
- What's new in Java 17

So, let's start with an overview of Java versioning.

Java versioning

There have been many changes made to the **Java versioning** scheme and schedule over its history. One first thing to note is that, at the very beginning, Java versioning used to follow a *1.x* scheme, with **1.3** essentially being the first widespread version.

Since **version 1.5**, however, the versioning scheme ditched the *1.x* prefix, so we had **Java 5**, **6**, and so on.

Another important point to make is about naming. The very first versions were called **JDKs** (short for **Java Development Kit** – more about this in a bit). Then, from versions **1.2** to **5**, the platform was named **J2SE** (for **Java 2 Standard Edition**). Since **Java 6**, at the time of writing, the platform is referred to as **Java SE** (for **Java Standard Edition**).

The most important thing to know about the JDK, a term that most of us are familiar with, is that until **Java 8**, the Java platform was distributed in two versions, the **Java Runtime Environment** (JRE) and the JDK. The JRE was basically a stripped-down version of the JDK, lacking all the development tools (such as the javac compiler). As said, since Java 8, only the JDK version is officially distributed.

In terms of release timelines, older Java releases used to have a long and non-uniform scheme, with major versions being released in intervals varying from between 1 and 3 years. Since **Java 9**, though, the platform's evolution has followed a 6-month release timeline for major versions.

One more point relates to **Long-Term Support** (LTS) releases. Roughly every 2 or 3 years, a version is considered LTS. This basically means a longer official support cycle (up to 10 years, depending on the vendor) with more features added (while non-LTS releases usually have fewer and simpler new features).

Last but not least, each major version (both LTS and non-LTS ones) also brings with it a set of minor versions, shipping patches, bug fixes, and security fixes.

In the following diagram, you can see the graphical representation of the support life cycle for some of the most important Java releases:

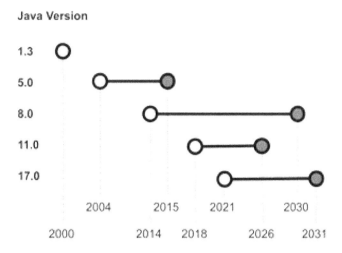

Figure 15.1 – Version support life cycle for some Java releases

Other than the version numbering, a further important consideration concerns the vendor ecosystem.

Vendor ecosystem

As many of you know, Java was released as a project by (the now defunct) Sun Microsystems. It was originally developed as a language for clients and what would later be called the **Internet of Things (IoT)**. Ironically, nowadays, it's rarely used in such scenarios and, conversely, very much used for server-side enterprise applications, which was likely not the first use case in mind when Java was designed.

In 2006, Sun released Java technology as open source under the GPL license. Sun later went out of business and was acquired by Oracle in 2010. With that transition, the Java ecosystem started to be governed mostly by Oracle itself.

Java releases are *certified* using the **Technology Compatibility Kit (TCK)**, which is a test suite used for testing the compatibility of Java distribution with the specifications included in a specific version. And talking of Java distributions, the most important project here is OpenJDK.

OpenJDK distributions

OpenJDK is the main source code repository from which many widespread JDK implementations have been derived, including the Oracle Java distribution.

We know that Oracle leads the open source development of Java within the OpenJDK community. OpenJDK is essentially the reference implementation of Java technology. Oracle ships the Oracle OpenJDK (which is free and not supported commercially) and the Oracle JDK (which is commercially supported under a paid subscription).

Many other vendors provide their own distributions, with small differences between them. All such distributions are created starting from the OpenJDK open source code base:

- **AdoptOpenJDK** is a multivendor project for distributing *vanilla* OpenJDK builds (`https://adoptopenjdk.net`).

- Red Hat provides its own build featuring support for the Red Hat Enterprise Linux operating system and some add-ons, such as support for the Shenandoah garbage collection implementation (`https://developers.redhat.com/products/openjdk`).

- Azul Technology builds a commercially supported implementation, including some proprietary garbage collection features (`https://www.azul.com/downloads`).

- AWS ships **Corretto**, an OpenJDK build designed to run on the AWS Cloud infrastructure (`https://aws.amazon.com/it/corretto`).

- IBM ships **OpenJ9**, originally developed for running on mainframe technology and now available, under the OpenJDK umbrella, for other architectures (`https://www.eclipse.org/openj9`).

- **GraalVM** is an interesting concept built on top of OpenJDK (we have already seen some of its features in *Chapter 7, Exploring Middleware and Frameworks*, when discussing the native compilation of Quarkus). GraalVM comes from the experience of Oracle Labs and brings a lot of different and interesting things to the Java technology, including a module for native compilation (as we mentioned before), and modules for `polyglot` usage, in order to run code written in **Python**, **JavaScript**, **Ruby**, and more.

These are the most commonly used Java distributions. The choice, unless you are looking for a very specific feature, is mostly dependent on circumstances, such as existing support contracts or commercial pricing. In the absence of specific needs, AdoptOpenJDK is usually a good place to start.

A recent ecosystem report built by **Snyk** (`https://snyk.io/jvm-ecosystem-report-2021`), shows that the builds of AdoptOpenJDK are the most popular by far (around 44%), followed by the different flavors (commercial and otherwise) of the Oracle distribution. Another important piece of news from the report is the growing adoption of **Java 11** and the move away from Java 8. However, we will see how the adoption of **Java 17** will grow in the upcoming months and years.

In this regard, let's see what's new in the latest version of Java, Java 17.

What's new in Java 17

Java 17 is an LTS release, meaning that, depending on the vendor, it will be supported for more than 5 years (up to 10, in some cases). It was released in September 2021.

Let's look at some of the new features introduced with this version.

Sealed classes

Sealed classes were introduced with Java 15, and the feature became officially supported with Java 17. They provide a way to declaratively define classes and interfaces while restricting which objects can extend it or implement such classes and interfaces.

This can be particularly useful in specific cases, such as if you are defining an API, as you can, at design time, control some aspects of the usage of APIs.

Here is a simple example:

```
public sealed class Payment permits Instant, Wire,
    CreditCard [...]
```

In this example, we declare a `Payment` class, and we define that only `Instant`, `Wire`, and `CreditCard` can extend it. In this particular example, we suppose these classes are in the same package as `Payment`, but it is possible to explicitly declare the full package if we wanted to place it somewhere else.

Also, the exact same syntax can be applied to interfaces:

```
public sealed interface Payment permits Instant, Wire,
    CreditCard [...]
```

This is the same behavior, just for interfaces, so the implementation is allowed only for the interfaces listed.

It's worth noticing that a compile-time error is raised if non-allowed operations (such as extending a class with a non-declared type) are performed. This will help the code to be more stable and testable.

Pattern matching for switch statements

This is a preview feature, meaning that it must be enabled (by passing a command-line parameter to the JVM) and is not officially completely supported (even if the exact boundaries of support are defined by each vendor).

This feature is about extending the behavior of the `switch` construct.

While there are many different potential use cases (and more will likely be refined and finalized in the upcoming releases), these three are the main ones:

- **Type checking**: The `switch` construct can behave like an `instanceof` operator, checking by type as in the following example:

```
[…]
switch (o) {
    case Instant i -> System.out.println("It is an
        instant payment");
    case Wire w    -> System.out.println("It is a wire
        transfer");
    case CreditCard c -> System.out.println("It is a
        credit card transaction");
    default -> System.out.println("It is another
        kind of payment");
    };
[…]
```

- **Null safety**: While, in the previous implementations, the `switch` expressions raised a `NullPointerException` if the object evaluated is `null`, with this new null safety feature, it is possible to explicitly check for the `null` case. In this example, the `switch` expression checks over a string variable, also checking the `null` case:

```
switch (s) {
    case "USD", "EUR" -> System.out.println("Supported
        currencies");
    case null    -> System.out.println("The String
        is null");
```

```
        default     -> System.out.println("Unsupported
            currencies");
    }
```

- **Refining patterns**: It is possible to use a syntax for expressing more than one condition in a `switch` branch. So, essentially, the following construct is allowed:

```
switch (o) {
    case Instant i && i.getAmount() > 100->
        System.out.println("It is an high value instant
            payment");
    case Instant i -> System.out.println("It is a
        generic instant payment");
    case Wire w     -> System.out.println("It is a wire
        transfer");
```

As you can see, this is a nice feature allowing for compact and readable code.

Strongly encapsulating JDK internals

Since Java 9, there's been a progressive effort to restrict access to the JDK internals. This is meant to discourage the direct utilization of classes residing in packages such as `sun.*`, `com.sun.*`, `jdk.*`, and more. The goal of this restriction is to reduce coupling to a specific JVM version (hence freeing the JVM developers up to evolve such classes, even introducing breaking changes if necessary) and enhance security.

To do so, the JDK progressively offered alternatives. Moreover, since Java 9 (and up to Java 16), source code using those internal classes and methods must be compiled by passing the `--illegal-access` parameter, which can be configured to permit, deny, or print warnings with details of usage.

In Java 17, this parameter is no longer usable. Instead, it is possible to use the `--add-open` parameter, which allows us to declare specific packages that can be used. It is a common opinion that even this possibility will progressively be denied in upcoming versions, to completely deny the explicit usage of JDK internals in custom code.

More changes in Java 17

A lot of other changes have been added to Java 17. Here are some highlights:

- **Support for the macOS/AArch64**: This allows the compilation and execution of Java code on Mac machines running on M1 chips.

- **Enhanced pseudo-random number generators**: This is a partial refactoring of utilities for pseudo-random number generation, including the deletion of duplicated code and the pluggability of different algorithms.

- **Foreign function and memory API**: This is an incubating set of features (which are still not stable and will be subject to further evolution) aimed at simplifying and securing access to resources (code and data) living outside the JVM. This means being able to access memory locations and call methods not managed or implemented in the JVM. To do so in previous versions, you were required to use **Java Native Interfaces** (**JNI**) classes, which are generally considered less secure (and more complex to use).

- **Context-specific deserialization filters**: As a part of an effort started some JVM versions ago, this is a way to define validation for code deserialization. Serialization and deserialization of classes are generally considered potential security issues, as specifically crafted payloads can execute arbitrary (and unsafe) operations. This feature allows the definition of filters to *prevalidate* the kind of code allowed in deserialization operations.

- **Deprecation of the applet API for removal**: Applets haven't been used for a long time, for many reasons, including performance and security issues. Moreover, most (if not all) of the modern browsers don't support them anymore. So, they are being deprecated and will be completely removed from the JDK.

- **Deprecation of the security manager for removal**: The security manager is an API primarily intended for usage along with applets. It was released in **Java 1.0**. It has been progressively abandoned, both due to complexity and performance issues and because applets are now less commonly used. So, it is now deprecated and will be removed in an upcoming version of the JDK.

- **Vector API**: This is a new API in the incubation phase (meaning it will be subject to changes and further evolution). It aims to define a new API for the computation of vectors. Other than being simple to use, this API is designed to compile code, specifically targeting available optimizations for supported CPU architectures, thereby boosting performance where possible.

While a number of other features have been added, modified, and removed, the preceding ones are the most important and impactful.

Summary

In this chapter, we have looked at some of the novelties introduced with the latest release of the Java platform (17).

We have had the opportunity to have a look at the Java versioning scheme and release schedule. We had a quick overview of the Java vendor ecosystem, a snapshot of what is an evolving situation at the time of writing. The same applies to the newest functionalities of the platform itself. While some features are notable by themselves, of course, many will be modified further in the near future.

This completes our journey into cloud-native architectures with Java. I hope I have provided some interesting insights and ideas, and I wish the best of luck to every reader in defining elegant and successful applications and having satisfying careers as software architects.

Further reading

- Oracle, *JDK 17 Release Notes* (`https://www.oracle.com/java/technologies/javase/17-relnote-issues.html`)

- Java Magazine, Mitch Wagner, *Is Java SE open source software? The short answer is 'yes.'* (`https://blogs.oracle.com/javamagazine/post/java-se-open-source-license`)

Index

N

U

Packt.com

Subscribe to our online digital library for full access to over 7,000 books and videos, as well as industry leading tools to help you plan your personal development and advance your career. For more information, please visit our website.

Why subscribe?

- Spend less time learning and more time coding with practical eBooks and Videos from over 4,000 industry professionals

- Improve your learning with Skill Plans built especially for you

- Get a free eBook or video every month

- Fully searchable for easy access to vital information

- Copy and paste, print, and bookmark content

Did you know that Packt offers eBook versions of every book published, with PDF and ePub files available? You can upgrade to the eBook version at packt.com and as a print book customer, you are entitled to a discount on the eBook copy. Get in touch with us at customercare@packtpub.com for more details.

At www.packt.com, you can also read a collection of free technical articles, sign up for a range of free newsletters, and receive exclusive discounts and offers on Packt books and eBooks.

Other Books You May Enjoy

If you enjoyed this book, you may be interested in these other books by Packt:

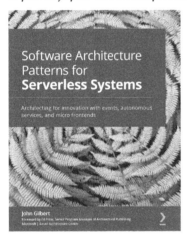

Software Architecture Patterns for Serverless Systems

John Gilbert

ISBN: 978-1-80020-703-5

- Explore architectural patterns to create anti-fragile systems that thrive with change
- Focus on DevOps practices that empower self-sufficient, full-stack teams
- Build enterprise-scale serverless systems
- Apply microservices principles to the frontend
- Discover how SOLID principles apply to software and database architecture
- Create event stream processors that power the event sourcing and CQRS pattern
- Deploy a multi-regional system, including regional health checks, latency-based routing, and replication
- Explore the Strangler pattern for migrating legacy systems

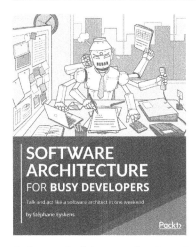

Software Architecture for Busy Developers

Stéphane Eyskens

ISBN: 978-1-80107-159-8

- Understand the roles and responsibilities of a software architect
- Explore enterprise architecture tools and frameworks such as The Open Group Architecture Framework (TOGAF) and ArchiMate
- Get to grips with key design patterns used in software development
- Explore the widely adopted Architecture Tradeoff Analysis Method (ATAM)
- Discover the benefits and drawbacks of monoliths, service-oriented architecture (SOA), and microservices
- Stay on top of trending architectures such as API-driven, serverless, and cloud native

Packt is searching for authors like you

If you're interested in becoming an author for Packt, please visit `authors.packtpub.com` and apply today. We have worked with thousands of developers and tech professionals, just like you, to help them share their insight with the global tech community. You can make a general application, apply for a specific hot topic that we are recruiting an author for, or submit your own idea.

Share your thoughts

Now you've finished *Hands-On Software Architecture with Java*, we'd love to hear your thoughts! Scan the QR code below to go straight to the Amazon review page for this book and share your feedback or leave a review on the site that you purchased it from.

https://packt.link/r/1-800-20730-1

Your review is important to us and the tech community and will help us make sure we're delivering excellent quality content.

www.ingramcontent.com/pod-product-compliance
Lightning Source LLC
Chambersburg PA
CBHW081453050326
40690CB00015B/2784